INCLUSIVE DEVELOPMENT OF SOCIETY

T0236594

PROCEEDINGS OF THE 6TH INTERNATIONAL CONFERENCE ON MANAGEMENT AND TECHNOLOGY IN KNOWLEDGE, SERVICE, TOURISM & HOSPITALITY (SERVE 2018), OCTOBER 6-7 AND DECEMBER 15-16, 2018, BALI, INDONESIA, AND DECEMBER 15-16, 2018, ROSTOV-ON-DON, RUSSIA

Inclusive Development of Society

Edited by

Ford Lumban Gaol

Bina Nusantara University, Jakarta, Indonesia

Natalia Filimonova

Vladimir State University, Russia

Irina Frolova

Southern Federal University, Russia

Ignatova Tatiana Vladimirovna

Russian Presidential Academy of National Economy and Public Administration, South Russia Institute of Management, Russia

CRC Press

Taylor & Francis Group

Boca Raton   London   New York

CRC Press is an imprint of the
Taylor & Francis Group, an **informa** business

A BALKEMA BOOK

CRC Press/Balkema is an imprint of the Taylor & Francis Group, an informa business

© 2020 Taylor & Francis Group, London, UK

Typeset by Integra Software Services Pvt. Ltd., Pondicherry, India

Library of Congress Cataloging-in-Publication Data

Applied for

Published by: CRC Press/Balkema
        Schipholweg 107C, 2316XC Leiden, The Netherlands
        e-mail: Pub.NL@taylorandfrancis.com
        www.routledge.com – www.taylorandfrancis.com

ISBN: 978-1-138-33476–2 (Hbk)
ISBN: 978-0-367-49751-4 (pbk)
ISBN: 978-0-429-44511-8 (eBook)
DOI: 10.1201/9780429445118 https://doi.org/10.1201/9780429445118

Inclusive Development of Society – Lumban Gaol (eds)
© 2020 Taylor & Francis Group, London, ISBN 978-1-138-33476-2

Table of contents

Preface

Managing Service, Education, and Knowledge Management in the Knowledge Economic Era are important factors that need to be there. Managing Service focuses on optimizing the service-intensive supply chains that are integrated into supply chain management as the intersection between the actual sales and the customer using Information Technology.

Information Technology is an important necessity in the globalization era, as any organization without IT is considered incomplete. Information management, service management and web intelligence are important components of IT, that are now playing a more important role in modern society. In short, information management is the collection and management of information from one or more sources and the distribution of that information to one or more audiences, which control the planning, structure and organization, controlling, processing, evaluating and reporting of information activities in order to meet client objectives and enable corporate functions in the delivery of information. Especially the education field benefits from not being left behind using IT systems in teaching and learning.

This book provides a platform for all researchers, enterprisers, and students to exchange new ideas and application experiences face to face, to establish business or research relations and to find global partners for future collaboration. The research topics cover every discipline in all fields of social science, economics, and arts. The conference topic areas especially focus on such disciplines as language, cultural studies, economics, behavior studies, political sciences, media and communication, psychology and human development.

The purpose of this book is twofold, presenting both theoretical research studies that provide a solid foundation for the development of new tools that explore the possibilities of developing tourism, hospitality and service industries in the Knowledge Economic Era, and empirical papers that advance our knowledge regarding the impact of information technologies in organizations' and institutions' practices.

We do hope that this book will have a positive impact on the social sciences including education, psychology, tourism and knowledge management areas.

Best regards,

Ford Lumban Gaol
Bina Nusantara University, Indonesia

Natalia Filimonova
Vladimir State University, Russia

Irina Frolova
Southern Federal University (Russia)

Ignatova Tatiana Vladimirovna
Russian Presidential Academy of National Economy and Public Administration,
South Russia Institute of Management, Russia

Inclusive Development of Society – Lumban Gaol (eds)
© 2020 Taylor & Francis Group, London, ISBN 978-1-138-33476-2

Editorial Board

Conference Chair:

- Vladimir Maslennikov - Vice-Rector for Research, Financial University, Russia
- Tatyana Rozanova - Vice-Rector for Continuing Education, Financial University, Russia
- Oleg Safronov - Head, Federal Agency for Tourism of the Russian Federation, Russia
- Wayne Marr - University of Alaska, USA
- A.J.W. Taylor - Victoria University of Wellington, New Zealand
- Nina Chaikovskaya - Director of Murom Institute (branch) Vladimir State University, Russia
- Natalia Filimonova - Head of Management and Marketing Department, Vladimir State University, Russia
- Maria de Lourdes Machado-Taylor - Taylor Center for Research in Higher Education Policies (CIPES), Portugal
- Anzor Saralidze - Rector of the Vladimir State University, Russia
- Inna K. Shevchenko - Rector ad interim of Southern Federal University, Russia

Program Committee Chair:

- Fonny Hutagalung - University of Malaya, Malaysia
- Ford Lumban Gaol - Bina Nusantara University, Indonesia
- Natalia Filimonova - Head of Management and Marketing Department, Vladimir State University, Russia
- Irina Frolova - Head of Accounting and Audit Department, Southern Federal University, Russia

Publication Chair:

- T.V. Matytsyna - Southern Federal University, Rostov-on-Don, Russia
- A.U. Polenova - English Language Department for Humanities, Southern Federal University, Rostov-on-Don, Russia
- T.G. Pogorelova - Accounting and Audit, Southern Federal University, Rostov-on-Don, Russia

Publicity Chair:

- N. Panchanatham - Department of Business Administration, Annamalai University, India
- T. Ramayah - Universiti Sains Malaysia, Malaysia

Program Committees:

- Yulia Gruzina - Financial University, Russia
- Mikhail Morozov - Financial University, Russia
- Marina Fedotova - Financial University, Russia
- Marina Nesterenko - perspektivy (Tourism: Practice, Issues and Prospects) Journal, Russia
- Natalia Filimonoa - Vladimir State University, Russia
- Natalia Morgunova - Vladimir State University, Russia
- Anastasia Ussuf - Vladimir State University, Russia
- Irina Zaytseva - Vladimir State University, Russia
- Svetlana Shtebner - Vladimir State University, Russia
- Julia Kochetova - Vladimir State University, Russia

- Olga Starostina - Vladimir State University, Russia
- Dr. Annibal Scavarda - Federal University of the State of Rio de Janeiro, Brazil
- Doctor Leonardo Baumworcel - Unigranrio University, Brazil
- Valentinas Navickas - Kaunas University of Technology, Lithuania
- Vaida Pilinkiene - Kaunas University of Technology, Lithuania
- Hsin Rau - Chung Yuan Christian University, Taiwan
- Fonny Dameaty Hutagalung - University of Malaya, Malaysia
- Panos M. Pardalos - University of Florida, USA
- Min-Young Lee - University of Kentucky, USA
- Igor K. Liseev - Financial University, Russia
- Alonso Perez-Soltero - Universidad de Sonora, Mexico
- Chew Fong Peng - University of Malaya, Malaysia
- Aishah Rosli - University of Malaya, Malaysia
- Zulkifli Md Isa - University of Malaya, Malaysia
- Kusmawati Hatta - University of Ar-Raniry-Aceh, Indonesia
- Zulkefli Mansor - University of Kebangsaan Malaysia, Malaysia
- Mohd Rushdan Abdul Razak - Open University, Malaysia
- Benfano Soewito - Bina Nusantara University, Indonesia
- Vladimir Yu. Kopytin - Faculty of Management. SFU, Russia
- Natalia Lebedeva - SFU, Russia
- Elena Mihalkina - Faculty of Economics, SFU, Russia
- Pavel Pavlov - Institute of Management in Economic, Ecological and Social Systems, SFU, Russia
- Tatiana Anopchenko - Faculty of Management, SFU, Russia
- Tatiana Ignatova - Cathedra of Economic Theory and Entrepreneurship, South-Russia Institute – Branch of Russian Academy of National Economy and Public Administration under President of RF
- Galazova Svetlana Sergeyevna - North Ossetian State University of Costa of Levanovich of Khetagurov, Russia

Inclusive Development of Society – Lumban Gaol (eds)
© 2020 Taylor & Francis Group, London, ISBN 978-1-138-33476-2

Development of a digital economy ecosystem in Russia

Irina B. Teslenko
Federal State-funded Educational Institution of Higher Education "Vladimir State University named after Alexander and Nikolay Stoletovs", Vladimir, The Russian Federation

Olga B. Digilina
RUDN University, Moscow, The Russian Federation

Nizami V. Abdullaev & Nadezhda V. Muravyeva
Federal State-funded Educational Institution of Higher Education "Vladimir State University named after Alexander and Nikolay Stoletovs", Vladimir, The Russian Federation

ABSTRACT: The article contains an analysis of the main directions of digital economy ecosystem development in the Russian Federation. The authors define ecosystem of digital business development as a number of companies that use intellectual, production, information, and other resources on a common basis. Because the digital economy ecosystem is the main condition of implementation of information and communication technologies in all spheres of people's lives, state regulation of cybersecurity, stimulation and support of promotion of information technology abroad, development of infrastructure of access and data storage, cashless payments, and all types of mass digital communication and services are necessary.

1 INTRODUCTION

The key component of the institutional environment of the digital economy is its ecosystem. In 1993, J. Moore began to use the term "ecosystem" with regard to business, transferring it from biology to economics. The Business Trends report published by Deloitte Consulting states that ecosystems are dynamic and jointly developing communities consisting of a variety of entities that create and receive new content during interaction and competition [7].

On the one hand, ecosystems are stable and durable, but, on the other hand, they are characterized by transformations, the constant internal dynamic processes. While some systems (e.g., software) are developed toward increasingly subtle individualization, other systems (e.g., technological infrastructure) on the contrary are combined, absorb smaller players, and create large firms that offer resources, information, and platforms to others.

Such a dynamic structure consisting of affiliates (small enterprises, corporations, universities, scientific organizations, nonprofit organizations, etc.), where cooperative and competitive processes are carried out, became known as the ecosystem. R.U. Ayres noted that the purpose of the ecosystem is to improve the company's interaction with its partners and increase competitive advantages in terms of the creation of innovative products, which will make it a leader in its industry. According to him companies need to implement innovations and build new relationships for a "healthy" ecosystem [17]. The term "ecosystem" has entered Russian business practice more recently [12].

According to the "Strategy for the Development of the Information Society in the Russian Federation for 2017–2030" (approved by RF Presidential Decree No. 203 dated May 9, 2017), the digital economy ecosystem is a partnership of organizations that provides constant interaction of their technological platforms, applied Internet services, analytical systems, and information systems of state authorities of the Russian Federation, organizations, and citizens. Such a definition is given in a dictionary of financial and legal terms.

By the ecosystem Sberbank means a network of organizations that are created around a single technological platform and use its services to make offers to customers and share access to them, and have plans to create a platform that will be a Russian analogue of Alibaba, Amazon, Facebook, and Google ecosystems.

The scientific literature provides the following definition of the business ecosystem. An ecosystem is a stable open system of various natural elements of the environment, between which there are regular constant processes of resource interchange, thus providing for the stable functioning at the corporate, regional, national, and integration levels [3].

In general, the significance of the use of the term "ecosystem" in the scientific literature is to focus on the establishment of close cooperation ties and interaction between organizations in order to actively create and implement innovations. In present-day conditions, foreign experts give preference to interaction and partnership.

Indeed, the digital economy ecosystem is a set of companies that share intellectual, production, information, and other resources and an access to them. A shared-use resource provides all participants with the access to the benefits of the scope of the ecosystem [1]. Along with cooperation and interaction, there is competition among organizations in the ecosystem, which is objective due to the characteristics of the digital economy.

According to the experts of the EDMA (the Electronics Developers and Manufacturers Association), the digital economy is characterized by global standardization of technologies, structures, and processes resulting in increased price competition and increased requirements on the range of activities. Companies that make easily scalable and replicable successful decisions do not share the market with the medium- and low-quality companies; they occupy it entirely. In each segment, a leader that controls 40% of the market takes more than 80% of the profit. A huge, ever-increasing gap emerges in the investment opportunities of the leader and other participants. There is market monopolization, emergence of oligopolistic structures, and the specialization of countries is fixed in the world market.

The ecosystem has a complex structure. It can be represented as a set of subsystems, such as technological (small innovative enterprises, start-ups, technology parks, business incubators, accelerators, clusters, associations, etc.), financial (development institutions, VC investors, VC funds, business angel investors, crowdfunding platforms, banks, etc.), science and education subsystem (universities, research institutes, research laboratories, research departments of large companies), information subsystem (information centers and portals, conferences, trade fairs, advisory agencies, etc.), retraining subsystem and innovation transfer subsystem (technology transfer centers, engineering centers, experimental machine design bureaus, patent offices, inspection and certification organizations, etc.), etc. Such an approach seems sound to the authors and really reflects the essence of any infrastructure.

2 STATEMENT

In this article, the digital economy ecosystem of Russia means the market segments where value added is created using digital (information) technologies. We can distinguish seven key components (hubs) in such ecosystem: the state and society, marketing and advertising, finance and trade, infrastructure and communications, media and entertainment, cyber security, and education and human resources [14]. So-called slicers are distinguished inside the components. There are 10 of them: development and design, analytics and data, AI and Big Data, hardware, business models, Internet of things, mobile, platforms, regulation, and start-ups and investments.

The purpose of our study was to determine the degree of development of the digital economy ecosystem in Russia at present, as well as the role of the state in the process of its creation.

If we take this approach to structuring the digital economy ecosystem, the following data show the state and development of its structural elements.

Currently, there are 3.773 billion active Internet users, 3.448 billion active mobile Internet users, and 2.907 billion active social media users in the world. In general, 50% of the world population uses the Internet and 46% uses the mobile Internet. In 2016, 2.3 million employees were employed in this field of information technology.

The index of digital literacy of the Russian population was 5.42 in 2016. Sixty-five percent of the Russian population are mobile Internet users. In Russia, 87.6 million people (71% of the population) use the Internet at least once a month, and 84.4 million people (69%) use the Internet at least once a week.

According to Mediascope, people in the 12- to 24 -year-old age group most often go on the Internet using a smartphone (89.8% of the population), while representatives of the 25- to 44 years age group of prefer to go on the Internet using a computer or laptop (75.4% of the population). However, 74.7% of the people in this age group go on the Internet using a smartphone.

Representatives of the 45+ age group choose a computer or laptop to get online (45.8% of the population). Residents of cities with a population of more than 100,000 often go on the Internet using a smartphone, whereas residents of cities of less than 100,000 also choose smartphones [14].The Russian statistics show the superiority of the mobile over the desktop: 65% versus 53%.

In 2016, marketing and advertising were estimated at 171 billion rubles, the Internet advertising market at 136 billion rubles, digital content at 63 billion rubles, and e-commerce at 1,238 billion rubles; in 2016, 36% of users made purchases in online stores, retail online exports amounted to $2 billion (including digital goods and services), infrastructure and software were estimated at 2,000 billion rubles (.RU is fifth among national domains in the world [ccTLD], .RU is the ninth among all domains in the world [ccTLD, gTLD, new gTLD], and .RF is the world's biggest IDN Domain).

In 2016, online retail was estimated at 706 billion rubles (an increase of 18% compared to 2015), online travel at 363 billion rubles (an increase of 15%), Internet services at 169 billion rubles (an increase of 15%), and online payments at 686 billion rubles (an increase of 17%) [14].

The top 100 online stores accounted for 68% of sales; at year-end 2016, every two of five purchases were made via mobile devices (including tickets and food). Curiously enough, there are many mobile orders in those categories of goods and services that are not considered to be of "Information Technology": DIY (Do It Yourself [14] is a kind of activity that includes self-production, repair, improvement of equipment, furniture, clothing, equipment, and other consumer goods), clothing, etc. 66% of smartphone owners use applications for payments and purchases. The most rapidly increasing category is food delivery (with small absolute volumes).

The Russian market volume M2M/IoT (Internet of things) as of the first half of 2016 reached 300 billion rubles, having increased from 225 billion rubles in the first half of 2015 [2]. The market of DPC (Data-Processing Centre) grows by 10–15% per year [14].

According to Gartner, there will be about 21 billion IoT-enabled devices in the world by 2020.The overall market potential of smart metering devices in Russia in the segment of private utility power and water consumption is more than 206 million smart meters or exceeds 400 billion rubles.

In the near future, the mobile economy will grow by 10.7% as a result of reducing costs, increasing efficiency, and providing new services through mobile devices and access [2]. According to the forecasts, by 2021, the contribution of the mobile economy to the economic growth will be 7.5% and the contribution to the country's GDP will be 4.7%.

Today, the mobile economy creates more than a million jobs, and by 2021 it will create one and a half million, solving more or less the problem of unemployment in the labor market. Owing to state investments in infrastructure and the availability of free Internet in Russia, the cost of data transmission is quite low.

In addition, the average cost of a smartphone in Russia is $168, and in other countries it is $241. Russia is distinguished by a high level of competence of IT specialists; many Russian developers are in the top 10 by revenue [4] at AppStore and Google Play app stores in other countries. Russia is in fifth place in the world in terms of application downloads [15].

As for the industry platforms, including in Russia, their activities are very important.They provide the following: worldwide access to resources without intermediaries, rent of resources (human resources, business models, technologies, IP blocks and finance), use of a volunteer model (Open Source Model is an Open Source Software), as well as on-demand sales through ecosystems [8].

The media and entertainment field is formed by digital content, games (about 72 million gamers are in Russia, which is about 65% of Internet users), social media, and books (the Russian e-book market grew 10-fold from 2013 to 2017) [5].

The number of vacancies for remote work and outsourcing has increased significantly. At the same time, the share of freelancers in the total number of employees in the Russian labor market is not more than 2% (up to 1.5 million people) [5].

As for start-ups and investments, there is a significant increase in indicators related to the purchase of start-ups in this field. There is also a growth of the mergers and acquisitions market. According to researchers, the number of start-ups in the world will grow, despite the increasing competition resulting from the growing technological revolution. According to Dow Jones analysts, their number has increased by 3.6 over the past two years.

Start-ups and small innovative enterprises have been increasingly coming up from the ideas of some individuals. In addition, no special production areas or a significant number of workers are required today. On crowdfunding platforms you can raise money, start a business, and become a world leader in the future to support your idea, as once happened in the companies that hold the top lines of the world ranking: Apple ($800 billion), Alphabet ($550 billion), and Microsoft ($420 billion) [5].

Corporate innovations are increasing and corporate investment is developing actively because of such structures as accelerators (e.g., IKEA, SAP, InspiRUSSIA, etc.) and corporate funds (e.g., Sistema VC, Sistema Asia Fund, etc.). In addition, the number of business angel investors and their investments [15], as well as the p2p network and the number of crowdfunding platforms are increasing. Among the most active are Venture Club, StartTrack, the Investment Club of the Skolkovo School of Management, the Investment Club of the Internet Initiatives Development Fun, SmartHub, and others.

In 2017, the Central Bank of Russia monitored the crowdfunding market for the first time. The Bank's assessment of the market capacity is the first step toward its regulation.

In total, the regulator analyzed the activities of the 10 largest crowdfunding platforms, among which were Planeta.ru, Boomstarter.ru, Kroogi.com, Thankyou.ru, Rusini.org, and Smipon.ru, and came to the conclusion that crowdfunding had a great growth potential.

According to the experts' forecasts, the volume of the crowdfunding market would have reached 4 billion rubles by the end of 2016, and then it would have grown by 15-30% annually (for comparison: the bank loans to individuals amounted to 7.2 trillion rubles in 2016) [13].

The development of the digital economy and its ecosystem makes it possible to predict that three quarters of Russians, 86.7 million people, will become Internet users by 2020 [15]. Until 2020, the labor market will remain a stable demand for IT specialists: in general, the demand for them is more than 350,000 people for the Russian Federation.

The goals and objectives of the digital economy development are defined until 2024 in the Digital Economy of the Russian Federation program within the frameworks of the following five basic directions [9]:

- Statutory regulation
- Personnel and education
- Formation of research competences and technical scopes
- Information infrastructure
- Information security

As for the digital economy ecosystem, the challenge is to ensure successful functioning of the following [9]:

- At least 10 leading companies (ecosystem operators) competitive on global markets
- At least 10 industry (industrial) digital platforms for the key subject fields of the economy (including digital health, digital education, and "smart city")
- At least 500 small and medium-sized enterprises in the field of digital technologies and platforms, as well as digital services

The following growing tendencies will contribute to the development of the ecosystem:

1. Growth of investments.
   IDC, the international research consultancy, conducted a survey of IT executives about future investments. Twenty-four percent of respondents noted that their organizations planned to increase expenditures for IT infrastructure by more than 10%, 21% by 5-10% in 2018.
   The reasons for such responses were the following: growth of the data volume (investments in equipment for their storage are needed); security vulnerability (in 2016, 31% of companies in Russia faced DDOS attacks, denial of service due to a hacker attack); insufficient network bandwidth (investments in network equipment are required); and development of wireless communications.
   According to IDC, the return on sales of cloud infrastructure components increased by 25.8% year-on-year and reached $12.3 billion in the second quarter of 2017. Moreover, the share of the infrastructure return for public clouds accounted for 33.5% of the market [18].

2. Improvement of the data storage system.
   According to IDC, the data volume in the world will increase 10-fold from 2013 to 2020 (up to 44 trillion gigabytes), so now the question arises: where to store these datasets?
   Western Digital (WD), the hardware manufacturer, believes that by 2020, 70% of all data will be placed on HDD (hard disk drives) that the company is developing actively. WD is engaged in disk capacity extension without increasing the media size. Relying on microwave-assisted magnetic recording (MAMR), the company is going to create HDDs of up to 40 TB. Samples of super-compatible hard disc drives will be available to corporate customers in 2019.
   IBM relies on flash memory. Its development is focused on reducing costs and accelerating the deployment of private clouds (with regard to the IBM FlashSystem 900).

3. Increase Internet speed.
   In the near future, we should expect an increase in the speed of data exchange between devices of the Internet of things (according to Gartner, their number will reach 20 billion by 2020).
   Cisco, which develops and sells network equipment, predicts that videos will account for 80% of all Internet traffic by 2021, compared to 67% in 2016. Next-generation mobile networks capable of processing more data in a shorter time will be needed.
   Companies have started to use 5G networks. Nokia teamed up with Amazon Web Services for joint developments in the field of IoT and 5G. A development center for the 5G network in Russia is planned [18].

4. Work in the artificial intelligence (AI) industry.
   The Gartner experts believe that the creation of systems that learn, adapt, and prepare to act autonomously would be the object of competition between technology suppliers until 2020.
   Artificial intelligence is beginning to be used to fight against cyber threats that, paradoxically, come from the same AI. The *Harvard Business Review* noted that the main hope for protection against attacks organized via AI was AI.

5. Development of the hyper-converged infrastructure market.
   Experts of HPE (Hewlett Packard Enterprise) came to the conclusion that technologies that efficiently use and process data can solve the problems of information volume growth. A hyper-converged infrastructure is capable of implementing it.
   A converged infrastructure is a type of infrastructure that represents a ready-made solution from a manufacturer and is designed to accelerate the infrastructure deployment. A hyper-converged infrastructure is an infrastructure in which computing power, storage, servers,

and networks are combined with software tools, and they are managed through a common administrative console. It is a software-defined technology, all components of which are integrated. For this reason, a single system administrator is sufficient to manage data stores and server hardware instead of a few IT specialists. This infrastructure is easily scaled: in order to increase capacity and performance, you need to add a new block. Instead of expanding capacity by increasing the number of disks, memory, or processors, performance is increased by adding new modules [6]. In 2017, sales of hyper-converged systems increased by 64.7% [18].

6. Increased popularity of cloud services (IaaS).

The IDC forecasts predicted that 40% of IT spending for hardware, software, and services would be concentrated around the cloud by the end of 2018.

In 2017, many companies migrated to the cloud. Start-ups also realized the advantages of the cloud, which become more flexible for their operation.

The market of tools for enterprise transition in the clouds is expanding. It is expected that 2018 will clear the way to cloud technologies for a record number of organizations.

3 CONCLUSIONS

Summarizing the foregoing, it can be noted that the ecosystem is developing and will be developed as the digital economy establishes. The ecosystem has a rather complex structure, the elements of which are in constant interaction and compete with each other.

The digital economy ecosystem has its own unique features: it easily scales resources, new goods, services, and new competencies; provides rapid promotion of basic technologies and entrepreneurial initiatives; significantly expands market coverage; and is not limited to the state boundaries [1].

At the same time, development of the digital economy and creation of the ecosystem require state regulation, primarily concerning issues of cyber security; stimulation and support for the promotion of information technology abroad, the development of access and data storage infrastructure, non-cash payments, and all types of mass digital communications and services [2]; and creation of comfortable conditions for IT companies to do business in Russia. Development of the appropriate tax legislation for the IT industry, regulatory documents related to big data, artificial intelligence, robotics, independent platforms, etc., is required [16].

State activities should be aimed at creating an institutional environment and appropriate ecosystem that would ensure alignment of interests of all stakeholders of the digital economy and promote conditions for ensuring the institutional equilibrium of the economic and social system.

REFERENCES

1. The Electronics Developers and Manufacturers Association (the EDMA) talked about the value of ecosystems for the digital economy development in Russia. Available at: http://arpe.ru/news/ARPE_rasskazala_o_tsennosti_ekosistem_dlya_razvitiya_tsifrovoy_ekonomiki_v_Rossii/
2. Business in the runet amounted to 1.500 billion rubles for 2016. Available at: http://www.tssonline.ru/newstext.php?news_id=116156.
3. golokhvastovd.v. ecosystem as a tool of balanced integration of interests of economic agents in the digital economy. Available at: http://izron.ru/articles/sovremennyy-vzglyad-na-problemy-ekonomiki-i-menedzhmenta-sbornik-nauchnykh-trudov-po-itogam-mezhduna/sektsiya-2-ekonomika-i-upravlenie-narodnym-khozyaystvom-spetsialnost-08-00-05/ekosistema-kak-instrument-sbalansirovannoy-integratsii-interesov-ekonomicheskikh-agentov-v-usloviyakh/
4. Investments in start-ups. Available at: http://smfanton.ru/nuzhno-znat/startup.html
5. Mobile economy contribution of the mobile economy to the GDP of Russia. Available at: http://vestnik-sviazy.ru dated august 13, 2017, http://www.vestnik-sviazy.ru/news/mobilnaya-ekonomika/

6. Something about converged (and hyper-converged) it infrastructure. Available at: https://habrahabr. ru/company/it-grad/blog/281813/

7. Review of new business trends or the labor market ecosystem. Available at: https://habrahabr.ru/com pany/mbaconsult/blog/295502/

8. Platform of industrial digital economy ecosystems. Available at: http://www.eurasiancommission. org/ru/act/dmi/workgroup/Documents/3.BC.pdf

9. The Digital Economy of the Russian Federation program. Decree of the Government of the Russian Federation No. 1632-r dated July 28, 2017. Available at: http://www.sbras.ru/files/news/docs/program ma_tsifrovaya_ekonomika.pdf, http://www.sbras.ru/ru/news/docs/government

10. Dictionary of financial and legal terms. Available at: https://www.consultant.ru/cons/cgi/online.cgi? req=jt;div=LAW, http://www.garant.ru/files/5/4/1110145/1110145.zip, http://www.garant.ru/prod ucts/ipo/prime/doc/71570570/.

11. http://www.consultant.ru/law/ref/ju_dict/word/jekosistema_cifrovoj_jekonomiki/

12. Formation of the University's business ecosystem: new challenges. Available at: http://kopnov.live journal.com/6597.html

13. The Central Bank monitored the crowdfunding market. Available at: http://web-payment.ru/newsi tem/1103/cbr-monitoring-crowdfunding/

14. Digital economy of Russia in 2016: Statistics and trends. Available at: http://www.iksmedia.ru/news/ 5401378-czifrovaya-ekonomika-rossii-v-2016.html

15. Digital economy of Russia in 2017: Analytics, figures and facts. Available at: https://www.shopolog. ru/metodichka/analytics/cifrovaya-ekonomika-rossii-2017-analitika-cifry-fakty/https://www.shopo log.ru/metodichka/analytics/cifrovaya-ekonomika-rossii-2017-analitika-cifry-fakty/

16. Digital economy of Russia: The development program. Available at: http://www.garantexpress.ru/ statji/zifrovaya-ekonomika-rossii-programma-razvitiya/. Digital economy was approved by the gov- ernment.. Available at: http://www.comnews.ru/node/108966, http://www.comnews.ru/node/ 108966#ixzz4yExXK1DN.

17. YakovlevaA.Yu. Factors and patterns of formation and development of innovation ecosystems. Ph.D. (Economics) Thesis, Higher School of Economics National Research University. M., 2012. Available at: https://search.rsl.ru/ru/record/01005472744. http://dlib.rsl.ru, http://libed.ru/knigi- nauka/98679-1-faktori-modeli-formirovaniya-razvitiya-innovacionnih-ekosistem.php.

18. 6 trends in IT infrastructure: forecast for 2018. Available at: https://habrahabr.ru/company/it-grad/ blog/341374/

Inclusive Development of Society – Lumban Gaol (eds)
© *2020 Taylor & Francis Group, London, ISBN 978-1-138-33476-2*

Identifying hot spot areas of substance abuse with a Geographical Information System in Malaysia: A case study

Y.F. Chan
Faculty of Education, Universiti Teknologi MARA, Malaysia

N.A. Adnan
Faculty of Architecture, Planning and Surveying, Universiti Teknologi MARA, Malaysia

ABSTRACT: Substance abuse such as illegal use of opiates and synthetics, hallucinogenic, stimulant, ketamine, and other drugs has received great attention from society as well as the government of Malaysia. Drug addiction field data compiled by the National Anti-Drug Agency in Malaysia indicated a high prevalence of opiate and amphetamine type stimulant (ATS) consumption in several high-risk states such as Kuala Lumpur, Selangor, Kelantan, and Pulau Pinang. However, not many studies have been conducted to understand the geospatial pattern of drug abuse in the high-prevalence areas in Malaysia. Therefore, the case study described in this article presented the use of a Geographical Information System (GIS) as an attempt to map the location of each drug client's data with the known high-prevalence areas. By using ArcGIS software, the researchers were able to identify the substance abuse hot spot areas in the two selected states of Kuala Lumpur and Selangor. The pattern of drug addiction can be analyzed and explained using GIS because it provides an accurate and comprehensive view of different attributes for drug data for each client according to district and sub-district. The geospatial data indicated the three hot spot areas in Kuala Lumpur are Cheras, Ampang, and Petaling Jaya. However, for the state of Selangor, the three hot spot areas identified are Klang, Ampang Jaya, and Damansara. With the identification of substance abuse hot spot areas, the relevant agencies such as the National Anti-Drug Agency and police officers will be able to accurately target the problematic areas for future prevention and awareness activities.

Keywords: Geographical Information System, hot spot areas, substance abuse

1   INTRODUCTION

Drugs are widely used in medicine, particularly to obtain relief from pain and stress (Gill et al., 2010). Although originally drugs were beneficial for many purposes, their abuse has caused adverse effects not only to the abusers themselves but also to society and the state as a whole. Substance abuse related to undesirable health effects and other social problems has been seen to be a major issue worldwide. The number of substance abusers is increasing and new psychoactive substances are emerging (UNODC, 2014). According to Devi et al. (2012), the World Health Report 2002 has expressed that the total burden of disease worldwide due to psychoactive substance abuse was 8.9% and included tobacco, alcohol, and illicit drugs.

This problem does not discriminate among gender, age, and socioeconomic status, despite many research studies correlating substance abuse with male gender (Choi et al., 2010; Maehira et al., 2013), low educational level (Chwarski et al., 2012), and low socioeconomic status (Rather et al., 2013) because some studies have proven otherwise (Humensky, 2010; Masood & Us, 2014). In addition there were many studies on substance abuse among adolescents (Walker et al., 2006; Yusoff et al., 2014), among whom were the highest substance abusers. The most common substances that are abused worldwide are opiates, cocaine, amphetamine, and cannabis, but the consumption patterns always change (Wu, 2010). These substances have different effects on theabusers according to the type of substance. However, the fact that their use generally leads to violence is very worrying to the government because substance abuse is synonymous with crime (Rusdi et al., 2008). Hence, monitoring the trend of substance abuse cases is an important aspect of preventing further serious problems from occurring. These trends could serve as an important incentive for the stakeholders involved to enhance prevention and awareness programs. Multivariate coupling with geographic information system mapping tools for many research areas has proven beneficial for understanding the circumstances of the problem, and proper planning could be suggested to overcome it. Therefore, this study investigated the use of a Geographic Information System (GIS) to identify hot spot areas of substance abuse in two high-risk states in Malaysia.

2 LITERATURE REVIEW

In advanced countries it has been quite some time since a GIS-based approach received great attention in the field of substance abuse (Latkin et al., 1998). A large geospatial database can be developed by integrating the geographic information with other substance abuse data. The database then can be linked to the GIS through physical features such as addresses and location coordinates (Mason et al., 2004). There are many analyses in GIS that can be used to integrate and analyze a huge amount data (spatial and nonspatial) from disparate sources (Geanuracos et al., 2007) as long as the data are linked to a geodatabase (Berke, 2010).

In fact, the advanced technology of GIS is proven through the understanding of the geographic distribution of a disease. In a study of substance abuse, Zhou et al. (2014) employed a geographic autocorrelation analysis and geographic scan statistics to characterize the geographic pattern of HIV and HCV infections among drug users at the township level. Brownstein et al. (2010) applied GIS to study the distribution of substance abuse and identify its hot spot areas, plus investigate the similarities between the locations that fall in the same cluster. Green and Pope (2008) also used GIS to compare the profiles of substance abuse treatment attendees from New Orleans to Houston based on the distribution of care centers for substance abuse clients. A study by Toriman et al. (2015) developed a map to display the distribution of drug addiction in Terengganu to identify hot spot areas of drug addiction and Buxton et al. (2008) also produced a map of needle distribution sites showing the limitation of primary distribution and assisted health authorities to assess the reaching of supplies in their locations. Through GIS, a disease mapping would make it easy to update information and identify hot spot areas (Srivastava et al., 2009; Kenu et al., 2014). In addition, Sanders et al. (2013) also used Getis-Ord Gi* to identify the hot spot areas for their study.

Through analysis of layers of substance abuse problems, risk areas could be determined. Mendoza et al. (2013) analyzed spatial consideration as a component of the risk assessment in Buffalo, New York. The influence of environmental factors on drug abuse, addiction, and seeking treatment has attracted many researchers to study this field using the GIS approach (McLafferty, 2008). Recent advanced applications of GIS in analyzing environmental factors are related to health issues in terms of identifying the distribution of diseases and the hot spot areas, which are very helpful in finding the proper solution to the problems (Botto et al., 2005; Shirayama et al., 2009; Seid et al., 2014).

3 METHODS

3.1 *Software*

ArcGIS is GIS software that was developed by Environmental System Research Institute (ESRI). It is suitable for working with maps and geographic information and uses a contextual tool for mapping and spatial reasoning (Esri, 2018). Hence, ArcGIS software version 10.4 has been proposed in this research to enable the researchers to explore data and share location-based insights with the stakeholders via apps, maps, and reports (Esri, 2018) and to produce hot spot and spatial distribution maps. This software provides tools and capabilities to allow any modification and manipulation of the spatial data and GIS information attribute productivity that focuses on data management, data development, modeling analysis, and map display.

3.2 *Data*

As substance abuse data are sensitive and confidential, only composite data from drug users in two high-risk states in Malaysia, namely Selangor and Wilayah Persekutuan Kuala Lumpur, were used to avoid the reveal of individual data. Earlier, the writers completed a research project of "Developing Geographical Information System to Identify High Risk Areas of Substance Abuse" in six different states in Malaysia (Kedah, Pulau Pinang, Kuala Lumpur, Selangor, Johor, Kelantan) with assistance and research grant from the National Anti-Drug Agency (*Agensi Antidadah Kebangsaan Malaysia* or AADK) for the period from January 2017 to June 2018. Hence, the researchers have established good collaboration with the Prevention Unit of AADK to obtain the database of substance abusers from the two selected states in Malaysia. For this study, the researchers only used the data from two different states to enter into the GIS system to identify hot spot areas of substance abuse. The major constraints of the study are due to accessibility to the data and accuracy of the data. First, the researchers have to rely on AADK to obtain this confidential data. Second, the accuracy of the data depends solely on the honesty of the clients in providing personal data when detected by the National Anti-Drug Agency in Malaysia. Furthermore, the drug addicts may not provide their actual address, and they will be moving from place to place. The data provided by the AADK are in Microsoft Excel data format with several types of attribute information such as drug client location (address), types of drug addicted, race, gender, and academic qualification entered into the ArcGIS software.

The temporal data in this study involves only data from January 2017 to June 2017 for the two selected states in Malaysia (Selangor and Wilayah Persekutuan Kuala Lumpur). All the data were converted into GIS data format (i.e. shapefile). The base map was defined in GIS and referred to a collection of GIS data or orthorectified imagery that forms the background setting for a map. The function of the base map was to provide background detail necessary to orient the location of the map. Typical GIS data and imagery that made up the layers for a base map such as streets, parcels, boundaries (country, county, city boundaries), shaded relief of a digital elevation model, waterways, and aerial or satellite imagery were used in this study. Base maps for this study were obtained from the local Department of Survey and Mapping Malaysia (JUPEM) for each district and sub-district of Peninsular Malaysia states.

3.3 *Processing the data*

The addresses of drug clients were added on to the ArcGIS map using the My Maps platform. The purpose of this process was to pinpoint the location based on the given addresses of drug clients. The drug clients' temporal patterns for the year 2017 according to districts and sub-districts were mapped using an interpolation model such as the kriging technique. Spatial interpolation is the process of assuming the spatial variation of attributes that includes spatially correlated components. In this study, kriging is used to estimate the value of drug clients from known data points. Then a reclassifying tool was used to reclassify the results derived from the interpolation technique. This tool was used to reclassify all the drug clients' data into several known categories such as Low, Moderate, and High drug client population and distribution. A hot spot map using kernel density was also performed for the known hot spot areas of drug addiction.

Figure 1. Spatial statistics tools for hot spot mapping.

3.4 *Results and data analysis*

The results from geospatial analysis are then gathered and used in the geospatial web-based drug client development. The web-based data were developed using a GIS ArcGIS-compatible platform. The web-based system was designed to enable an authority such as the AADK to easily access the results of geospatial analysis such as the high concentrations of drug clients for each state according to districts and sub-districts in Peninsular Malaysia, the drug client hot spot areas, and relevant maps according to attribute data as stated in the data collection part. Several specific tools are then proposed according to analysis and data viewing and reporting system as needed by the AADK.

The production of hot spot maps takes place through the ArcGIS software. A spatial statistics tool is used to produce a hot spot map. In the spatial statistics tool, there is a hot spot analysis (Getis-Ord Gi*) under the mapping clusters toolset. The data obtained from the AADK are processed through hot spot analysis (Getis-Ord Gi*). That means there are four hot spot maps that represent two maps of the drug addicted hot spot. These hot spot maps are then compared with the interpolation map to analyze the relationship between drug addicted rates. Figure 1 shows the spatial statistics tools for hot spot analysis.

4 RESULTS

GIS analysis in the two selected states of high-risk areas, namely Kuala Lumpur and Selangor, has identified several hot spot areas as illustrated in Figures 2, 3, and 4. Geographically, Selangor state is divided into 10 districts. Most of the districts have reported drug clients of between 1 and 100 persons. The highest districts of Klang, Ampang Jaya, and Damansara have reported 101– 200 clients. However, in Kuala Lumpur, there are seven districts in this metropolitan city of the country. As this is the metropolitan city, it is not surprising to identify an average of 501–600 clients in almost every district in the whole state. The data imply we could easily find 501–600 drug addicts in the cities of Bandar Kuala Lumpur, Ampang, Batu, Setapak, Hulu Kelang, Petaling, and Cheras.

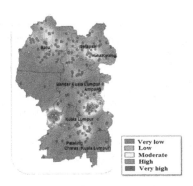

Figure 2. Hot spot areas in Kuala Lumpur (Cheras, Ampang, Petaling Jaya).

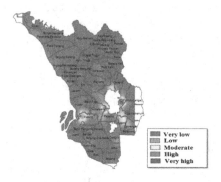

Figure 3. Hot spot areas in Selangor (Klang, Ampang Jaya, Damansara).

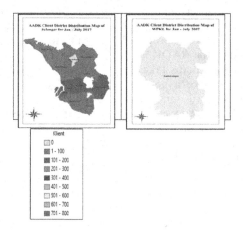

Figure 4. Drug addict distribution areas in Kuala Lumpur and Selangor.

5 DISCUSSION

The identification of the distribution of high-risk areas in this study will be very helpful for the relevant authorities to take action. For example, police officers will have a better idea of where to monitor the drug-related activities and drug addicts. The National Anti-Drug Agency will be able to plan for more relevant substance abuse prevention activities in the particular districts. The government will also be able to determine where to set up the Cure and Care Clinics in the country.

According to Sorensen (1997), many law enforcement departments have conducted crime analysis using maps, including detecting crimes caused by substance abusers. For example, Chikumba (2009) used GIS to produce a map helping drug officers obtain information about environmental factors and location of existing facilities and acting as an information tool through the integration of spatial and drug logistic data. Brouwer et al. (2006) agreed that environmental surroundings and geographic boundaries affect the scene of substance abuse, and in their study they used GIS to analyze the effect of environmental factors such as migration, neighborhood characteristics, and proximity that contribute to substance abuse.

Certain research studies applied GIS to analyze the distance of the treatment center. In a study by Guerrero et al. (2011), the GIS approach was used to map the distribution of Latino residents and calculate the distance between a census block and facilities offering services. Guerrero et al. (2013) used GIS to determine the travel distance from substance abuse outpatients to treatment facilities with services. They also analyzed the data to identify the

12

proximity of hot spot areas and clusters of census tracts with high-density populations to the treatment services using network analysis and autocorrelation analysis. In another study, Hunt et al. (2008) developed a deterrence model based on GIS compromising the information of dealer locations and distribution of dealers in the market and the relationship between dealers and location and crimes, thus identifying the target of enforcement and planning suitable strategies. This model suggested reducing the violence related to street dealing. Through developing GIS content, all the related information assists decision makers to establish proper planning in overcoming the problem as studied by Geanuracos et al. (2007), who developed a database of demographic, health, and crime data to access the transition of HIV disease. In addition, a study by Davidson et al. (2011) also used the GIS approach by developing a database and generating a map containing collected data to improve needle exchange service delivery. Further, Zhu et al. (2006) found that there is a significant link between hot spot areas of substance abuse and criminal problems. Besides these, there were a few studies on substance abuse that attempted to identify the risk areas of substance abuse such as research conducted by Mendoza et al. (2013) that used GIS to determine the risk areas. The results showed that there were risk and neighborhood factors associated with substance abuse treatment. The risk factors can be determined using GIS through ecological parameters analyses (Daash et al., 2009). Because drugs have been considered the numbor one enemy in the country, the use of GIS will to some extent be able to provide solutions to overcome the related problems.

6 CONCLUSION

This analysis revealed that mostly "very high" hot spot areas are located in the capital city for each of the states such as WP Kuala Lumpur (Ampang, Cheras) and Selangor (Klang, Damansara, Ampang, Jaya). This specific data analysis related to drug abuse can be used by local authorities to objectively focus on these areas to intensify drug prevention programs such as *Perangi Dadah Habis-habisan* (PDH) or Fighting the Drugs initiatives to reduce the number of drug abuse cases in these six high-prevalence areas. Hence, the authors believe that the geospatial data obtained in this study will benefit many parties if the data are used wisely to address substance abuse issues and challenges. As an effort to eradicate substance abuse problems, GIS can be modeled to help police, officers of the National Anti-Drug Agency, and stakeholders involved in identifying hot spot and risk areas, thus assisting in planning solutions for the problems.

ACKNOWLEDGMENTS

This study was funded by the External Research Grant from *Agensi Antidadah Kebangsaan Malaysia* and the Institute of Research Management and innovation, Universiti Teknologi MARA (Kod Projek:100-IRMI/GOV 16/6/2 (011/2017).

REFERENCES

Berke, E. M. 2010. Geographic information systems (GIS): Recognizing the importance of place in primary care research and practice. *The Journal of the American Board of Family Medicine*, 23(1):9–12.

Botto, C., Escalona, E., Vivas-Martinez, S., Behm, V., Delgado, L., & Coronel, P. 2005. Geographical patterns of onchocerciasis in southern Venezuela: Relationships between environment and infection prevalence. *Parassitologia*, 47(1):145–150.

Brouwer, K. C., Weeks, J. R., Lozada, R., & Strathdee, A. 2006. Integrating GIS into the study of contextual factors affecting injection drug use along the Mexico/US Border. In: Y. F. Thomas, D. Richardson, & I. Cheung (Eds.), *Geography and Drug Addiction* (pp. 27–42). Amsterdam: Springer.

Brownstein, J. S., Green, T. C., Cassidy, T. A., & Butler, S. F. 2010. Geographic informations systems and pharmacoepidemiology: Using spatial cluster detection to monitor local patterns of prescription opioid abuse. *Pharmacoepidemiology Drug Safety*, 19(6):627–637.

Buxton, J. A., Preston, E. C., Mak, S., Harvard, S., & Barley, J. 2008.Harm reduction strategies and services communities, more than just needles: An evidence-informed approach to enhancing harm reduction supply distribution in British Columbia. *Harm Reduction Journal*, 5:37. https://doi.org/ 10.1186/14777517-5-37

Chawarski, M. C., Vicknasingam, B., Mazlan, M., & Schottenfeld, R. S. 2012. Lifetime ATS use and increased HIV risk among not-in-treatment opiate injectors in Malaysia. *Drug and Alcohol Dependence*, 124(1–2):177–180.

Chikumba, P. A. 2009. Application of geographic information system (GIS) in drug logistics management information system (LMIS) at district level in Malawi: Opportunities and challenges. In *The International Conference on Einfrastructure and E-Services for Developing Countries*, 38 (pp. 105–115). Berlin: Springer.

Choi, P., Kavasery, R., Desai, M. M., Govindasamy, S., Kamarulzaman, A., & Altice, F. L. 2010. Prevalence and correlates of community re-entry challenges faced by HIV-infected male prisoners in Malaysia. *International Journal of STD and AIDS*, 21(6):416–423.

Daash, A., Srivastava, A., Nagpal, B. N., Saxena, R., & Gupta, S. K. 2009. Geographical information system (GIS) in decisión support to control malaria: A case study of Koraput district in Orissa. *India Journal Vector Borne Diseases*, 46(1):72–74.

Davidson, P. J., Scholar, S., & Howe, M. 2011. A GIS-based methodology for improving needle exchange service delivery. *International Journal of Drug Policy*, 22(2):140–144.

Devi, J. P., Azriani, A. B., Mohd, Z. W., Ariff, M. N. M., & Hashimah, A. N. 2012. The effectiveness of methadone maintenance therapy among opiate-dependants registered with Hospital Raja Perempuan Zainab II Kota Bharu, Kelantan. *Malaysian Journal of Medical Sciences*, 19(4):17–22.

Esri. 2018. *Work Smarter with ArcGIS*. Retrieved from http://www.esri.com/arcgis/about-arcgis.

Geanuracos, C. G., Cunningham, S. D., Weiss, G., Forte, D., Reid, L. M. H., & Ellen, J. M. 2007. Use of geographic information systems for planning HIV prevention interventions for high-risk youths. *American Journal of Public Health*, 97(11):1974–1981.

Gill, J. S., Rashid, R. A., Hui, K. O., & Jawan, R. 2010. History of illicit drug use in Malaysia: A review. *International Journal of Addiction Sciences*, 1(1):1–6.

Green, T. C., & Pope, C. 2008. Using a GIS framework to assess hurricane recovery needs of substance abuse center clients in Katrina- and Rita-affected areas. In Y. F. Thomas D. Richardson, & I. Cheung (Eds.), *Geography and Drug Addiction* (pp. 369–393). Amsterdam: Springer.

Guerrero, E. G., Pan, K. B., Curtis, A., & Lizano, E. L. 2011. Availability of substance abuse treatment services in Spanish: A GIS analysis of Latino communities in Los Angeles County, California. *Substance Abuse Treatment, Prevention, and Policy*, 6:21. https://doi.org/10.1186/1747-597X-6-21

Guerrero, E. G., Kao, D., & Perron, B. E. 2013. Travel distance to outpatient substance use disorder treatment facilities for Spanish-speaking clients. *International Journal Drug Policy*, 24(1):38–45.

Humensky, J. L. 2010. Are adolescents with high socioeconomic status more likely to engage in alcohol and illicit drug use in early adulthood? *Substance Abuse Treatment, Prevention, and Policy*, 5(1):1–10.

Hunt, E. D., Sumner, M., Scholten, T. J., & Frabutt, J. M. 2008. Using GIS to identify drug markets and reduce drug-related violence a data-driven strategy to implement a focused deterrence model and understand the elements of drug markets. In: Y. F. Thomas, D. Richardson, & I. Cheung (Eds.), *Geography and Drug Addiction* (pp. 395–413). Amsterdam: Springer.

Kenu, E., Ganu, V., Calys-Tagoe, B. N. L., Yiran, G. A. B., Lartey, M., & Richard, A. 2014. Application of geographical information system (GIS) technology in the control of Buruli ulcer in Ghana. BMC. *Public Health*, 14:724. https://doi.org/10.1186/1471-2458-14-724

Latkin, C., Glass, G. E., & Duncan T. 1998. Using geographic information systems to assess spatial patterns of drug use, selection bias and attrition among a sample of injection drug users. *Drug and Alcohol Dependence*, 50(2):167–175.

Maehira, Y., Chowdhury, E. I., Reza, M., et al. 2013. Factors associated with relapse into drug use among male and female attendees of a three-month drug detoxification–rehabilitation programme in Dhaka, Bangladesh: A prospective cohort study. *Harm Reduction Journal*, 10(1):1–13.

Mason, M., Cheung, I., & Walker, L. 2004. Substance use, social networks, and the geography of urban adolescents. *Substance Use and Misuse*, 39(10–12):1751–1777.

Masood, S. & Us Sahar, N. 2014. An exploratory research on the role of family in youth's drug addiction. *Health Psychology and Behavioral Medicine*, 2(1):820–883.

McLafferty, S. 2008. Placing substance abuse: Geographical perspectives on substance use and addiction. In Y. F. Thomas D. Richardson, & I. Cheung (Eds.), *Geography and Drug Addiction* (pp. 1–16). Amsterdam: Springer.

Mendoza, N. S., Conrow, L., Baldwin, A., & Booth, J. 2013. Using GIS to describe risk and neighbourhood-level factors associated with substance abuse treatment outcomes. *Journal of Community Psychology*, 41(7): 799–810.

Rather, Y. H., Bashir, W., Sheikh, A. A., Amin, M., & Zahgeer, Y. A. 2013. Socio-demographic and clinical profile of substance abusers attending a regional drug de-addiction centre in chronic conflict area: Kashmir, India. *Malaysian Journal of Medical Sciences*, 20(3): 31–38.

Rusdi, A. R., Noor, Z. M. H. R., Muhammad, M. A. Z., & Mohamad, H. H. 2008. A fifty-year challenge in managing drug. *Journal of the University of Malaya Medical Centre*, 11(1): 3–6.

Sanders, L. J., Aguilar, G. D., & Bacon, C. J. 2013. A spatial analysis of the geographic distribution of musculoskeletal and general practice healthcare clinics in Auckland, New Zealand. *Applied Geography*, 44: 69–78.

Sorensen, S. L. 1997. Smart mapping for law enforcement settings: Integrating GIS and GPS for dynamic, near-real time applications and analyses. In W. David & M. E. Tom(Eds.), *Crime Mapping and Crime Prevention* (pp. 349–378). New York: Criminal Justice Press.

Srivastava, A., Nagpal, B. N., Joshi, P. L., Paliwal, J. C., & Dash, A. P. 2009. Identification of malaria hot spots for focused intervention in tribal state of India: A GIS based approach. *International Journal of Health Geographics*, 8: 30. https://doi.org/ 10.1186/1476-072X–8–30

Toriman, M. E., Abdullah, S. N. F., Azizan, I. A., Kamarudin, M. K. A., Umar, R., & Mohamad, N. 2015. Spatial and temporal assessment on drug addiction using multivariate analysis and GIS. *Malaysian Journal of Analytical Sciences*, 19(6): 1361–1373.

UNODC. 2014. *World Drug Report, 2014*. Vienna: United Nations. Retrieved from: http://www.unodc.org/documents/wdr2014/World_Drug_Report_2014_web.pdf

Wu, L. T. 2010. Substance abuse and rehabilitation: Responding to the global burden of diseases attributable to substance abuse. *Substance Abuse and Rehabilitation*, 1: 5–11.

Yusoff, F., Sahril, N., Rasidi, N. M., Zaki, N. A. M., Muhamad, N., & Ahmad, N. 2014. Illicit drug use among school-going adolescents in Malaysia. *Asia-Pacific Journal of Public Health*, 26(5S): 100S–107S.

Zhou, Y. B., Liang, S., Wang, Q. X., et al. 2014. The geographic distribution patterns of HIV-, HCV- and co-infections among drug users in a national methadone maintenance treatment program in Southwest China. *BMC Infectious Diseases*, 14: 134. https://doi.org/10.1186/1471-2334-14-134

Zhu, L., Gorman, D.M., & Horel, S. 2006. Hierarchical Bayesian spatial models for alcohol availability, drug "hot spots" and violent crime. *International Journal of Health Geographics*, 5: 54. https:///doi.org/10.1186/1476

Inclusive Development of Society – Lumban Gaol (eds)
© 2020 Taylor & Francis Group, London, ISBN 978-1-138-33476-2

Impact of the external environment risk factors on organizations engaged in transport infrastructure construction

N.V. Kapustina, D.A. Macheret & Z.P. Mejokh
Russian University of Transport (RUT - MIIT), Moscow, Russian Federation

I.V. Frolova
Southern Federal University, Rostov-on-Don, Russian Federation

A.P. Kolyadin
Pyatigorsk State University, Pyatigorsk, Russian Federation

ABSTRACT: Searching for and substantiation of mechanisms and tools for transition to a sustainable development of domestic construction companies, involved in the transport infrastructure construction, under conditions of the transport infrastructure upgrade, predefines the relevance of the issue under study. The purpose of the study is to develop a methodological tool for assessing the impact of external risk factors on the activity of organizations, involved in the transport infrastructure construction, with due consideration of specific features of the current business environment. During the study, logical, statistical, sociological and simulation methods were applied, in particular: a) expert evaluation method; b) correlation analysis; c) regression analysis; d) comparison method. The obtained results of assessment of external risk factors for organizations, involved in the transport infrastructure construction, have shown that even those construction companies, which have sufficient potential at present, in the nearest future may be exposed to unfavorable external environment risk factors, if they do not increase investments in fixed assets and will not implement technological innovations. Identify external risk factors of the business environment for organizations; select indicators, characterizing external factors of the business environment of organizations; Identify key external risk factors for the business environment of organizations, involved in the transport infrastructure construction was developed during the study. Therefore, searching for tools, facilitating transition to sustainable development and enhancement of business efficiency of their operations, apparently becomes a priority task for these companies, allowing them to create an environment, favorable for continued development based on continuous improvements in engineering and technology, the use of innovative raw materials and quality improvement as well as for development of the competence of employees and organizations in general. The methodology proposed for assessing external risk factors of organizations, involved in the transport infrastructure construction, can be used to formulate various organization development vectors.

1 INTRODUCTION

The relevance of the search for and justification of mechanisms and tools for transition to the sustainable development of domestic construction companies, involved in the transport infrastructure construction, is defined by new business environment. The specific feature of today's business environment is the upgrade of the existing transport infrastructure both in our country and globally, in particular, due to the development of more high-tech and high-speed transport.

The radical change in the business environment of companies, involved in the transport infrastructure construction, is caused by the emergence of a new technological order in the transport system development, which resulted in the transformation of the transport economy and management mechanisms, including management mechanisms of construction companies, involved in this sector. Therefore, the present and future development of the Russian transport infrastructure depends on the pace of integration of construction companies in the modern format of the economy.

Issues, related to the development of theories and methodologies for assessing the impact of external risk factors on the business activity of organizations, were discussed in publications released by such authors as V. Avdiysky and V. Bezdenezhnykh (2013), V. Kossov (2009), M. Krui, D. Galai, R. Mark (2011), B. Fliveborg, N. Bruzelius, V. Rothengatter (2014), A. Shapkin (2005), N. Kapustina (2014), R. Fedosova (2014), R. Kachalova (2012), Prince Boateng (2014), Robert W. Poole, Jr. and Peter Samuel (2011).

A lot of publications was devoted to the transport infrastructure development and efficiency assurance - in particular, those issued by A.V. Ryshkov (2010), A.Yu. Beloglazov (2010), K. V. Zakharova (2010), I. Kirtsner (2010), B.M. Lapidus (2008), Yu.V. Yelizaryeva (2008), M. Yu. Sokolova (2016, 2017), B.A. Lyovin (2016), Macheret D.A. (2015), I.M. Mogilevkin (2010), A.A. Walters (2004), R. Vogel (1964), S.M. Grant-Mueller (2001), E. Quinet (1998).

During the study, logical, statistical, sociological, modeling methods were used, in particular: a) expert evaluation method; b) correlation analysis; c) regression analysis; d) comparison method.

Taking into account the above-mentioned circumstances, the objective of the study is to develop methodological tools for assessing the impact of external risk factors on business activities of organizations, involved in the transport infrastructure construction, taking into account specific features of the current business environment.

The following working hypothesis was assumed: development and economic efficiency of organizations, involved in the transport infrastructure construction, depend on a number of external environment factors, basically associated with the implementation of innovations and indicators characterizing living standard of the population.

This working hypothesis was put forward since citizens are the main users of transport infrastructure facilities, which use these facilities both in commercial and personal purposes, and the growth in the scope of construction of new infrastructure transport facilities and the refurbishment of obsolete facilities is associated with the need to implement innovative management & service methods in the transport sector. Thus, the pace of growth in the transport infrastructure construction depends on quantitative and qualitative indicators, which characterize citizens and the innovative activity of entrepreneurs in the country.

Stability, available potential for sustainable growth and efficiency improvement, active implementation of advanced innovative production and management technologies are the key to the sustainable success of organizations, involved in the transport infrastructure construction in the current business framework.

To assess external risk factors for organizations, involved in the transport infrastructure construction, an attempt was made to develop a model for integral assessment of external risk factors for organizations, involved in the transport infrastructure construction.

External business environment risk factors for organizations, involved in the transport infrastructure construction, were assessed by quantitative evaluation methods. In terms of structure and composition, the external environment of organizations, involved in the transport infrastructure construction, is rather heterogeneous and numerous and can include a great number of factors and components, which, at a different degree, with different characteristics and frequency, affect economic and organizational efficiency of these organizations.

2 RESEARCH METHODOLOGY

In accordance with a system approach, a methodical approach to evaluating external environment risk factors for enterprises, involved in the transport infrastructure construction, was developed during the study.

Identifying external risk factors, affecting the state of the business environment in the external environment, is a very important task in the assessment of external risk factors. The correlation analysis method was applied to analyze external risk factors of the business environment. The task was to assess the degree of influence of various external risk factors on enterprises, involved in the transport infrastructure construction. Political, economic, socio-demographic, technological, information, innovative and environmental issues, affecting the activity of organizations, involved in the transport infrastructure construction, were analyzed.

The assessment of external risk factors for organizations, involved in the transport infrastructure construction, includes the resolving of the following tasks:

- Applying the expert evaluation method, identify external risk factors of the business environment for organizations, involved in the transport infrastructure construction;
- Applying the expert evaluation method, select indicators, characterizing external factors of the business environment of organizations, involved in the transport infrastructure construction;
- Select the resulting indicator;
- Acquire data on previously defined indicators and provide initial data processing;
- Develop a correlation model for analyzing data obtained by a correlation analysis, and identify the dependencies;
- Identify key external risk factors for the business environment of organizations, involved in the transport infrastructure construction;
- Construct a regression model and check it for representativeness;
- Analyze the regression model.

3 METHODS OF TESTING AND ANALYSIS OF RESULTS

Having defined external risk factors of the business environment for organizations, involved in the transport infrastructure construction, experts selected characterizing indicators. Risk factor groups, which significant affect the efficiency of operations of organizations, involved in the transport infrastructure construction, are as follows:

Socio-demographic risk factor group;
- Economic risk factor group;
- Technological risk factor group;
- Political risk factor group;
- Innovative risk factor group;
- Environmental risk factor group.

To assess socio-demographic external risk factors, such indicators as the population employment level; unemployment rate; number of personnel engaged in research; number of units in organizations, implementing technological innovations; number of post-graduates with defended thesis; population economic activity level; minimum wage rate; number of citizens, whose income is below the subsistence minimum level; population morbidity; occupational injuries.

The indicators, characterizing external economic risk factors of the business environment, include the IEF global competitiveness index; investments in intellectual properties; investments in fixed assets; R&D expenses; capital-to-labor ratio variation indices; capital productivity ratio variation indices; education expenses (from all sources of funding to all levels of education); increase in high-performance jobs; coefficient of fixed assets renewal in the Russian Federation by types of the economic activity.

As for the technological risk factors, affecting the efficiency of operations of organizations, involved in the transport infrastructure construction, the following indicators were suggested for use: global innovation index; innovative activity of organizations; specific weight of organizations, implementing technological innovations, in the total number of organizations surveyed; specific weight of organizations, implementing managerial innovations; technological innovation costs; number of issued patents for industrial designs.

To evaluate political risk factors, the following indicators were selected: current intellectual property protection laws (normative and legal acts, regulating intellectual and innovative activities at the federal level); state administration quality index; federal target programs to support innovations.

Innovative risk factors, affecting operations of organizations, involved in the transport infrastructure construction, are as follows: state expenditures on education; publishing activity in the Web of Science international database; participation of organizations in joint innovation projects; national share in the international market for high-tech products; technical achievements; new Russian technologies (technical achievements), transferred to foreign organizations; exports of innovative goods and services.

Key indicators, characterizing environmental risk factors of the surveyed organizations, are as follows: environmental protection expenditures; ratio of environmental protection expenditures to GDP; production and consumption waste generation; funds (claims) and fines, recovered to compensation for damage caused due to environmental law violations.

As a resultant indicator, the following indicator was chosen: commissioning of fixed assets of the transport infrastructure in the Russian Federation. Table 1 presents data on the resultant indicator for 2005-2016.

The statistical data show that the commissioning of fixed assets in the Russian transport infrastructure tended to decline in 2010, 2013 and 2015; then, in 2016, there was a rise, but it failed to reach the 2012 level, when the maximum scope of commissioning of fixed assets in the Russian transport and communication infrastructure was registered.

Regarding socially demographic external risk factors for organizations, involved in the transport infrastructure construction and upgrade, it should be noted that the external environment is gradually becoming more favorable, but the pace of change is rather insignificant. E.g., unemployment is steadily declining, however, population's cash incomes tend to decrease slightly. Data on technological risk factors indicate the growth in the scope of own-produced innovative products and developed advanced production technologies. The growth of these indicators confirms the efficiency of growing costs for technological innovations. However, it is worth noting a decrease in the patent issuance in 2016, a decline in the share of organizations, implementing managerial and technological innovations, as well as a drop in the global innovation index and innovative activity of organizations. Data, characterizing economic risk factors, affecting business activities of construction companies, show that existing trends are generally positive, but the pace of growth is still rather insignificant. Such indicators as investments in fixed assets and the R&D costs grow most steadily. However, other indicators, e.g., capital productivity and capital intensity variation indices, tend to decline.

With regard to political risk factors of the business environment for organizations, involved in the transport infrastructure construction, it should be noted that the situation is not entirely

Table 1. Commissioning of fixed assets in the RF transport and communication infrastructure, billion rubles (Federal State Statistics Service).

Year	Indicator value	Rate of growth, %
2005	630.3	-
2006	695.6	110,36
2007	813.2	116,91
2008	1169.9	143,86
2009	1374.8	117,51
2010	1084.2	78,86
2011	2117.6	195,31
2012	2641.2	124,73
2013	2228.2	84,36
2014	2319.9	104,12
2015	1909.5	82,31
2016	2544.2	133,24

favorable. The number of economy development institutions, established by the state, has dropped nearly 18-fold times to zero and remains almost at the same level to date. It's worth noting that the scope of legal documents, regulating intellectual property rights (regulations, governing intellectual activity at the federal and regional levels), has grown several times over the period of the study, whilst federal target innovation support programs remain at the same level for several years.

The situation with environmental risk factors, affecting business environment quality in organizations, involved in the transport infrastructure construction, should also be assessed as not very favorable. The environmental protection expenses, expressed in percentage of GDP, are on decline, although the total environmental protection expenses are on rise. Production and consumption waste volumes as well as funds (claims) and penalties, levied to compensate for damage caused by nature protection legislation violations, tend to grow. This fact confirms a growing number of situations and the necessity of improving the environmental legislation.

All data collected on the business environment indicators were processed in special software product.

The analysis of pair correlation results showed that the rate of fixed assets commissioning in Russian transport and communication infrastructure, is at the highest level, which is associated with the increase (decrease) in the following factors:

- Population life expectancy (0.909);
- Technological innovation expenses (0.905);
- Investments in fixed assets (0.926).

From the correlation analysis results it follows that the reduction in these factors as well as unstable dynamics of these factors, adversely affecting the commissioning of fixed assets of the transport and communications infrastructure, are the main risk factors, affecting quality of the existing business environment in companies, involved in the transport infrastructure construction.

As a result, it should be noted that the working hypothesis, assuming that the development and economic efficiency of organizations, involved in the transport infrastructure construction, is affected by a number of external environment factors, basically associated with the implementation of innovations and indicators, characterizing living standards of the population, was put forward.

To confirming the suggested working hypothesis, a regression model was developed.

The analysis of risk factors, affecting the development and economic efficiency of organizations, involved in the transport infrastructure construction, with the use of this regression model, allowed us the main external environment risk factors and construct a real model, expressed by the formula (1):

$$Y = -874.34 + 13.343X_1 + 0.407X_2 + 0.131X_3, \tag{1}$$

Where X_1– population life expectancy;
X_2 – technological innovation expenses;
X_3 – investments in fixed assets.

The regression model, characterizing mutual influence of the above-entioned risk factors for organizations, involved in the transport infrastructure construction, includes three key factors which characterize risk environment of the companies under study, and confirms that the growth of these factors positively impacts on the development and economic efficiency of the organizations surveyed.

At the same time, the growth of all the above-mentioned indicators should positively influence the commissioning of fixed assets of the Russian transport and communication infrastructure, which confirms the working hypothesis suggested.

4 CONCLUSIONS

Our analysis results confirms that currently the main risk factors do not adversely affect the development and economic efficiency of the organizations, involved in the Russian transport infrastructure construction. However in the future the situation may change and, if the trend for key indicators changes, their efficiency will decline. However, it's possible that new risk factors, more strongly affecting activities of organizations, involved in the transport infrastructure construction, may emerge.

Thus, the assessment of external risk factors for organizations, involved in the transport infrastructure construction, shows that business environment today and in the nearest future is not very favorable even for construction companies with a current sound potential, unless they build up investments in fixed assets and implement technological innovations. Therefore, these companies apparently have to search for tools, facilitating the transition to sustainable development and the improvement of the economic efficiency of the company operations, which allow the creation of an environment, favorable for a continuous development based on steady improvements in engineering and technology, as well as for the use of innovative feedstock and materials, quality improvement, development of the competence of employees and organization as a whole.

REFERENCES

Avdiyskiy, V.I. & Bezdenezhnykh, V.M. 2013. *Risks of Business Entities: Theoretical Fundamentals, Analysis, Forecasting and Management Methodology: Teaching Manual.* Moscow: Alfa-M: NITs INFRA-M.

Balabanov, I.T. 1996. *Risk Management.* Moscow: Finansy i Statistika.

Grant-Muller, S.M., Mackie, P.J., Nellthorp, J., Pearman, A.D. (2001). *Economic appraisal of European transport projects the state-of-the-art revisited.* Transport Reviews, 21(2), 237–261.

Fliveborg B., Bruzelius N., Rothengatter V. 2014. *Megaprojects and Risks: Anatomy of Ambitions* — M.: Alpina Publisher.

Kossov, V.V. 2009. *Fundamentals of Innovation Management.* Moscow: Magistr.

Kapustina, N.V. & Fedosova, R.N. 2014. *Managing the Development of an Innovative Organization Based on Risk Management.* Voprosy Ekonomiki I Prava, no. 9: 46–49.

Kachalov R.M. (2012) *Economic Risk Management: Theoretical Foundations and Applications: monograph.* Petersburg: Nestor-history, 2012, 248.

Kirtsner I. 2010. *Competition and entrepreneurship / translation from English.* - Chelyabinsk: Sotsium. - XIV+272p.

Lapidus B.M., Macheret D.A. 2015. *Modern problems of development and railway transport's reform //* Vestnik VNIIZhT. 6. pp.3–8.

Lyovin B.A. 2016/17. *Creation the Russian University of transport is a stage in the evolution of industry education//*Transport strategy - XXI century. 35 (4), p.84

Macheret D.A., Ryshkov A.V., Beloglazov A.Yu., Zakharov K.V. 2010. *Macroeconomic assessment of transport infrastructure development//*Vestnik VNIIZhT. 5. pp.3–10.

Mogilevkin I.M. 2010. *Global infrastructure: a mechanism for moving to the future/*IMEMO RAN. Moscow: Magistr, - 317p. (rus).

Sokolov M.Yu. 2017. *A new stage in the development of industry transport education //* Transport strategy - XXI century. No 35 (4). pp.80–81 (rus).

Shapkin A.S. 2005. *Theory of Risk and Modeling of Risk Situations.* - Moscow: Dashkov and Co publisher,– pp.30.

Prince Boateng. 2014. *A dynamic systems approach to risk assessment in megaproject /* A thesis submitted in partial fulfilment of the requirements for the degree of Doctor of Philosophy / Royal Academy of Engineering Centre of Excellence in Sustainable Building Design, School of Energy, Geoscience, Infrastructure and Society, Heriot-Watt University, Edinburgh, UK.

Quinet E. *Principles of transport economy.* Economica, Paris, 1998.

Robert W. Poole, Jr. and Peter Samuel. 2011. *Transportation Mega-Projects and Risk/*Reason Foundation Policy Brief 97.

Vogel R.W. 1964. *Railroads and American Economic Growth: Essays in Econometric History.* John Hopkins University Press.

Walters A.A. 2004. *Excessive consumption (overload). In Ekonomicheskaya teoriya.* Moscow: INFRA-M, pp.157–166.

Inclusive Development of Society – Lumban Gaol (eds)
© 2020 Taylor & Francis Group, London, ISBN 978-1-138-33476-2

Assessment of a sustainable development potential of printing companies in the digital economy environment

R.N. Baykina (Fedosova) & A.L. Lisovsky
Financial University under the Government of the Russian Federation, Moscow, Russian Federation

A.A. Yussuf
Moscow Witte University, Moscow, Russian Federation

ABSTRACT: This article substantiates the necessity of a sustainable development potential of printing companies in the digital economy environment. A methodology for assessing the sustainable development level of companies, which was tested on a group of printing companies demonstrating their potential for sustainable development, has been suggested. The methodology developed for identifying steady enterprises with a sustainable development potential was tested based on the balance sheet data of printing companies. The methodology suggested for analyzing and evaluating the level of the company's sustainable development potential allows identification of stable enterprises with high chances for providing effective operations in the business environment characterized by dynamic technological changes. It was revealed that for the transition to sustainable development, the printing industry enterprises should achieve high resource potential indicators.The results of this study may lay the foundation for further studies of urgent issues related to the analysis and evaluation of a sustainable development potential of printing companies.

1 INTRODUCTION

Today, printing companies are facing a typical number of problems, related to both management and production. Unstable financial positions, technological backwardness of production, and often a lack of understanding of business features of printing work customers are some of the main problems. All of these destabilize a company's operations and undermines its competitiveness. This dramatic situation is aggravated by global digitalization. A failure to be in tune with the need for today's printing companies to use their potential in full therefore forces them to struggle for survival instead of following the sustainable development path.

A successful transition of modern organizations from a socioeconomic development to a sustainable development is undoubtedly determined by the external and internal environments in which they operate. If the business environment is common to all enterprises of the printing industry, the internal environment is rather company specific, adaptive, and more turbulent. In the authors' opinion, the quality of tangible and intangible factors of the internal environment predefines the availability and level of the company's development potential. In analyzing and diagnosing the sustainable development potential, we suggest using a resultant approach rather than a nonproviding one (Aksenov, 2017).

2 RESEARCH METHODOLOGY

To assess the internal environment (tangible and intangible factors), a methodology for assessing the sustainable development level of companies, which was tested on a group of printing companies demonstrating their potential for sustainable development, has been suggested.

It is advisable to evaluate the internal environment each year. Therefore, companies that have not been included in a group of steadily developing enterprises based on the results of multiyear assessment of their operations may be recognized as successful based on one-year operating results. Thus it is very important to identify this state in a timely manner and ensure a target use of management tools that allows the positive trend to be strengthened.

The authors of this study suggest the following step-by-step procedure for analyzing and diagnosing the sustainable development potential level of companies under digital transformation conditions:

- Identify indicators and their ratios for characterizing the sustainable development potential (Table 1).
- Define and characterize sustainable development potential levels of industrial enterprises (Table 2).
- Calculate indicator correlations.
- Interpret the results obtained.
- Issue recommendations for developing tools for providing sustainable development of the enterprise.

A company's balance sheet is an indispensable document, being a source of information for analyzing the sustainable development potential of the company. It reflects resulting

Table 1. Ratios for diagnosing the level of sustainable development potential of a company[a].

Item	Reference ratio	Legend
1	$(C+StFI) \geq AP$	C – cash StFI – short-term financial investments AP – accounts payable
2	$StR \geq StL$	StR – short-term receivables StL – short-term liabilities
3	$(SsA+LtR+OCA) \geq$ $(LtL+DI+PfLC)$	SsA – slow-selling assets LtR – long-term receivables OCA – other current assets LtL – long-term liabilities DI – deferred income PfLC – provisions for liabilities and charges
4	$NcA \leq (SE+P)$	NcA – noncurrent assets SE – stockholder equity P – provisions

Compiled by the authors.

[a] For calculations, data from the Company's Balance Sheet Form 1 for the year under study are used.

Table 2. Description of the company sustainable development potential levels.

Level	Attributes
High	All inequalities match reference values
Above average	Partial correspondence (3 of 4)
Average	Partial correspondence (2 of 4)
Low	Partial correspondence (1 of 4)
No development potential	Full mismatch of all inequations to reference values

Compiled by the authors.

indicators of the evaluation of available consolidated local potentials of the enterprise (economic, social, and environmental components) that are optimally used in the company operations, as well as hidden unused opportunities, which under the impact of the digital environment will be able to build up the company's competence in the information technologies (Kuntsman, 2016) and ensure its transition to sustainable development in the digital economy environment.

The potential level for the company's sustainable development is defined by the number of inequalities, which correspond to the reference values presented in Table 1. Table 2 describes parameters for evaluating the availability of the sustainable development potential.

3 STUDY RESULTS

The methodology developed for identifying steady enterprises with a sustainable development potential was tested based on the balance sheet data (For honest business 2018) of 31 printing companies.

To classify groups of companies by their sustainable development potential level, balance sheets of these companies for 2015 were analyzed. The results of verification of all four inequalities of the methodology are presented in Table 3.

Table 3.   Company sustainable development potential evaluation.

Item	Company	Calculation stages			
		1	2	3	4
1	Goznak, JSC	<	>	<	>
2	JSC SPA Kripten	>	>	<	<
3	Aliot, LLC	<	>	<	<
4	NOVAKARD, JSC	<	<	>	>
5	Polygraph-Zaschita SPB	<	>	<	>
6	CONCERN ZNAK, JSC	>	>	<	<
7	ORENCART	<	<	<	>
8	FIRST PRINTED YARD	>	>	>	<
9	FLEXOZNAK	>	>	>	<
10	KT, JSC	<	<	>	<
11	SPA NEOPRINT	<	<	>	<
12	FORMAT, JSC	>	<	>	<
13	ORGA ZELENOGRAD, JSC	>	>	<	<
14	GHP DIRECT RUS	>	<	<	<
15	NOVOKARD	<	<	>	<
16	OREN-CART	<	<	>	>
17	TYUMEN PRINTING HOUSE, OJSC	>	<	>	<
18	N.T. GRAPH	<	<	>	>
19	PLASTIC ON LINE	<	<	<	>
20	EUROCOPY PRINTING HOUSE - 2SPB	>	>	<	<
21	PPD GT	>	>	<	<
22	KBI, CJSC	>	<	>	<
23	ACARD	>	>	>	<
24	SIBPRO, CJSC	>	>	<	<
25	NKS	<	>	<	<
26	LIT, CJSC	>	>	>	<
27	SPRINTEX	<	>	<	<
28	SIBZNAK LTD	<	>	<	<
29	NEOPRINT	<	<	>	<
30	FIRM POLY-CART	>	<	>	<
31	CENTRAL PRINTING HOUSE	<	<	>	<

As a result of the study, it was revealed that in 2015 only 4 (FIRST PRINTED YARD, FLEXOZNAK, ACARD, LIT) of the 31 companies surveyed had the greatest potential for sustainable development (high level). A group of outsiders (lack of potential for sustainable development) includes ORENCART and PLASTIC ON LINE.

Such companies as Aliot, KT, SPA NEOPRINT, NOVOKARD, NEOPRINT, CENTRAL PRINTING HOUSE, SPRINTEX, SIBZNAK LTD, NKS, and GHP DIRECT RUS (approximately 30% of total printing enterprises surveyed) show average performance indicators (only two ratios meet the criteria).

Such companies as SPA Kripten, CONCERN ZNAK, ORGA ZELENOGRAD, FORMAT, TYUMEN PRINTING HOUSE, EUROCOPY PRINTING HOUSE - 2SPB, PPD GT, KBI, SIBPRO, and FIRM POLY-CART are near the leaders. Analysis of balance sheets of these companies showed the inconsistency between slowly selling assets and long-term liabilities (identified for six companies) or between short-term receivables and short-term liabilities (identified for four companies).

Thus, we can conclude that in 2015, the development vector of only four leading enterprises corresponded to the modern digital business environment requirements. For the transition to sustainable development, the printing industry enterprises should achieve high resource potential indicators.

4 CONCLUSIONS

The methodology suggested for analyzing and evaluating the level of the company's sustainable development potential allows identification of stable enterprises with high chances for providing effective operations in the business environment characterized by dynamic technological changes. However, the steady success of a modern organization depends on the choice and adaptation of relevant mechanisms and strategic tools for ensuring sustainable development of printing enterprises, rather than on their existing potential.

The availability of sustainable growth potential, and active implementation of advanced digital production and management technologies are the key to the sustainable success of printing industry companies in today's digital environment.

Timely identification of the sustainable development potential level seems to contribute to the development and implementation of an effective enterprise development strategy in an economy that is based on digital technologies, information, and knowledge. Regular assessment of the resource potential allows verification of the adequacy of the selected management tools with the aim to strengthen positive trends and take corrective actions, if required, in a timely manner.

REFERENCES

Aksenov, P.V. 2017. *Ensuring Sustainable Development of Industrial Enterprise on the Basis of Strategic Competitive Advantages*. Dissertation, Financial University, Moscow.

Analysis of the World's Industry Development Experience and Approaches to the Digital Transformation of the Industry in the EAEC Member States. Information and analytical report. Available at: file:///G:/ЦИФРОВАЯ%20ЭКОНОМИКА/ЦИФРОВАЯ%20ТРАНСФОРМАЦИЯ%20ПРОМЫШЛЕННОСТИ%2013.02.2017.pdf (accessed 23 December 2017).

Chuprov, S.V. 2016. Information resource supply of Finns stability of industrial enterprise: Analysis, limits, forecast. *Proceedings of the 17th All-Russian Symposium*. Moscow, April 12–13,2016 /(pp. 160–163). Under the editorship of corresponding member. Corr. Ran G. B. Kleiner. Moscow: CEMI RAS.

Connect where it counts. Mapping your transformation into a digital economy with GCI 2016 Available at: http://www.huawei.com/minisite/gci/files/gci_2016_whitepaper_en.pdf?v=20170420

For honest business (portal about companies and business in Russia) Available at: https://zachestnyi biznes.ru/ (accessed February 2018).

Kuntsman, A.A. 2016. Transformation of internal and external business environment in the digital economy. *Management of Economic Systems* 11 (93). Available at: https://cyberleninka.ru/article/n/transformatsiya-vnutrenney-i-vneshney-sredy-biznesa-v-usloviyah-tsifrovoy-ekonomiki

Management sustainable development. 2015. In A.V. Trachuk (ed.), SPb. "Publishing house "the Real economy"".

Official site of Federal State Statistics Service 2017. Available at: http://www.gks.ru/ (accessed March 11, 2018).

Russian Printing Industry: Condition, Trends and Prospects for Development. Industry Report, 2016. Available at: www.fapmc.ru/mobile/activities/reports/2016/poligrafiya.html (accessed May 11, 2018).

Survey of the printing services market in Russia, 2016. Available at: https://www.openbusiness.ru/biz/.../obzor-rynka-poligraficheskikh-uslug-v-rossii/ (accessed May 29, 2018).

Sustainable development in Russia. 2013. In S. Bobylev and R. Flights (ed.), Berlin and Saint-Petersburg.

Tsygankov, A.P. From cleaner production to sustainable development. Available at: http://ruscp.ru/ru/aboutus/63-otk (accessed March 11, 2018).

Inclusive Development of Society – Lumban Gaol (eds)
© 2020 Taylor & Francis Group, London, ISBN 978-1-138-33476-2

Consideration of risk management of internet project promotion

Natalia M. Fomenko
National Research Nuclear University MEPhI, Volgodonsk Technical Institute, Volgodonsk, Russian Federation

ABSTRACT: The 21st century is the age of informatization and superiority of automated information technologies. The introduction of new information technology introduction and application of the global internet network affect all areas of individual organizational activity and the economy as a whole. In such circumstances the issue of consideration of risk management of internet project promotion is quite relevant. The objective of this study was to examine models of interaction between participants in the network market and simulate the assessment of the company internet project risk. The modeling process is based on the example of a group of business risks. The article divides the process into groups of risks and individual risks in groups. It estimates the value of losses by expertise (severity of the consequences) for each risk. The value of the loss is modeled by generating random numbers. The relative probability of each individual risk is calculated by a paired comparison method. The total losses are determined based on the simulation of risk values and the probability of occurrence.

1 INTRODUCTION

It has become clear that in the 21st century the competitiveness of an organization largely depends on its ability to respond in a timely manner to changes in the external environment.

Russia is increasingly drawn into the process of globalization under modern conditions. Today most innovative projects are implemented in a network format with the help of information and communication technologies.

One of the features of economic development in advanced countries in the early 21st century is the transition from an industrial economy to a post-industrial one. It happens in the formation of a single global information space based on the Internet. Information processes and virtualization of all spheres of the economy penetrate into all processes of the modern organization, and information management markets are created. As a result of these changes the virtual world begins to function along with the real environment of the organization.

According to the journal *E-commerce*, there were more than 3.8 billion Internet users in the world on January 1, 2018 despite the fact that the first website was launched in August 1991. Sales in e-commerce in 2017 reached $2.29 trillion, and by the end of this year, it is projected to grow to $2.77 trillion. Sales on the Internet should be 8.8% of global retail sales in 2018.

According to some estimates, e-commerce already accounts for more than 1% of Russia GDP, and in recent years Russian lawmakers have been increasingly paying attention to the regulation of online purchases, indirectly confirming their importance for the country's economy.

It becomes obvious that the development of global information and communication technologies led to the formation of a new global electronic environment, the so-called network (or digital) economy, which in turn led to the emergence of new forms of operation and development of organization management systems, including e-business.

The interaction of participants in the market business environment with the use of information and communication technologies, the network market, e-business as well as the risk-management assessment of Internet projects is reflected in the works of Russian scientists: V. Lipaeva (2005),

Alexunin (2006), I. Lukasevich (2006), A. Dolzhenko (2007), V. V. Bugorsky (2008), E. Efimov (2010), K. Adamadziev (2011), V. Dolyavotovsky (2012), G. Hubayeva (2012), and others.

However, the ambiguity of the risk-management assessment of the application of information and communication technologies and the network environment is revealed in the analyzed works, which, in turn does not allow correct estimation of their advantages, new possibilities, and efficiency of application.

E-business is growing rapidly in the economy, that penetrating all areas of human activity very quickly. Currently, e-business is in a stage of intensive growth, which will persist for several years.

The e-business market depends on the participants in electronic relationships. Currently the following models of participant interaction in the network market are distinguished:

B2B (*business-to-business*): e-business systems in which commercial entities (information catalogs, corporate information and broker sites, Internet shop, electronic trading platform, electronic exchanges, auctions, electronic communities) act as subjects of processes.

B2C (*business-to-customer*): e-business systems in which a legal entity acts as the seller, and the buyer is a private (physical) person (electronic intermediation, an online sales and service department, a virtual auction).

B2G (*business-to-government*): e-business systems in which legal entities (enterprises, organizations) and state institutions act as parties to business relations. Special cases: business-to-administration, B2A; consumer-to-administration, C2A; administration-to-administration, A2A.

C2G (*consumer-to-government*) and ***G2G*** (*government-to-government*): e-business models that cover the management of business relations between the state and citizens as well as citizens and state bodies among themselves. They are called upon to make the government easily accessible to the population of the country and provide citizens with access to government documents and selected representatives in government bodies. These models can contain elements of e-business, for example, for collecting taxes, registering vehicles, registering patents, issuing the necessary information, etc.

C2C (*consumer-to-consumer*): e-business systems between individuals where the site acts as an intermediary. Some private individuals can put up for sale items for purchase by other individuals at the electronic auction.

C2B (*customer-to-business*): When consumers make their choice by offering their prices for various goods and services offered by enterprises through the Internet. The seller uses the data of the current demand for the final price decision. The C2B site acts as a broker intermediary that provides a search of the seller according to the price formed by the buyer offers.

E2E (*exchange-to-exchange*): e-business that arose after the widespread use of Internet exchanges. If an application cannot be operated on this exchange it is automatically transferred to another exchange. If there is no desired product or service on that exchange it is transferred further until it is satisfied. Special cases: C2E (consumer-to-exchange); B2E (business-to-exchange).

The interaction of the electronic market subjects can be represented structurally in Figure 1.

The Internet makes it easier for companies to develop new techniques of generating revenue by creating added value to existing products and services or by developing the basis for the production of new products and services.

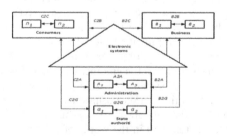

Figure 1.   Structure of the network market subject interaction.

However, like any type of commercial activity, e-business is accompanied by risks. It is obvious that the level of risk situations in e-business far exceeds the risks in traditional business as it is carried out in a highly dynamic and rather unstable environment of the Internet space.

2 METHODOLOGY

Investment activity in any project involves a certain risk and the project can lead to failure. Therefore, when analyzing an investment project risk factors should be taken into account. We should identify as many of types of risks as possible and try to minimize the overall risk of the project. It is necessary to estimate how much the expected income will compensate the perceived risk to make the right investment decision. However, the complexity of this process is that the assessment of the risk of investment is less formal than other methods of evaluation. Nevertheless, risk analysis is necessary and extremely important for investment expertise.

Various methods are used to assess threats and vulnerabilities in domestic and foreign practice. They are based on expert assessments, statistics, and consideration of factors that affect the levels of threats and vulnerabilities.

The most widespread approach is the approach, the main essence of which is to take into account the various factors that affect the levels of threats and vulnerabilities. It allows abstracting from minor technical details, taking into account not only software and technical aspects.

The analysis of problems of information systems risk assessment allows us to draw the following conclusions:

- The creation and implementation of almost any information system is realized under conditions of significant uncertainty that manifests itself in the form of incompleteness or inaccuracy of information about the parameters of the system and the conditions for its implementation.
- Uncertainty accompanies all stages of the information system life cycle, starting with design, implementation, subsequent operation, and modernization.
- It is advisable to perform risk analysis and assessment in at least two stages: design and operation of the information systems.

It is obvious when implementing web-oriented information systems that there is a set of specific risks when the system functions in a network environment.

The initial stage of the project is characterized by setting goals and formulating a requirement. Its main properties are incompleteness and inaccuracy, and the design of systems is subject to change. The analysis stage is based on the share of uncertainty associated with the decisions made on the architecture of the system.

The design phase is associated with the uncertainty in providing functionality and quality characteristics of the system (performance, reliability, availability, integrity, adaptability, integrability, etc.).

The implementation phase forms a relationship between the uncertainty in the quality of the system deployment, the level of user training, and the impact of the implementation of the developed information systems on the organization's business processes.

The uncertainty accompanying the process of design of the information system can lead to the creation of unfavorable situations that hamper the achievement of goals in the subsequent operation of information systems.

Existing methods of identifying risk management, as a rule, are based on the use of checklists followed by an analysis of the decisions made. To do this, apply information on previously developed projects and involve experienced developers. The main areas of risk in these conditions are the aspects that relate to support, technical, and programmatic issues.

The drawbacks of such techniques are the difficulty in working with checklists, which can include up to several hundred positions, and the need to attract the help of experienced developers (domain experts).

The following methods can be used for risk analysis:

- "Brainstorm". This method assumes that ideas for risk analysis are presented without discussion and evaluation.
- Methods of expert evaluation. The Delphi method is a vivid example.
- Sensitivity-based and probabilistic analysis.
- Simulation modeling.
- Utility theory.
- Analysis with the help of decision trees, and so on.

The disadvantages of the proposed options for risk analysis lie in the absence of a unified methodology for determining which allows integration of qualitative and quantitative approaches to their risks.

When planning responses for leading risks and threats, a responsible executor is identified for their reduction, and activities and key indicators are planned.

In the general case of risk management the following steps are to be performed at each stage of the lifecycle of the information systems:

Step 1. Forming a list of possible risks
Step 2. Ranking of risks by the degree of their influence
Step 3. Forming actions to reduce the impact of the most dangerous threats
Step 4. Assessing the response to risks
Step 5. If the iteration is complete stop control; otherwise go to step 2.

Identifying the risks of web-based information systems is expedient to perform at two main stages: at the stage of its creation and the stage of its operation.

Analysis of the literature made it possible to identify the following risk groups from the whole possible set of web-oriented information systems: *project risks* is at the stage of information systems creation (*PR*); *technological risks* (TR), *business risks* (BR), and *juridical risks* (JR) are at the stage of operation (Table 1).

Table 1. Risk management of the enterprise Internet project promotion.

Risk group	Types of risks
Project risks	Nonconformity of the purpose of information systems project with the organization goals (objectives)
	The complexity (size) of the information systems project
	Ensuring the required functional characteristics of the information systems
	Ensuring the required application performance
	Ensuring the required reliability and security of information systems services
	Ensuring the required availability of information systems services
	Use in the project of new technologies (software–hardware, web technologies, computing methods, etc.)
	Sustainability of system architecture (design solutions) to possible changes in requirements
	Competence of the customer in the information systems field
	Developer competence
	Other project risks
Technological risks	Decreased site traffic
	Destruction of the information assets of the site
	Loss of site availability
	Other technological risks including project miscalculations
Business risks	Additional costs and complexity of website promotion
	Loss of profit as a result of production downtime
	The damage from idle time in employees' work

(*Continued*)

Table 1. (*Continued*)

Risk group	Types of risks
	Decrease in business reputation
	Loss of profit due to disruption of obligations
	Damage from unforeseen expenses
	The damage from the company stock value fall
	Loss of profits from out-of-time decisions
	Losses related to the suboptimal functioning of the business due to erroneous decisions
	Other business risks including project miscalculations
Juridical risks	Administrative/criminal liability due to the use of counterfeit software
	Idle time due to the seizure of computers by law enforcement agencies during the audit
	Changes of design work conditions by the customer
	Other juridical risks including project miscalculations

Regulation of the impact of risks can be implemented through various organizational and economic activities. For example, risk management can reduce their level by improving the technology of production and sale, improving product quality control, improving the processability and sustainability of production, and allocating additional funds for the creation of additional reserves.

Using a cooperative system of production and sales, it is expedient to redistribute risks between all business partners through corporate property responsibility.

Considering the Internet project risks can be implemented, for example, by optimizing the information system architecture. Losses related to business risks represent the inability to implement some information system functions and they appear because of technical risks. For instance, the risk of not entering information into the database may arise due to the technical risk of "disruption of the communication channel." On the other hand, business risk can be parried by the appropriate process organization and/or architectural solution.

It becomes obvious that the cost of the risks of the Internet project information system used is characterized by the cost of business risks, the likelihood of technical risks, and the matrix of correspondence between them. In this case, the correspondence matrix is determined by the information system architecture. Therefore, it is possible to calculate all business risks that have the opportunity to appear in the system and determine their value. Next, we should determine the technical risks based on the selected solutions and determine the possibility of occurrence of these risks and the degree of business risk impact (build a compliance matrix). Then, we should choose the best option for its construction, changing the architecture and technical characteristics of the system in several iterations.

The proposal to create (or use on a contract basis) an information and analytical service at the enterprise cannot be rejected. According to some authors, the greatest effect of e-business will be achieved through the implementation of a set of analytical methods. At the same time analytical support of business integration on the Internet can include

- Analysis of the company internal resources
- Analytical rationale of optimal business model choice for the Internet
- Project support in the process of its creation
- Project support at the first stage of its work on the web
- Internet business forecasting and justification of the reengineering need of the current model in the future.

3 RESULTS

Considering the foregoing, we believe that simulation modeling is the most acceptable option for the Internet project risk assessment. This is due to the fact that the risks of the project implementation can be considered as a random manifestation of various factors, the impact of which can be assessed on the basis of stochastic (probabilistic) models of Internet project implementation [14, 54, 76, 213].

The modeling process is performed on the example of a business risk group in the following sequence:

- Breakdown of the analyzed process into groups of risks and individual risks in groups
- Expert assessment of the cost loss value (severity of consequences) for each risk: minimum (min.), most probable (m.p.), and maximum (max) values (with denomination of thousands)
- Simulation of loss price values, based on the characteristics defined earlier, by generating random numbers, for example, by a discrete distribution (when the values of random numbers and the corresponding occurrence probabilities are specified)
- Calculation of the relative strength (size or probability) of each individual risk by paired comparisons and subsequent evaluation of the corresponding element of the priority matrix eigenvector normalized to unity (the procedure of determining the eigenvectors of matrices is approximated by calculating the geometric mean in the hierarchy analysis method)
- Calculation of total possible losses based on the modeled risk value and its probability of occurrence
- Calculation of statistical characteristics of simulated total risk values and formulation of conclusions

First, we determine the characteristics of the possible cost loss on the example of project business risks (Table 2), where by the cost of risk losses we mean the possible losses of resources in case of unfavorable events and deviations from the planned strategy, as well as loss of business operation profit.

The second stage is the modeling of the values of the cost losses using a discrete distribution. The results of business risk simulation are shown in Table 3.

The next stage is the construction of paired comparison matrix for business risk group indicators with a dimension of 10×10. We fill in the paired comparison matrix with numerical characteristics and calculate the priority vector using the hierarchy analysis method (Table 4).

Table 2.   Estimating of possible cost of business project risk losses.

Risks	Value	Cost loss (thousand rubles)		
		min	m.p.	Max
Additional costs and complexity of website promotion	b_1	10	50	80
Loss of profit as a result of production downtime	b_2	600	800	950
The damage from idle time in employees' work	b_3	18	30,0	39
Decrease in business reputation	b_4	85,0	100	120
Loss of profit due to disruption of obligations	b_5	140	200	280
Damage from unforeseen expenses	b_6	55	70	90
The damage from the company stock value fall	b_7	375	400	420
Loss of profits from out-of-time decisions	b_8	75	90	102
Losses related to the suboptimal functioning of the business due to erroneous decisions	b_9	240	300	330
Other business risks including project miscalculations	b_{10}	120	150	165

Table 3. Results of business risk modeling.

Criterion	b_1	b_2	b_3	b_4	b_5	b_6	b_7	b_8	b_9	b_{10}
Mean	48,8	796,5	30,1	99,9	204,8	70,7	398,6	89,5	296,6	148,1
Standard error	1,5	7,0	0,4	0,7	3,0	0,8	0,9	0,5	1,7	0,9
Standard deviation	15,0	69,7	4,4	7,2	30,4	8,4	8,8	5,5	16,6	9,2
Coefficient of variation	0,3	0,1	0,1	0,1	0,1	0,1	0,0	0,1	0,1	0,1
Sample variance	224,8	4861,4	19,7	52,5	922,2	70,2	77,8	30,1	277,2	84,3
Excess	1,0	0,8	0,6	0,9	0,5	0,4	0,9	-0,1	2,3	0,9
Asymmetry	-0,5	-0,3	-0,2	0,2	0,5	0,4	0,1	0,3	-1,0	-0,9
Interval	70,0	350,0	21,0	35,0	140,0	35,0	45,0	24,0	90,0	45,0
Minimum	10,0	600,0	18,0	85,0	140,0	55,0	375,0	78,0	240,0	120,0
Maximum	80,0	950,0	39,0	120,0	280,0	90,0	420,0	102,0	330,0	165,0

Table 4. Paired comparison matrix.

	b_1	b_2	b_3	b_4	b_5	b_6	b_7	b_8	b_9	b_{10}	Priority vector	Normalized vector
b_1	1	7	6	7	5	7	8	7	7	7	5,560	0,370
b_2	1/7	1	5	6	7	5	6	7	5	5	3,309	0,220
b_3	1/6	1/5	1	3	4	3	2	3	4	2	1,500	0,100
b_4	1/7	1/6	1/3	1	2	2	3	2	2	3	1,013	0,067
b_5	1/5	1/7	1/4	1/2	1	4	5	3	4	5	1,157	0,077
b_6	1/7	1/5	1/3	1/2	1/4	1	3	2	4	3	0,782	0,052
b_7	1/8	1/6	1/5	1/3	1/5	1/3	1	2	2	1	0,454	0,030
b_8	1/7	1/7	1/3	1/2	1/3	1/2	1/2	1	2	2	0,508	0,034
b_9	1/7	1/5	1/4	1/2	1/4	1/4	1/2	1/2	1	1/2	0,350	0,023
b_{10}	1/7	1/5	1/5	1/3	1/5	1/3	1	1/2	2	1	0,408	0,027
Total											15,041	1,000

The values of the normalized vector are taken as the relative strength, magnitude, or probability of each individual risk occurrence ($p_{b,i}$).

Total business risk (BR) of the Internet project will be defined as

$$BR = \sum_i p_{b,i} b_i^m$$

Then the total risk of the Internet project will be determined as

$$SR = \sum_i p_{b,i} b_i^m + \sum_i p_{t,i} t_i^m + \sum_i p_{j,i} j_i^m,$$

where $p_{b,i}$, $p_{t,i}$, and $p_{j,i}$ is the probability of individual business risk occurrence, technological risks, and juridical risks respectively; and b_i^m, t_i^m, and j_i^m are modeled values of the cost of possible losses of individual business risks, technological risks, and juridical risks respectively.

The statistics of the simulated total business risk value BR are given in Table 5, and the histogram and integral percentages in Figures 2 and 3.

It can be seen from the data that the final modeled distribution of business risks (BR) is close to the normal distribution. This allows us to assess the possible business risk of an Internet project as follows:

- A probability of 0.683 BR will be in the range of 248.25 ± 16.85 thousand rubles
- A probability of 0.954 BR will be in the range of 248.25 ± 33.7 thousand rubles

Table 5. Results of business risk (BR) simulation.

Criterion	Value	Criterion	Value
Mean	248.2508	Excess	0.608739
Standard error	1.685464	Asymmetry	−0.44949
Standard deviation	16.85464	Interval	81.14618
Coefficient of variation	0.0679	Minimum	200.9884
Sample variance	284.079	Maximum	282.1346

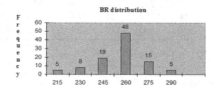

Figure 2. Distribution of simulated business risk (BR) values.

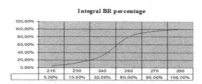

Figure 3. Integral business risk (BR) distribution percentage.

4 CONCLUSION

The proposed methodology for assessing the risk management of the Internet project of the enterprise is based on a process and statistical approach for determining the cost of losses (severity of consequences) for each risk and calculating the relative strength (magnitude or probability) of each individual risk by paired comparisons in the hierarchy analysis method. The main advantage of the methodology is that it makes it possible to obtain the values of risk assessments in monetary terms, which in turn makes it possible to use them directly in calculating the economic effect of the project.

REFERENCES

Adamadziev, K. R., Adamadzieva, A. K., Magomedgadzhiev, Sh. M., Gadzhiev, N. K., & Omarova, E. Sh. 2011. *Network Economy*. – Makhachkala: Publishing and polygraphic center of Dagestan State University.

Alexunin, V. A., & Rodigina V. V. 2005. E-commerce and marketing.

Dolyatovsky, V. A. 2012. *System Analysis in the Management of the Firm*. Guide-Saarbrücken: LAP Lambert Academic Publishing.

Dolzhenko, A. I. 2007. Fuzzy productivity models for risk assessment of information systems projects/problems of the federal and regional Economy: Scholarly notes (10: 83–89). Rostov-on-Don: RSEU RINH.

Fomenko, N. M. 2015. *Information and Communication Technologies in Organization Management*. Rostov-on-Don: SKNTS VS SFU.

Khubaev, G. N., & Streltsova, E. D. 2005. Quantitative quality assessment of decision support tools. Economic and Organizational Problems of Design and Application of Information Systems: Materials of VII scientific and practical conference (pp. 214–222). Rostov-on-Don.

Krepkov, I. M., Efimov, E. N., & Fomenko, N. M. 2010. Analysis and consideration of risks of enterprise Internet-project promotion. *MPEI Bulletin*, 2.

Lipaev, V. V. 2005. Analysis and reduction of software project risks. JetInfo online, 1 (140).

Lukasevich, I. Ya. 2006. Simulation of investment risks. Corporate Management.

Markswebb. Available at: http://markswebb.ru/e-commerce/e-commerce-user-index/

Statistics of the Internet 2018: Sites, blogs, domains, e-commerce: Interesting figures and facts from around the world. Available at: https://sdvv.ru/articles/elektronnaya-kommertsiya/statistika-interneta-2018-sayty-blogi-domeny-elektronnaya-kommertsiya-interesnye-tsifry-i-fakty-so-v/(29.06.2018).

Inclusive Development of Society – Lumban Gaol (eds)

Supply chain efficiency analysis in innovative company based on MCC-3C methodology

A.Y. Podchufarov
Higher School of Economics, Moscow, Russia

V.I. Ponomarev
RUDN University, Moscow, Russia

R.V. Senkov
Higher School of Economics, Moscow, Russia

ABSTRACT: The expansion of three-circuit Matrix of Core Competencies (MCC-3C) methodology aimed at improving supply chain (SC) efficiency in an innovative company is examined. It is demonstrated that analysis of production and development factor groups as part of the MCC-3C model, when analysing the supply chain structure, makes it possible to distinguish the criteria for decision-making on allocation and optimising the utilisation of competencies involved in the company's activities. Approaches for determining the applicability limits of qualitative theories and formation of quantitative assessments are proposed that allow justifying the structure of internal SC and positioning in external ones. An example of the MCC-3C methodology application in the SC design of an international company producing final hi-tech products is provided.

1 INTRODUCTION

The resulting (basic target) indicators (BTI) of the activities of economic entities are largely due to the adopted structure of the target indicators (TI) of internal subsystems and the effectiveness of their achievement (Podchufarov 2013, Samoylov & Podchufarov 2016). Supply Chain Management (SCM), irrespective of the breadth of interpretation of this concept, is an important factor for success in a market economy. In this regard, the credibility of the assessment of the significance and effectiveness of internal and external elements of SCM largely determines the rationality of the involvement and use of resources, both in the implementation of operational and strategic development plans.

The methodological basis for the development of a balanced SC structure can be the provisions of the theory of system interaction, which, based on the principles of the system approach, provides an effective tool for finding the best ways to achieve the resulting target indicators. Over the past few years, this direction has been actively developed by specialists of the Faculty of World Economy and World Politics at the Higher School of Economics, Moscow based on the Department of foreign economic association "Avtopromimport" "State and Corporate Management Systems".

When formulating justified solutions in the field of socio-economic interaction, a significant role is played by the availability of adequate models of the objects under consideration. The complex nature of the field of research has determined the development of a wide range of approaches to its analysis and forecasting. However, despite the high interest in this subject, there is still lack of practice-oriented methods that combine the depth of analysis with the visibility and convenience of their application in the stages of preparation and adoption of managerial decisions. To date, one can note the prevalence of two approaches in this field:

– the use of detailed mathematical models that require specialized knowledge in the theory of exact sciences and modern software-analytical tools;
– decision making on the basis of qualitative estimates and generalized characteristics of the object under study with minimal involvement of the mathematical apparatus.

Typical examples of the first approach, as a rule, include VAR-modeling (Vector AutoRegression), system dynamics model, DSGE models, etc. Examples of the second approach include the structural and strategic analysis methods (the five-factor model, the diamond model of competitive advantage, the value chain of M. Porter), the methods of expert evaluations (evaluation of competitive strength by A.A. Thompson and A. J. Strickland), matrix methods (BCG, GE/McKinsey, ADL) and many others.

Relevant situation is in SC sphere. The specific nature of SCM activities has determined the range of methods that take into account its features. As a rule, more sophisticated methods are provided by integral approaches aimed to analyze performance measurements and overall efficiency of SC based on relatevly complicated methodologies: system dynamics, DEA, etc. However, the same problems can be solved by methods that used rather simple mathematics: SCOR model, methods based on balanced scorecard and benchmarking methodology, etc. Comprehensive reviews of the SCM concept and respective methods can be found in Goedhals-Gerber (2010), Pettersson (2008), Sillanpää (2010).

There is also a range of methods aimed to analyze specific SC problems: cost and cost-benefit analysis emphasized on assessing logistics costs and ways to reduce them; ABC-analysis – a method of classifying resources by significance; XYZ-analysis – a method of classifying resources by characteristic trends in consumption; RFM-analysis – a method of customer segmentation based on the nature of consumption of the products produced by the company. The underlined methods are practice-oriented, are based on relatively simple mathematical models, give clear and easily interpreted results.

However, the above mentioned approaches do not fully bind SC elements with overall competitiveness of the company, which makes it difficult to obtain an integral assessment of their influence on the company's activities as a whole and take into account the significance of every single of them in achieving it.

2 MCC-3C METHODOLOGY

For a comprehensive analysis of the activities of market relations in modern conditions, a methodology for assessing and managing competitiveness based on the three-circuit MCC-3C model, proposed by the authors of the article, is widely used at present. In general, competitiveness in the approach has three circuits for consideration, reflecting the interconnection of its components. In the first circuit, consumer quality indicators are analyzed, in the second – the competitiveness factors and in the third – the core competencies (CC). In the first circuit, the optimization problem is solved based on the requirements for external consistency, in the second – internal consistency and in the third – consistency in the subsystems of the life cycle (Podchufarov 2018).

Within the framework of the approach, the object under study is represented in the form of a system of interconnected segments, refer to Figure 1, and is characterized, on the one hand, by an integrated competitiveness indicator, determined by the joint influence of the competitiveness factors included in the segments under consideration, on the other hand, by indicators of the position of the object on the market.

As a result, the mechanism of the iterative verification (sequential refinement) of the received estimates is laid in the methodology, which ensures the convenience of using MCC-3C models when developing algorithms and coordinating management decisions. Analysis of the effectiveness of the use of competitiveness factors and their significance in achieving the resulting (target) indicators is carried out on the basis of Matrices of Core Competencies (Podchufarov 2013).

The model considers the structure and the assessments of the competitiveness factors linearized close to the equilibrium trajectory, where each factor is described by its comparative indicator (CI) and the significance indicator (SI):

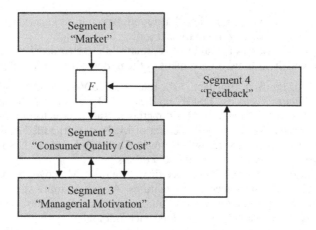

Figure 1.   MCC-3C model segments and their interaction.

– *CI* is a nondimensional value that determines the level of the current comparative state of the factor with respect to the competitor in question, the valuation of which is assumed to be 1;
– *SI* is a nondimensional value that determines the degree of influence (lag/advance) of the considered factor value on the basic target indicator.

The *CI* and *SI* are defined by assessments of variables or functional dependencies in the planning interval under consideration. Calculation of resulting indicators is made in accordance with the economic interpretation of this concept and the rules applied in the Control Systems theory to the transfer functions.

The generalizing indicator K_i for the *i*-th factor in static case (the coefficient of the transfer function) and generalizing indicator K_n for the *n* sequential elements are determined by the formulas:

$$K_i = 1 - (1 - CI_i) \times SI_i, \tag{1}$$

$$K_n = \prod_{i=1}^{n} K_i, \tag{2}$$

where *n* – total number of factors.

For example, if specified factor for the object under investigation has *CI* = 1.2 and *SI* = 0.5 when it means that we assess it by 20% better than the same factor of our competitor (which equals to 1). From (1) and (2) follows that it defines 10% (20% × 0.5) increase in K_n (respect to case *CI* = 1).

The indicator of competitiveness (*CMPI*) in case of direct circuit:

$$CMPI = K_{MR} \times K_{CQ/C} \times K_{MM}, \tag{3}$$

where K_{MR} – coefficient of the transfer function for Segment 1, $K_{CQ/C}$ – for Segment 2, K_{MM} – for Segment 3.

The proposed approach in many ways makes it possible to combine the depth of the mathematical apparatus of the model with an intuitive interpretation of the intermediate and final results. To increase the reliability, methods of harmonizing evaluations with statistical results of the analysis of significance of factors can be used based on the VAR analysis of the relevant time series.

Taking into account the dynamics of factors and their cross-influences, it is possible to generate dynamic MCC-3C models in which the interaction with internal subsystems is represented as an oriented graph. In such a model, the competitiveness indicator of the subsystem of a lower level determines the *CI* of the corresponding factor in the higher-level subsystem, and its SI is an estimate of the cross-influence between the specified subsystems (Podchufarov 2018).

The decomposition of the basic system into internal subsystems and the methodology for assessing the cross-influence of factors proposed by the MCC-3C approach allow us to construct an "end-to-end" scheme for the formation of competitiveness, to identify its most significant elements and to set requirements for the values of their indicators. The implementation of a set of inter-system interaction measures eliminates sources of imbalance in the systems under consideration, which are the cause of unsustainable or inefficient development, and allows the formation of proposals aimed at their further improvement. However, practice shows that the results acceptable for application in some cases can be obtained on the basis of simplified MCC-3C models that consider the first two interaction circuits and do not include the analytical instrument of cross-influence accounting.

It is important to note that the proposed approach allows linking indicators of various functional directions determining the achievement of BTIs and taking into account the potential of core competencies structured in accordance with the MCC-3C.

3   MCC-3C IMPLEMENTATIONS IN SUPPLY CHAIN MANAGEMENT

Considering SCM as one of the important components of management, which in many cases encompasses the complete circuit of the life cycle of the products, it is logical to adopt the prevailing opinion about the high relevance of the task related to the choice of the methodological basis of the SCM analytical apparatus as an element of the overall management system.

Using the MCC-3C models in SCM makes it possible to identify problem factors and to form the structure of cause-effect relationships that determine the achievement of the resulting indicators. Analysis of the effectiveness of SC allows to identify the criteria that justify the decision-making about business models of participation in external SC and the transfer of internal SC elements to the external interaction circuit.

As noted above, the application of the MCC-3C approach allows us to determine the competitiveness factors and obtain their numerical estimate. Also, the functionality of MCC-3C model makes it possible to evaluate groups that characterize certain areas of activity. Analysis of the importance of groups of factors and their correlation with the CC structure forms the basis for the preparation of conclusions about the feasibility of developing certain competencies in the internal or external circuits of SC.

Introducing the concept of significance indicator of the k-th group of factors (SIG_k):

$$SIG_k = \sum_{i \in F_k} (SI_i), \tag{4}$$

where F_k – the subset of factors in the k-th group, and the significance indicator of the k-th group of factors in the two-dimensional state-space with coordinates SIG^D – significance indicator of the development factors and SIG^P – significance indicator of the production factors:

$$SIG_k^{(D;P)} = \left(SIG_k^D; SIG_k^P\right), \tag{5}$$

$$SIG_k^D = \sum_{i \in D_k} (SI_i), \quad SIG_k^P = \sum_{i \in P_k} (SI_i),$$

where D_k and P_k – the subsets of factors belonging to the k-th group and referring to "development" and "production" respectively ($D_k \cup P_k = F_k$), it is possible to form an array of data characterizing the groups of the object under investigation. It is convenient to project them on the coordinate plane with the X axis corresponding to the SIG^P and the Y axis

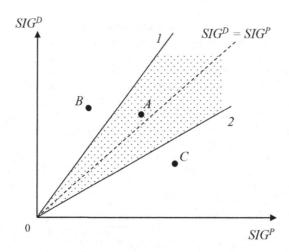

Figure 2. Factor groups representation in the area of specialization.

corresponding to the SIG^D, referred to as the "area of specialization", Figure 2. Groups on the figure are represented by points in the first quadrant of the plane.

Line $SIG^D = SIG^P$ divides the allocated quadrant into two equal parts. In the case where the indicators of the groups lie above this line (for example, groups A and B), the significance of the production factors for them is higher than the significance of the development factors. Consequently, for the indicators of the groups lying below the line (for example, group C) the situation is reversed, and the significance of the factors of production will be lower than the significance of the development factors. In practice, it is convenient to single out a "sector" of values, in which these indicators are comparable, that is, their ratio is close to one. The given region is bounded by arrows 1 and 2 (highlighted by dotted lines).

Of particular interest is the analysis of significance indicators of the resultant groups of factors characterizing the objects in question as a whole (resulting significance indicator of the group, SIG_{res}). This indicator is calculated using formula (4), where the whole set of factors formed by MCC-3C model for the object under investigation is considered.

In the case the SIG_{res} areas on the figure fall outside the shaded segment, decisions about the inclusion of SC elements in the external or internal contour, as a rule, with acceptable reliability can be justified by qualitative estimates, and in many cases it is sufficient to use the provisions of the theory of transaction costs (Coase 1988). For example, in the case of an arrangement above arrow 1 (point B), it is logical to transfer to the external contour the links that based on R&D activities, and transfer the production of the main components to the internal contour. Otherwise, when objects lie below arrow 2 (point C), it is advisable to include the elements responsible for the production of components of the final product in the external contour, while the R&D activities are carried out in-house.

In the case when SIG_{res} is placed in the allocated segment, between arrows 1 and 2, in order to work out a decision on the structure of SC, a more detailed analysis of the group indicators, as to their mutual location in the area of specialization and relationship with the core competencies (Podchufarov & Senkov 2017) is required.

Particular attention should be paid to the situations when the companies under consideration are characterized by the SIG_{res} placed within the allocated segment, and belong to a class of high-tech enterprises focused on producing end products. As a rule, they are characterized by active interaction of production units and R&D, and in many cases small- and medium-term specialization.

Below is an example of an analysis of the SC structure based on the MCC-3C model for a high-tech company (an international machinery holding), whose feature is the active interaction of R&D and production-related units both at the development stages and support in the manufacturing of the final products, specializing in medium-series production.

The use of the MCC-3C model in the analysis of the activity of this enterprise made it possible to identify the factors of development and production and obtain estimates of the comparative and the significance indicators (an extract for aggregated groups of factors is provided), refer to Table 1.

In the case under consideration, the estimates of the significance of the resulting factors of production (1.5) and development (1.7) are comparable, which corresponds to the shaded segment of the area of specialization on Figure 2. An analysis of the position of the group indicators in this area presented in the table and their relationship to the core competencies (Podchufarov & Senkov 2017) made it possible to formulate the requirements for the SC structure of the enterprise under study aimed at balanced utilization of its own CCs, significance indicators limits, as well as the development of cooperative ties and strategic partnerships in justified areas of activity, refer to Figure 3. Radius of each circle on Figure 3 is proportional to the utilization of respective CC. Factors that form groups placed inside circles are treated as activities developed inside the company. To design efficient SC other factors are more likely to handle on "outsource" basis.

Taking into account the specifics of the enterprise, determined by the big role of R&D and the limited serial production of end products, additional requirements were determined for SCM processes in terms of their flexibility and adaptability to changes. Special attention was

Table 1. Indicators of the factors of the MCC-3C model for international machinery holding (extract).

Consumer quality factors	CI	SI	K
Production			
Components of own production	1.10	0.15	1.015
Components of external manufacturers	0.80	0.15	0.970
Assembling the final products	1.10	0.10	1.010
Ensuring product lifecycle	0.50	0.15	0.925
Other	0.85	0.15	0.978
Development			
Components of own production	1.30	0.10	1.030
Components of external manufacturers	0.80	0.20	0.960
Integrated solutions (complexes)	1.05	0.10	1.005
Processing and management algorithms	0.90	0.20	0.980
Technological processes	0.80	0.20	0.960
Other	0.85	0.20	0.970

Cost factors	CI	SI	K
Production			
Components of own production	1.20	0.20	1.040
Components of external manufacturers	1.10	0.20	1.020
Assembling the final products	1.20	0.20	1.040
Ensuring product lifecycle	1.10	0.10	1.010
Other	1.05	0.10	1.005
Development			
Components of own production	1.30	0.20	1.060
Components of external manufacturers	1.10	0.10	1.010
Integrated solutions (complexes)	1.10	0.10	1.010
Processing and management algorithms	1.20	0.10	1.020
Technological processes	1.10	0.10	1.010
Other	1.05	0.10	1.005

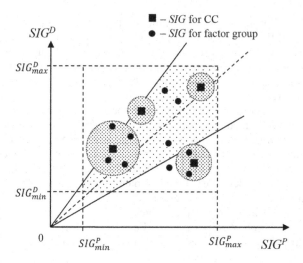

Figure 3. Analysis of core competencies in the area of specialization for international machinery holding (extract).

paid to SC in the R&D sphere, which is due to the need for deep immersion in the subject area when placing orders in the external circuit for components of newly developed products and quality control of their execution. Also, the features of the external cooperation factors were taken into account. In the case examined, the cost of the purchased products was largely determined by the conditions of the relations between the counterparties at the product development stage and the indicators of mutual interest in a long-term partnership, which is characteristic of the specific nature of the company under study.

Situations similar to those considered in the terminology of the XYZ analysis are characterized by the nomenclature of the category Z, which significantly exceeds the number of values corresponding to large-series production enterprises. As a consequence, in many cases, the top management of the enterprise assumes the processes for its formation and control over execution, with institutional support from the competent collegial bodies, including representatives of owners and counterparties, and, in the presence of state orders, regulating and controlling authorities. The units that support SCM processes associated with the nomenclature corresponding to categories X and Y generally have a negligible amount of output in terms of money.

Features of the case are quite common in the production of critical components and products for long periods of use. Examples of sectors of the economy that are consumers of this class of products include shipbuilding, machine-tool construction, aviation, space, chemical, power industry, and others.

4 CONCLUSIONS

A high level of competition in the modern market dictates the need to build effective supply chains. A typical trend in such conditions is the management's desire to increase the reliability of analytical information for making decisions about the development of its internal core competencies and the transfer of non-core SC elements to the external contour within the framework of strategic partnerships and external cooperation.

The extension of the MCC-3C approach presented in the article includes a methodology based on an analysis of the significance of development and production factors, which makes it possible to identify the composition and obtain quantitative assessments of the criteria that justify the development of requirements for the SC structure. The class of companies in which the specified groups of factors have comparable indicators of significance is investigated. Special attention is paid to enterprises specializing in medium- and small-series production of high-tech end products. The essential features of the SC of such companies are indicated.

In general, the principles of the system approach, the theory of interaction of systems and the MCC-3C models developed on their basis allow to do quantitative assessments of the influence of the SC structure on the achievement of the BTI of the subjects of economic relations being investigated, to set requirements for SCM processes and to develop an informed decision on their internal or external implementation. At the same time, the joint use of the proposed approaches with specialized methods of analysis and forecasting, according to the authors, can provide the maximum positive effect when planning the company's activities in the SCM field.

REFERENCES

Coase, R. 1988. *The Firm, the Market, and the Law*. University of Chicago Press.
Goedhals-Gerber, L. 2010. *The Measurement of Supply Chain Efficiency: Theoretical Considerations and Practical Criteria*. Stellenbosch University.
Pettersson, A. 2008. *Measurements of efficiency in a Supply chain*. Lulea University of Technology.
Podchufarov, A.Y. 2013. Base approaches for quality improvements of sectoral management in defense industry of Russia. *Government and corporate management systems in defense industry*. NRU HSE: 8–13.
Podchufarov, A.Y. 2018. Development sustainability and balance of competitiveness factors. *International economic symposium – 2018, April 19-21, 2018*. St.Petersburg University Press: 250–250.
Podchufarov, A.Y. & Senkov, R.V. 2017. Theory and practice of competitiveness management using the case of leading Russian and foreign companies. *Book of abstracts of International scientific forum "Public administration", 19-20 June 2017*. North-West Institute of Management: 67–72.
Samoylov, V.I. & Podchufarov, A.Y. 2016. Competitiveness assessment methodology of industrial holding in conditions of complex import substitution. *News of the Tula state university. Technical sciences*. 10: 349–354.
Sillanpää, I. 2010. *Supply chain performance measurement in the manufacturing industry*. University of Oulu.

Inclusive Development of Society – Lumban Gaol (eds)
© 2020 Taylor & Francis Group, London, ISBN 978-1-138-33476-2

Integrating risk-management practices and risk behaviour to sustain R&D project performance

N. Zainal Abidin, M.S.M. Ariff, N. Zaidin, N.Z. Salleh & R.Md. Nor
Azman Hashim International Business School, Universiti Teknologi Malaysia, Johor, Malaysia

N. Ishak
Lembaga Tabung Haji, Kuala Lumpur, Malaysia

ABSTRACT: The significant effect of risk-management practices (RMPs) on project performance (PP) is well acknowledged in many fields of project management, for example, construction, information technology (IT), and software development. The issue of risk behaviour (RB) arises because it is recognized as having a direct influence on individual orientation towards risk. Past research has addressed RMP and RB separately, making it complicated to judge the success of a project from a risk-management perspective. Thus, to sustain research and development (R&D) PP, these two components must be integrated for better prediction of PP. This research proposes a model to predict R&D PP by integrating RMP and RB. A self-administered questionnaire was employed to collect the required 265-participant sample size from R&D project leaders of a research university. The findings reveal that RB of R&D project leaders is positively and highly correlated with RMP, and significantly affects R&D PP. In addition, RB was found to exert a slightly higher effect on R&D PP. This study concludes that the inclusion of RB in the model of RMP and PP leads to better performance of R&D projects.

Keywords: risk-management practices, risk behaviour, R&D project, project performance

1 INTRODUCTION

In many instances, risk-management practice (RMP) improves project performance (PP) via systematic identification, appraisal and management of project-related risk (Epstein, 2002). Past studies have demonstrated that RMP has a positive effect on PP, particularly in the industries of construction (Kululanga & Kuotcha, 2010; Lee & Ali, 2012), information technology (IT) and software development (Ahlan & Arshad, 2012; Saleem & Abideen, 2011), accountancy (Association of Chartered Certified Accountants [ACCA], 2012) and transportation (Caltrans, 2012). However, there is a lack of studies on RMP in research and development (R&D) projects. R&D projects are usually very costly and are exposed to significantly higher risk than investing in tangible assets (Abdel-Khalik, 2014), therefore, this research is crucial given that it aims to explore RMP in R&D projects to examine how RMP effects R&D PP.

Recent trends in RMPs consider risk behaviour (RB) one of the indicators of PP. Human factors or error contribute to the failure of RMP implementation, and consequently affect PP. This has been demonstrated in past research in the industries of construction (Kululanga & Kuotcha, 2010); finance (Ashby, 2010); business (Stan-Maduka, 2010); and accountancy (ACCA, 2012); yet, no solution for combatting the effect of human factors and error has been proposed. The principal cause of this failure is due to individual behaviour rather than to technical failure. Therefore, the correlation between RB and RMP, and the effects on R&D PP should be examined to benefit the success of R&D projects. However, current models of RMP and PP do not integrate RB, despite the importance of this factor having been highlighted in past studies. Thus, this research aims to fill this the gap in the literature by integrating RB into the current model of RMP and PP to predict R&D PP. This new model is expected to provide a comprehensive framework for predicting PP from a risk-management perspective.

Specifically, this study aims to examine the combined effects of RMP and RB on R&D PP. The study objectives are to find answers to the following research questions: (i) What is the correlation between RB and RMP?; (ii) What is the effect of RMP on R&D PP?; (iii) What is the effect of RB on R&D PP?; (iv) What is the combined effect of RMP and RB on R&D PP?. This study provides a new model of RMP and RB in predicting PP in relation to R&D projects.

2 LITERATURE REVIEW

Risk management is defined as the 'coordinated activities to direct and control organization with regard to risk' (International Organization for Standardization [ISO]), and has been describe as being about 'predicting the unpredictable' (Lee & Ali, 2012). Risk management is also defined as a process of identifying, evaluating, and managing risk. Indeed, the ultimate goal of applying risk management is to minimize risk.

Rohrmann (2005) defines RB as the actual behaviour of people when confronted with risks. It is characterized by the degree of risk associated with decision making. Factors of culture, personality, habit, and motivation are categorized under RB (Van Winsen et al., 2011), and influence individuals' behaviour towards risks.

Process theory has been utilized to examine RMP in successful project management. In fact, the formal process of managing risk is based on process theory (Kwak & Stoddard, 2004). Process theory of RMP includes practices of context establishment, and the identification, analysis, evaluation, treatment, monitoring, and review of risk (Kululanga & Kuotcha, 2010).

RB is connected to behavioural theory. This theory is defined as the belief that a manager's level of success is based on the way they act or behave (Cherry, 2010). In behavioural theory, organizations are treated as a coalition of individuals (i.e., top management, staff, and shareholders). This attempts to predict organizational behaviour and decision-making processes. In addition, it can predict and understand what managers or other people do and will do rather than what they should do. In behavioural theory, management will take new action to manage a problem when a unit's performance is below a given reference point. Thus, the RB of employees positively affects RMP (Cherry, 2010). Further, failure to manage behaviour leads to the failure of project. Hence, RB is predicted to have significant effect on project failure or success (Ashby, 2010; Cherry, 2010).

In the present research, 14 studies from the period 2010–2015 were examined in relation to RMP, and it was found that all these studies utilized identification of risk and analysis of risk in measuring the RMP. This demonstrates that both these practices of RMP are crucial. Twelve studies highlighted response to and treatment of risk in relation to RMP, eight applied risk monitoring to measure RMP, five employed risk communication to measure RMP, and four employed risk evaluation of RMP. Only three of the 14 studies applied context establishment to measure RMP in the management of various projects.

These practices RMP are widely practiced in the projects of various industries, for example, in construction (Kululanga & Kuotcha, 2010; Lee & Ali, 2012); transportation (Caltrans, 2012); IT and software development (Ahlan & Arshad, 2012; Saleem & Abideen, 2011); as well as in other industries (Junior & Carvalho, 2013; Rasid, Golshan, Ismail & Ahmad, 2012). Given that R&D is related to project management, the RMPs of these industries were examined in relation to their suitability in the context of R&D projects. The present study combines some practices of RMP due to the similarities between the practices. Thus, for this study, the following were identified as the practices of RMP: a) risk identification (RI); b) risk analysis combined with risk evaluation to form risk analysis and evaluation (RAE); c) risk response

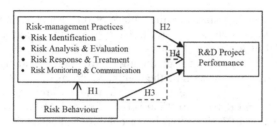

Figure 1. Research framework.

and treatment (RRT); d) risk monitoring combined with risk communication to form risk monitoring and communication (RMC).

Based on process theory and behavioural theory, as well as on findings from previous research, the framework of this study is developed as presented Figure 1.

According to behavioural theory, managers' behaviour or acts can be predict and understood. Thus, if a person demonstrates appropriate behaviour towards RMPs, then it can be understood that they are willing to implement risk management in their project. According to Cherry (2010), RB addresses a person's belief about an object or project, and it is crucial in determining their perspectives on it. In addition, the success rate of a project depends on how individuals behave towards risks and implementation of RMP. This indicates that RB positively affects RMP. Thus, H1 is proposed:

H1: RB of an R&D project leader positively correlates with RMP.

Individuals, groups, departments, and organizations can implement RMP to mitigate risk. A project will be positively affected by applying RMP, and these practices can reduce risk and the incidence of risk events. Given that risk management has a positive effect on projects (Ashby, 2010; Junior & Carvalho, 2013; Kululanga & Kuotcha, 2010; Saleem & Abideen, 2011; Stan-Maduka, 2010; Wang, Lin & Huang, 2010), risk management is important for ensuring a project's success. Past research has indicated that RMP positively affects PP in relation to cost (Rasid et al., 2012; Saleem & Abideen, 2011); cost and time (Caltrans, 2012); time and quality (Junior & Carvalho, 2013); and cost, time and quality (Ahlan & Arshad, 2012; Lee & Ali, 2012). Thus, H2 is proposed:

H2: RMP of an R&D project leader positively and significantly affects R&D PP.

The failure to manage behaviour leads to the failure of project. Most studies argue that although behaviour error leads to project failure indirectly, it is undeniable that behaviour influences PP (Ashby, 2010). In behavioural theory, a manager's level of success depends on their actions or behaviours (Cherry, 2010). Thus, if the behaviour of a manager is positive, then the behaviour will also have a positive effect on the success of the R&D project. Thus, H3 is proposed:

H3: RB of an R&D project leader positively and significantly affects R&D PP.

The effects of RMP and RB on PP have been discussed in past research. The present study proposes a new research model by assessing the combined effects of RMP and RB on R&D PP. Thus, H4 is proposed:

H4: RMP and RB of an R&D project leader positively and significantly affect R&D PP.

3 METHODOLOGY

The primary data were obtained through a self-administered questionnaire. A total of 35 questionnaire were developed: demographic (five items); RMP (20 items were adapted from past studies). It consists of RI (identification of risk), RAE (evaluation and assessment of risks), RRT (actions for managing R&D risks), RMC (monitoring of risk-mitigation actions) and RB (level of responsibility in managing R&D risks). All these items were reconstructed to suit the context of R&D projects to allow the project leaders to respond based on their experience in managing risk in relation to R&D projects. The proposed questionnaire was examined by two internal experts of risk management for validation. Upon the completion of the questionnaire design, the researchers conducted a pilot study on 15 project leaders to ensure the reliability of the questionnaire. The pilot study found the Cronbach's alpha for RMP was 0.825; for R&D performance was 0.832; and was 0.864 for RB.

The respondents to the study were all R&D project leaders who have received grants for, and completed, research projects under university research grants from different faculties and research alliances of the university under study. The respondent list was provided by the Office of Research Management Center of the university. The population consists of 850 project leaders for Tier 1 and Tier 2 university research grants from 2011 to 2014. According to Krejcie and Morgan (1970), the sample size for this study should be 265 project leaders. However, only R&D project leaders who have completed their research projects and successfully managed risks associated with the projects were selected to ensure the suitability of the respondents to the study. These R&D project leaders were chosen using the stratified random sampling, and based on their research alliances, the participants are attached to the two types of grants (Tier 1 and Tier 2).

The researchers conducted exploratory factor analysis (EFA) to determine the number of common factors that influence a set of measures in this study. The Kaiser–Meyer–Olkin (KMO) measure of sampling adequacy for RMP is 0.901; for R&D PP is 0.77; and for RB is 0.771; all scores are supported by Bartlett's test of sphericity of 0.00. In the first round of EFA for RMP, 20 items were loaded into three components. However, three items were removed because there was a factor loading of more than 0.3 for all three components. The second round of EFA also extracted three components with an eigenvalue >1.0, and all 17 items with a factor loading of >0.5 were retained, with total variance explained of 69.155%. Thus, some changes were made for RMP based on the EFA results, that is, a) RI; b) RAE was combined with RRT to form risk analysis and treatment (RAT); and c) RMC. For R&D PP, only one component was extracted, with total variance explained of 66.698%. All items of R&D PP were accepted, with a factor loading of >0.5. For RB, one item was extracted because the factor loading was <0.5, and the other four items were retained. The total variance explained for RB was 70.971%, with only one component extracted.

All Cronbach's alpha scores for RMP, R&D PP and RB were >0.8. The scores for RI, RAT, and RMC of RMP were 0.898, 0.914, and 0.836, respectively. The scores for R&D PP and RB were 0.873 and 0.844, respectively. Thus, RMP, RB and R&D PP were reliable for further analysis.

4 RESULTS

Research question 1 aims to determine the correlation between RB and RMP. Pearson correlation analysis was performed to examine H1. The correlation was 0.698 (Sig. 0.000), which means there is a high correlation between RB and RMP. The correlation between RMC of RMP and RB was 0.619 (Sig. 0.000), which was the highest correlation, followed by RI with a correlation of 0.588 (Sig. 0.000) and RAT with a correlation of 0.575 (Sig. 0.000). Thus, H1 is supported.

Research question 2 aims to determine the effect of RMP on R&D PP. The multiple regression result R^2, 42.7% (Sig. 0.000) of the RMP was explained by the variation of RI, RAT, and RMC. As presented in Table 1, the highest beta (β) among all RMP components against R&D PP was RAT $(\beta\ 0.295,\ t\ 4.820,\ Sig.\ 0.000,\ VIF\ 1.71)$. This means that RAT of RMP had the highest effect on R&D PP. The second highest β was for RI $(\beta\ 0.262,\ t\ 3.898,\ Sig.\ 0.000,\ VIF\ 2.065)$. The lowest β was for RMC $(\beta\ 0.221,\ t\ 3.837,\ Sig.\ 0.000,\ VIF\ 1.507)$. Thus, H2 is supported.

Research question 3 aims to determine the effect of RB on R&D PP. As presented in Table 2, RB positively affects R&D PP $(\beta\ 0.655,\ t\ 14.049,\ Sig.\ 0.000,\ VIF\ 1.000)$. Thus, it can be concluded that the RB of R&D project leaders positively and significantly affects R&D PP. Further, R^2 of 42.9% of the variation in R&D PP was explained by RB. Thus, H3 is supported.

Table 1. Effect of RMP on R&D PP.

Model	Unstandardized coeff. B	Std. error	Standard-ized coeff. β	T	Sig.	Collinearity statistics Tolerance	VIF
1 (Constant)	.673	.233		2.885	.004		
RI	.313	.080	.262	3.898	.000	.484	2.065
RAT	.342	.071	.295	4.820	.000	.584	1.711
RMC	.179	.047	.221	3.837	.000	.664	1.507
F statistic	64.855		R^2		.427		

a. Dependent variable: R&D PP

Table 2. Effect of RB on R&D PP.

Model	Unstandardized coeff B	Std. error	Standard-ized coeff. β	t	Sig.	Collinearity statistics Tolerance	VIF
1 (Constant)	.917	.204		4.495	.000		
RB	.706	.050	.655	14.049	.000	1.000	1.000
F stat	64.855		R^2	.429			

a. Dependent variable: R&D PP.

Table 3. Effect of RMP and RB on R&D PP.

Model	Unstandardized coeff. B	Std. error	Standard-ized coeff. β	T	Sig.	Collinearity statistics Tolerance	VIF
1 (Constant)	.250	.218		1.147	.252		
RMP	492	.078	.382	6.282	.000	.513	1.948
RB	.419	.066	.388	6.393	.000	.513	1.948
F statistic	132.862		R^2		.504		

a. Dependent variable: R&D PP

Research question 4 aims to examine the combined effect of RMP and RB on R&D PP. As presented in Table 3, R&D PP is affected by RMP and RB, with R^2 of 50.4%. The highest β was for RB (β 0.388, t 6.393, Sig. 0.000, VIF 1.948) compared with RMP (β 0.382, t 6.282, Sig. 0.000, VIF 1.948). That is, RB has a higher effect on R&D PP than RMP. Thus, H4 is supported.

5 DISCUSSION

This study found that RB is positively correlated to RMP. The correlation of RI, RAT, and RMC of RMP with RB is consistent with past research (ACCA, 2012; Ahlan & Arshad, 2012; Ashby, 2010; Navare, 2003; Stan-Maduka, 2010). This demonstrates that to ensure RMP is effectively practiced, R&D project leaders must have appropriate behaviour related to risk. Thus, all R&D project leaders must know their role and responsibilities in relation to risk-management implementation, and have a strong belief that implementing RMPs will contribute to the higher success rate of their project. Behaviours related to risk management should be a part of organizational culture (Stan-Maduka, 2010) to ensure the successful completion of R&D projects.

The results of this study demonstrate that RMPs influence R&D PP. This finding is also consistent with past research (Ahlan & Arshad, 2012; Lee & Ali, 2012). This demonstrates that practicing risk management will contribute to the success of R&D projects in relation to efficient use of the R&D budget, meeting all project milestones, and attaining good PP.

This research has demonstrated that RB positively and significantly affects R&D PP, which is consistent with past research (Navare, 2003; Morrow, 2009). This finding demonstrates that the RB of R&D project leaders influences the on-time completion of research projects, adherence to research procedures and budget allocation, and meeting the project's performance targets. The RB of project leaders allows them to manage and resolve risks associated with R&D project implementation, thus contributing positively to project success.

Based on the R^2 scores (see Tables 1, 2 and 3), the variance in R&D PP (50.4%) is better explained when RMP and RB are combined. That is, the variation in R&D PP is higher when RMP and RB are combined to predict R&D PP. Thus, RMP and RB combined is more important for predicting R&D PP than either RMP or RB separately. RB has a slightly higher effect on R&D PP than RMP. This is consistent with Navare (2003), who highlights that RMP cannot work alone without appropriate RB. As noted by Ashby (2010) and Stan-Maduka (2010), human factors affect the application of RMP and PP. The present study also finds that ineffective culture and behaviour of project leaders in relation to risk may contribute to the failure of R&D projects.

6 CONCLUSION

Both RMP and RB are crucial in ensuring a higher success rate of R&D projects. However, measures of RMP and RB should be further enhanced to better predict their influence on PP. For example, constructs of RMP should be expanded to cover organizational contexts, and risk-management tools and techniques. The 'belief' and 'culture' components of RB should be more specific in relation to their application in the R&D context. The personality, motivations and habits of project leaders' RB should be explored in measuring RB. To generalize the association between RMP, RB and R&D PP, future studies should examine the leaders of all R&D grants, particularly national and international R&D projects with high risk profiles.

ACKNOWLEDGEMENTS

The authors would like to acknowledge the Ministry of Education (PY/2013/00656) and Universiti Teknologi Malaysia for providing financial support for this research.

REFERENCES

Abdel-Khalik, A. R. (2014). CEO risk preference and investing in R&D, *ABACUS*, *50*(3), 245–278.
Ahlan, A. R. & Arshad, Y. (2012). Information technology risk management: The case of the IIUM. *Journal of Information Systems Research and Innovation*, 58–67.
Ahmad, A. (2014). *Examining risk behavior and risk management practices in oil and gas construction industry* (Unpublished master's thesis), UTM.
Ashby, S. (2010). The 2007–09 Financial Crisis: Learning the risk management lessons. Retrieved from http://www.nottingham.ac.uk/business/businesscentres/crbfs/documents/researchreports/paper65.pdf.
Association of Chartered Certified Accountants (ACCA). (2012). *The accountants for business rules for risk management: Culture, behaviour and the role*. London.

Banholzer, W. F. & Vosejpka, L. J. (2011). Risk taking and effective R&D management. *Annual Review of Chemical and Biomolecular Engineering, 2*, 173–188.

Caltrans Office of Statewide Project Management Improvement. (2012). *Project risk management handbook: Threats and opportunities* (2nd ed.). Sacramento, CA: Caltrans. Retrieved from http://www.dot.ca.gov/hq/projmgmt/guidance_prmhb.htm

Epstein, M. (2002). *Risk management of innovative R&D projects.*

International Organization for Standardization (ISO). *MS ISO 31000:2010: Risk management—principles and guidelines.* Retrieved from https://www.iso.org/iso-31000-risk-management.html

Junior, R. R. & Carvalho, M. (2013). Understanding the impact of project risk management on project performance: An empirical study. *Journal of Technology Management & Innovation, 8*, 6–6.

Krejcie, R. V. & Morgan, D. W. (1970) Determining sample size for research activities. *Educational and Psychological Measurement, 30*(3), 607–610.

Kululanga, G. & Kuotcha, W. (2010). Measuring project risk management process for construction contractors with statement indicators linked to numerical scores. *Engineering, Construction and Architectural Management, 17*(4), 336–351.

Lee, C. S. & Ali, A. S. (2012). Implementation of risk management in the Malaysian construction industry. *Journal of Surveying, Construction and Property, 3*(1).

Navare, J. (2003). Process or behaviour: Which is the risk and which is to be managed? *Managerial Finance, 29*, 5–6.

Onaran, Y. & Faux, Z. (2015). Ex-Lehman CEO Fuld says 27,000 employees were risk managers. *Bloomberg Business.* Retrieved from http://www.bloomberg.com/news/articles/2015-05-28/ex-lehman-ceo-fuld-saysall-27-000-employees-were-risk-managers

Project Management Institute (PMI). (2000). *A guide to the project management book of knowledge (PMBOK) 2000 edition.* Newtown Square, PA: PMI.

Rasid, S. Z. A., Golshan, N. M., Ismail, W. K. W. & Ahmad, F. S. (2012). Risk management, performance measurement and organizational performance: A conceptual framework. In *3rd International Conference on Business and Economic Research proceeding* (pp. 1702–1715).

Research Management Centre (RMC). (2014). *Research & development facts and figures 2013.*

Saleem, S. & Abideen, Z. U. (2011). Do effective risk management affect organizational performance? *European Journal of Business and Management, 3*(3), 258–267.

Stan-Maduka, E. (2010). The impact of risk management practice on the development of African businesses. *World Journal of Entrepreneurship, Management and Sustainable Development, 6*(3), 213–219.

Tilk, D. (2011). *PwC: Project success through project risk management.* California: PricewaterhouseCoopers LLP.

Wang, J., Lin, W. & Huang, Y. H. (2010). A Performance-oriented risk management framework for innovative R&D projects. *Technovation, 30*(11), 601–611.

Inclusive Development of Society – Lumban Gaol (eds)

Risk-management practices for effective management of risk in research universities

M.S.M. Ariff, N. Zaidin, N.Z. Salleh, R. Md. Nor & M.N. Som
Azman Hashim International Business School, Universiti Teknologi Malaysia, Johor, Malaysia

N. Ishak
Lembaga Tabung Haji, Kuala Lumpur, Malaysia

ABSTRACT: This study proposes risk-management practices (RMPs) that are suitable for effective risk management in research universities (RUs) in Malaysia. Based on process theory and risk-management implementation in various industries, five practices of risk management—risk governance and management systems (RGMS), risk identification (RI), risk analysis and evaluation (RAE), risk-mitigation strategy and control (RMSC), and risk monitoring and communication (RMC)—are proposed to manage risks effectively in the RU setting. The proposed RMPs are a result of two stages of instrument development, as well as factor analyses that were employed to determine suitable practices of risk management in the RU setting. Self-administered questionnaires were used to collect data, with 288 completed questionnaires collected from the senior administrators of RUs. Exploratory and confirmatory factor analysis (CFA) validated that the five RMPs should be systematically applied to ensure all risks are effectively mitigated and managed. This study provides specific RMPs that capture appropriate methods for effective risk-management implementation in the RU setting.

Keywords: risk management, risk-management practices, research universities

1 INTRODUCTION

Risk-management practices (RMPs) in for-profit organizations are well addressed, particularly in areas such as construction, business and finance. However, in higher-education institutions, RMP appears to be significantly less developed than in much of the corporate world (Tufano, 2011). In higher education, particularly in the context of research universities (RUs), very few institutions have implemented risk management using an integrated approach to their quality-assurance regime or strategic-planning framework (Brewer & Walker, 2011). In Malaysia, five universities—Universiti Teknologi Malaysia, Universiti Sains Malaysia, Universiti Kebangsaan Malaysia, Universiti Putra Malaysia and Universiti Malaya—have been granted RU status (Ariff, Zakuan, Tajudin & Ismail, 2015). These universities are facing a host of risks, and are still searching for suitable effective risk-management models to implement to mitigate risk.

While higher-education institutions increasingly recognize that the effective management of risk is important for them, their focus has been on preventing risk from occurring and managing risk after the event (Brewer & Walker, 2011). In research on risk management, it has found that different sets of RMPs have been used and applied in different industries; therefore, there is no single standard or universal RMP available for risk management in organizations. Until recently, there were no studies examining RMP specifically in the RU setting. Most risk-management studies in higher-education services focus on the identification of measures, prevention, and post-risk management (Ariff et al., 2015). Therefore, universities need to strive to establish a concrete foundation in

RMP to ensure they are parallel with other industries, and are not left behind on contemporary knowledge, approaches, and practices of risk management. This limitation of RMP measures in the university environment highlights the need to determine RMPs that are workable in the RU setting. Further, it is important to produce RMPs that are not only suitable for universities to manage risk, but more importantly, contribute to university performance. Thus, this research aims to find a suitable RMP scheme for effective management of risk in the RU setting.

The goal of this study is to examine and propose appropriate practices of risk management in the RU setting. The study is significant because it provides suitable RMPs that can contribute to RU performance. It is hoped that this study will stimulate innovative ideas in managing risk in RUs worldwide.

2 LITERATURE REVIEW

Risk is defined as the 'effect of uncertainty on objectives [that] aids decision making by taking account of uncertainty and its effect on achieving objectives and assessing the need for any action' (International Organization for Standardization [ISO]). Thus, risk management is an important management tool that is used to eliminate potential problems in an organization.

Past research has confirmed that the formalization of processes based on process theory includes the process of managing risks (Kwak & Stoddard, 2004). The term 'process' refers to the steps and procedures in a given timeframe or situation that achieve a certain outcome or reaction (Ariff et al., 2015). Applying process theory to RMP involves determining the processes or activities involved in the practice of risk management. The variables of risk-management processes consist of the activities of communicating, consulting, establishing context, identifying, analysing, evaluating, treating, monitoring, and reviewing risk (Kululanga & Kuotcha, 2010). These practices are considered fundamental to risk management, and importantly, are all derived from process theory.

Based on past research conducted between 2010 and 2015, the application of 18 RMPs in various industries (e.g., construction, information technology, software development, business, and accountancy) were reviewed. The outcome of this review revealed that RMPs include the following activities: (i) identification of risk (16/18); (ii) risk analysis (12/18); (iii) risk evaluation (8/18); (iv) risk assessment, which combined risk analysis and risk evaluation, including risk prioritization (10/18); (v) risk response/treatment/mitigation/control (11/18); (vi) risk monitoring/reporting/ review (10/18); risk communication/consultation (1/18) (Ariff et al., 2015; Banholzer & Vosejpka, 2011; Brewer & Walker, 2011; Lee & Ali, 2012; Saleem & Abideen, 2011; Wang, Lin & Huang, 2010; Yazid, Hussin & Daud, 2011; Zigic & Hadzic, 2012). This review finds the following implications: (i) different sets of RMPs are applied in different industries, thus RMPs specifically for RUs should be explored; (ii) two studies are related to research and development, but no study has been conducted specifically on RMPs in RUs. Further, while conventional RMPs are implemented throughout the industries noted here, developing specific RMPs for RUs requires additional RMPs. These additional RMPs have been identified based on Committee of Sponsoring Organizations (COSO) and the ISO's ISO 31000:2009/MS and ISO 31000:2010. These additional RMPs are (i) risk governance; (ii) risk context; (iii) tools and technology; (vi) continuous improvement of risk management. All these RMPs were further evaluated (as described in the methodology section) to form a construct of RMPs specific to the RU setting.

3 METHODOLOGY

The two stages of the research are presented in the flowchart of this study (Figure 1). In Stage 1, the researchers attempted to establish RMPs for RUs. The following seven practices of risk management are proposed as a result of the activities described in the flowchart: governance of risk, identification of risk, analysis and evaluation of risk, response/treatment/mitigation of risk, monitoring and communication of risk, tools and technology, continuous improvement.

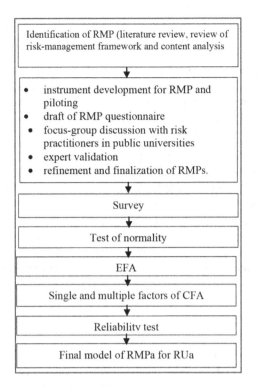

Figure 1. Development and validation of RMP.

a. review and use of process theory to determine RMPs in risk-management implementation
b. past research on RMP contributing to the identification of four RMPs for RUs
c. review of risk-management frameworks for managing risk, risk in the higher-education industry, and demand for effective risk management, contributing to the identification of three additional RMPs for RUs
d. content analysis of risk-management implementation in selected universities in the United Kingdom, the United States, Australia, Canada, and Singapore—this analysis was based on information related to risk-management implementation in the studied universities' websites, with the aim of examining and identifying RMPs used by universities.

Stage 2 of the research involved conducting focus-group sessions with risk practitioners and experts to finalize the initial proposed RMPs, as well as to verify the developed questionnaire. The focus-group sessions included discussion with directors and staff of Malaysia's public universities who are involved in the risk-management implementation. In this stage, content validation was also conducted with risk-management consultants and experts who are directly involved in risk implementation. The outcome of the Stage 2 activities was the identification of the final five practices of risk management: risk governance and management systems (RGMS); risk identification (RI); risk analysis and evaluation (RAE); risk-mitigation strategy and control (RMSC); and risk monitoring and communication (RMC), and a revision of the questionnaire used to measure RMP. It is important to note that four of the five RMPs proposed are a combination of different RMPs that contribute to the creation of a different set of RMPs specifically suitable for managing risk in RUs.

A pilot project to ensure the reliability of the questionnaire was conducted with 30 administrators of the studied universities. The results of the pilot confirmed the appropriateness of the proposed RMPs and the items used in the questionnaire. Thus, the RMPs and the questionnaires were finalized. The questionnaire featured the following 26 items:

a. RGMS (A1: risk-management committee; A2: clear roles and responsibilities; A3: risk-management policy and guidelines; R4: risk-management meetings; A5: use of system/risk dashboard)

b. RI (B1: comprehensive and systematic RI; B2: involvement of staff in RI; B3: risks related to R&D&C identified; B4: risks related to financial sustainability; B5: risks related to strategic direction; B6: new-risk identification).

c. RAE (C1: financial implication of risks; C2: quantitative and qualitative measures of likelihood; C3: risk rating; C4: effectiveness of existing controls; C5: risk list and prioritization).

d. RMSC (D1: risk-mitigation strategy; D2: risk target; D3: risk treatment/additional control; D4: positive and negative action to mitigate risk; D5: risk treatment/mitigation plan)

e. RMC (E1: monitoring of risk; E2: review of risk treatment effectiveness; E3: risk-management report and profile update; E4: improvement of risk-management activities; and E5: risk communication).

The senior administrators involved in risk-management implementation of the RUs served as the population of the study. It was determined to be necessary to gather 370 to achieve a 5% confident level given that the total number of questionnaires was 37. Two hundred and eighty-eight completed questionnaires were collected from the respondents, which constitutes a 77.8% response rate.

4 RESULTS

Exploratory factor analysis (EFA) and confirmatory factor analysis (CFA) were performed to test the validity of the RMP instrument. As presented in Table 1, the Kaiser–Meyer–Olkin (KMO) test for RMP is 0.970, supported by Bartlett's test for sphericity of 0.000, showing the adequacy of samples for EFA. Based on the EFA result, all these practices and items of RMP were retained given that the rotated component yielded five components and that the factor loading scores were ≥0.5 for all the items. The total variance explained for RMP was 67.557%.

For the single factor analysis of CFA (see Table 2), the results indicated that all indices values for the five practices of RMP met the parameters, thus all the variables of this study achieved satisfactory results. All factor loadings for each of the items of RGMS, RI, RAE, RMSC and RMC of RMP were >0.6.

The results of composite reliability for the five constructs of RMP were >0.7, supporting the measurement model of the study (Hair, Black, Babin & Anderson, 2010). For the discriminant validity, the average variance extracted (AVE) values of >0.7 for all practices of RMP are greater than square of correlation amongst the constructs, showing that the discriminant validity is supported.

For construct validity, the fitness indices of measurement model were assessed after setting the free parameter. As presented in Figure 2, all the indices values for the final measurement model of RMP were within the required levels, which demonstrates the measurement model is satisfactory. Further, all AVE values are >0.5, supporting the convergent validity of the measurement model of RMP.

The results of the composite reliability for the five constructs of RMP were >0.7 (RGMS: 0.925; RI: 0.941; RAE: 0.924; RMSC: 0.945; and RMC: 0.943), supporting the measurement model of the study. For discriminant validity, the AVE values of >0.7 (RGMS: 0.713; RI: 0.726; RAE: 0.709; RMSC: 0.776; RMC: 0.769) for all practices of RMP were greater than the square of correlation among the constructs, showing that discriminant validity is supported. All factor loadings for the items of RMP were >0.6.

For the reliability of the instrument, the Cronbach's alpha scores for RGMS, RI, RAE, RMSC, and RMC of RMP were 0.868, 0.852, 0.848, 0.894, and 0.899, respectively. Thus, based on the content validity, the EFA, single factor and multiple factors of CFA, as well as the construct and convergent validity of the study, it can be stated that the proposed RMPs are valid and reliable for effective risk-management implementation in RUs.

Table 1. EFA results for the RMPs.

	Component				
	1	2	3	4	5
A1:	.218	.212	.734	.206	.053
A2:	.329	.265	.700	.207	.129
A3:	.206	.290	.588	.302	.150
A4:	.294	.207	.700	.170	.189
A5:	.341	.264	.536	.289	.113
B1:	.262	.200	.268	.761	.227
B2:	.289	.170	.278	.710	.241
B3:	.274	.211	.363	.522	.281
B4:	.265	.199	.130	.765	.201
B5:	.281	.335	.157	.666	.248
B6:	.261	.295	.240	.695	.201
C1:	.373	.543	.201	.142	.206
C2:	.307	.661	.269	.138	.060
C3:	.218	.683	.081	.334	.124
C4:	.278	.560	.201	.293	.224
C5:	.258	.689	.233	.179	.198
D1:	.690	.213	.297	.177	.210
D2:	.622	.349	.269	.154	.247
D3:	.648	.290	.188	.317	.134
D4:	.636	.345	.141	.247	.070
D5:	.560	.357	.181	.255	.135
E1:	.010	.273	.205	.410	.609
E2:	.193	.270	.271	.236	.686
E3:	.247	.222	.203	.206	.655
E4:	.318	.171	.234	.079	.687
E5:	.153	.255	.287	.093	.724

Total variance explained	67.557%.
KMO test	0.970

Bartlett's test of sphericity	
Approx. chi-squared	10940.991
df	10
Sig.	.000

Table 2. Single factor analysis of CFA.

Name of cat	Name of index	Indices values				
		RGMS	RI	RAE	RMSC	RMC
Absolute fit	RMSEA	0.076	0.055	0.048	0.040	0.098
	GFI	0.995	0.978	0.982	0.991	0.971
Incremental fit	AGFI	0.946	0.962	0.971	0.998	0.928
	TLI	0.987	0.991	0.994	0.996	0.980
	CFI	0.986	0.993	0.995	0.998	0.971
Parsimonious fit	Chi-squared/df	2.976	2.031	1.795	1.555	4.342

5 DISCUSSION

The findings demonstrate some similarities and differences between the RMPs found in this study and those found in other industries. Specifically, RI, RAE, RMSC, and RMC are widely recognized as 'must-do' activities in risk-management implementation (Ariff et al., 2015; Lee &

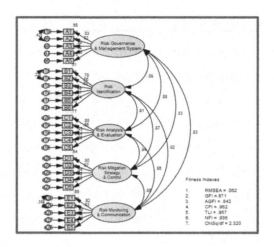

Figure 2. The final model of RMP for RU.

Ali, 2012; Saleem & Abideen, 2011; Yazid et al., 2011; Zigic & Hadzic, 2012). RGMS is critical and should be emphasized (Ariff et al., 2015; ISO), particularly in the early stage of risk-management implementation. Although RAE consists of different activities, this RMP should be combined because they can be considered a continuous process of analysing risk.

Risk strategy identifies the response that should be implemented to mitigate the identified risk (Wang et al., 2010) before determining additional controls to mitigate the identified risk; thus, they can be run simultaneously. Risk prioritization is a result of RAE; thus, it can be considered a part of the RAE process.

For practical purposes, it is suggested that the following RMPs be integrated and implemented by RUs to ensure effective risk management:

a. RGMS—The items that describe this RMP call for RUs to establish appropriate risk governance and tools to manage risk effectively.
b. RI—This RMP is similar to those implemented in other industries and is the most important step in managing risk. Without identifying risks, they cannot be managed. This suggest that without RI, all risks are reactively managed, resulting in ineffective risk-management implementation.
c. RAE—This RMP is loaded under the analysis and evaluation of risk that can be managed simultaneously, particularly in the service industry. Although risk evaluation requires an independent party to rate the identified risk by considering the effectiveness of the current controls for avoiding the occurrence of risks or for minimizing the effects of risks if they occur, in reality, particularly in the early stage of risk implementation, combining these is highly recommended.
d. RMSC—This RMP consists of activities related to risk mitigation. It is suggested that implementing a strategy of risk mitigation and the control of risk be combined. When managers determine the risk-mitigation strategy that should be executed, the additional control of the risks must also be addressed. This then ensures the risk-mitigation strategy is easy to develop.
e. RMC—This RMP involves activities that are essential for the involvement of all RU staff in risk management to ensure the identified risks are systematically monitored and communicated. This RMP calls for relevant authorities in the university, usually those in the risk-management unit or those sitting on risk-management committees at the faculty and university levels to continuously manage risk, including by identifying new risks and reprofiling current risks.

6 CONCLUSION

As stated, different sets of RMPs are used and applied for different industries; therefore, there is no single standard or universal RMP that can manage risks in all types of organizations. Based on the EFA and CFA results of this study, it was found that the constructs of RMPs for RUs are RGMS, RI, RAE, RMSC, and RMC. Thus, this study contributes significantly to the research on risk management by providing specific RMPs for RUs, particularly in the early stage of risk-management implementation.

The management of the RUs should consider the proposed RMPs for effective risk-management implementation. The risk-management framework and policy guidelines of RUs should encompass these RMPs. In addition, RUs should apply the RMPs systematically to improve performance by effectively managing risks with the aim of sustaining the RU's status and increasing its ranking in the global education market.

The limitations of this study include the inability to reach samples who are actively involved in risk-management implementation in the RUs under investigation. Thus, future research should find a systematic way of examining the responses of this group in the future. Further, this research did not focus on specific risks facing the universities. Future research should focus on identifying specific risks encountered by universities so that risk-mitigation strategies and additional controls to mitigate risks can be properly identified and planned.

ACKNOWLEDGEMENTS

The authors would like to acknowledge the Ministry of Education (PY/2013/00656) and Universiti Teknologi Malaysia for providing financial support for this research.

REFERENCES

Ariff, M. S. M., Zakuan, N., Tajudin, M. N. M. & Ismail, K. (2015). A conceptual model of risk-management practices and organizational performance for Malaysia's research universities. *The Role of Service in the Tourism and Hospitality*. 153–178.

Banholzer, W. F. & Vosejpka, L. J. (2011). Risk taking and effective R&D management. *Annual Review of Chemical and Biomolecular Engineering, 2*, 173–188.

Baumgartner, H. & Homburg, C. (1996). Applications of structural equation modeling in marketing and consumer research: A review. *International Journal of Research in Marketing, 13*(2), 139–161.

Brewer, A. & Walker, I. (2011). Risk management in a university environment. *Journal of Business Continuity & Emergency Planning, 5*(2), 161–172.

Hair, J. F., Black, W. C., Babin, B. J. & Anderson, R. E. (2010). *Multivariate data analysis*. Maxwell Macmillan International Editions.

International Organization for Standardization (ISO). *MS ISO 31000:2010: Risk management—principles and guidelines*. Retrieved from https://www.iso.org/iso-31000-risk-management.html

Kululanga, G. & Kuotcha, W. (2010). Measuring project risk management process for construction contractors with statement indicators linked to numerical scores. *Engineering, Construction and Architectural Management, 17*(4), 336–351.

Kwak, Y. H. & Stoddard, J. (2004). Project risk management: Lessons learned from software development environment. *Technovation, 24*(11), 915–920.

Lee, C. S. & Ali, A. S. (2012). Implementation of RM in the Malaysian construction industry. *Journal of Surveying, Construction and Property, 3*(1).

Mohr, L. B. (1982). *Explaining organizational behavior*. San Francisco, CA: Jossey Bass.

Saleem, S. & Abideen, Z. U. (2011). Do effective risk management affect organizational performance. *European Journal of Business and Management, 3*(3), 258–267.

Tufano, P. (2011). Managing risk in higher education. *Forum Futures*, pp. 54–58.

Wang, J., Lin, W. & Huang, Y. H. (2010). A performance-oriented RM framework for innovative R&D projects. *Technovation, 30*(11), 601–611.

Wilson, C., Negoi, R. & Bhatnagar, A. (2010). University RM. *Internal Auditor, 67*(4), 65–68.

Yazid, A. S., Hussin, M. R. & Daud, W. N. W. (2011). An examination of enterprise risk management (ERM) practices among the government-linked companies (GLCs) in Malaysia. *International Business Research, 4*(4), 94–103.

Zigic, D. & Hadzic, M. (2012). The process of risk management in financial business. *Journal of Applied Science, 9*(2), 33–40.

Inclusive Development of Society – Lumban Gaol (eds)

Assessing viral-advertising pass-on behaviour of online consumers: The consumer attitude perspective

N.N.A. Mahat, M.S.M. Ariff, N. Zaidin, N.Z. Salleh & R. Md. Nor
Azman Hashim International Business School, Universiti Teknologi Malaysia, Johor, Malaysia

N. Ishak
Lembaga Tabung Haji, Kuala Lumpur, Malaysia

ABSTRACT: This paper assesses the viral-advertising pass-on behaviour (VAPB) of online cosmetic consumers on social media (Instagram) from the consumer-attitude perspective. The construct of online consumers' attitudes are as follows: attitude towards Instagram of social media (AI), attitude towards advertising on Instagram of social media (AAI), attitude towards advertising in general (AAG), attitude towards advertised brand (AAB), and attitude towards advertising messages of the advertised product on Instagram (AAM). AAM is a new attitudinal factor added to the conceptualization of VAPB to predict VAPB comprehensively. An online survey using a self-administered questionnaire was employed. Three hundred and ten questionnaires were sent to the targeted users of Instagram. Two hundred and eighty completed and usable questionnaires were used, representing a response rate of 90%. The result of the multiple regression analysis shows that all the five attitudinal factors of online consumers significantly affect VAPB. This study provides a useful model of predicting VAPB from the consumer-attitude perspective. AAM is significant in measuring Instagram users' VAPB. It is also useful in designing persuasive advertising on social media.

Keywords: consumer attitudes, viral-advertising pass-on behaviour, social media, Instagram

1 INTRODUCTION

Instagram is one of the fastest growing social-media platforms with a growth rate of 19 to 47% from 2015 to 2017 (Wong, 2017). As one of the most popular platforms of social media, Instagram deserves attention from researchers and e-marketers to examine users' attitudes towards viral-advertising pass-on behaviour (VAPB). VAPB describes users' behaviour in relation to passing on advertising messages, therefore, understanding what stimulates users to pass on messages based on their attitudes in relation to VAPB has attracted a great deal of research attention.

Cosmetics is a lucrative industry that draws a great deal of attention in marketing research (Hayley, 2016) and makes wide use of online advertising, particularly on Instagram. One of the biggest industry players, MAC cosmetics (a brand of makeup and skin care) adopted Instagram as a trendy promotional channel. In fact, MAC cosmetics was ranked first, with 17.2 million Instagram followers (Wong, 2017), thus demonstrating that Instagram is the most popular social-network site and an important online means for MAC to connect with its customers. This trend has attracted researchers to study online consumers' VAPB in the cosmetics industry, particularly with MAC.

The current literature acknowledges the importance of understanding consumer attitudes in measuring the VAPB of online consumers (Nur Hidayah et al., 2016). The present study includes the following constructs in relation to VAPB: attitude towards Instagram of social media (AI) (Chu, 2011; Nur Hidayah et al., 2016); attitude towards advertising on Instagram of social media (AAI) (Chu, 2011; Kateler, 2012; Nur Hidayah et al., 2016); attitude towards advertising in general (AAG) (Chu, 2011); attitude towards advertised brand (AAB) (Chu, 2011; Kateler, 2012; Tan, Kwek & Li, 2013). These measures of consumer attitudes are well addressed in the conceptualization of VAPB in the literature; however, consumer attitude towards advertising messages (AAM) is less emphasized in the literature. Advertising messages on social media relay what a product is intended to be used for, and the superiority of one brand's offerings over its competitors' offerings. Advertising messages on social media provide useful marketing information, making it easier for consumers to exchange information and make information go viral (Chu, 2011; Nur Hidayah et al., 2016; Paquette, 2013). Users of social media who have confidence in these messages (Mahmood, 2016; Soyoen, Huh & Faber, 2014) may cause the messages to go viral among their connections by sharing and disseminating the marketing information and by receiving feedback about the product before they decide to purchase (Nur Hidayah et al., 2016). Thus, including this construct of consumer attitudes in measuring consumers' VAPB on social media, specifically Instagram, is of great importance.

This research aims to investigate the current conceptualization of VAPB of online consumers in relation to the following issues: (i) addressing and testing a comprehensive model of consumer attitudes that affect the VAPB; (ii) answering whether AAM and all the other tested constructs of consumer attitudes positively affect the VAPB of Instagram users, specifically among consumers of MAC cosmetics. In doing so, this study adopts a comprehensive examination of consumer attitudes to determine users' VAPB on Instagram. This research extends the current literature on VAPB in the context of the cosmetics industry, and enriches the literature by adding the dimension of AAM in the conceptualization of VAPB.

2 LITERATURE REVIEW

In predicting the behavioural intentions of social-media users in passing on advertising messages based on a consumer-attitude perspective, the theory of planned behaviour (TPB) (Fishbein & Ajzen, 1975) appears to be the most widely adopted theory in the research. TPB postulates that the primary determinants of a specific behaviour are the underlying intentions governed by three determining factors: attitude towards the behaviour; the subjective norms; and perceived behavioural control (Fishbein & Ajzen, 1975). The first component of TPB reflects the perception an individual has towards the potential consequences of behaviour. The formed perception, whether positive or negative, and the salient beliefs an individual might hold determine an individual's general attitude towards the behaviour. It shows that when a user of social media forms favourable AAM, they might engage in VAPB.

Previous studies on consumer attitudes affecting VAPB can be grouped into three types: (i) studies that use of attitude towards social media, attitude towards advertising on social media, and attitude towards advertising in general as independent variables of VAPB (Chu, 2011; Nur Hidayah et al., 2016); (ii) studies that use the additional dimension of attitude towards the advertised brands to predict VAPB (Kateler, 2012; Nur Hidayah et al., 2016); (iii) studies that highlight the importance of AAM as a new variable for predicting VAPB on social media (Mahmood, 2016; Soyoen et al., 2014). Based on the model that uses attitudes towards the advertisement, a consumer forms various feelings (affects) and judgements (cognitions) as the result of exposure to an advertisement, which, in turn, affects the consumer's attitude towards the advertisement and their attitude towards the brand (Schiffman & Kanuk, 2014). It reflects the perception, whether positive or negative, the individual holds towards the potential consequences of behaviour. In the context of advertising on social media, when an individual is exposed to an advertising message, this perception, in conjunction with the salient beliefs the individual might hold determine the individual's attitude towards the advertisement and the brand (Ketelar, 2012; Nur Hidayah et al., 2016; Schiffman & Kanuk, 2014). Thus, it can be concluded that assessing consumer attitudes in relation to VAPB on social media (for this study, Instagram) should include the following constructs: AI, AAI, AAG, AAB, and AAM.

It has been found that users of social media treat the use of Instagram as part of their daily routine, thus contributing to users' favourable attitude towards the networking site (Johnston, Chen & Hauman, 2013). When a favourable attitude is formed towards a social-media site (i.e., Instagram in this case), users are willing to participate actively on the networking site to provide or pass on product-related information (Chu, 2011; Nur Hidayah et al., 2016). In the context of VAPB, when users are exposed to an advertising message on social media, they are willing to viral the message among their peers. This kind of attitude drives users to engage in VAPB. Hence, users' AI influences their behavioural intention to viral the advertising messages posted on Instagram. Thus, this study proposes the following hypothesis:

H1: *Users AI positively affects their VAPB.*

Attitude towards advertising on social media refers to users' attitude towards advertisements, the brand, and their purchase intentions (Chu, 2011), as well as to how users or followers of a social-media platform react to advertisements appearing in the media. If users of social media react positively to the advertising messages in their purchase decision to buy a product, or prefer to read marketing information in the media, then a favourable attitude towards the advertisement on social media is formed. This attitude is affected by a variety of factors, for example, the users' demographic characteristics; the quality of the information provided on the social-media site; and the users' ability to access the media and information efficiently. Based on previous research (Chu, 2011; Nur Hidayah et al., 2016), this study hypothesizes that AAI has a positive effect on VAPB on Instagram. Thus, this study proposes the following hypothesis:

H2: *Users AAI positively affects their VAPB.*

Individuals' AAG is influenced by how they react to any particular advertisement. In general advertising, the attitude can influence consumer attitudes towards advertisements, the brand, and purchase intention (Chu, 2011), as well as towards VAPB. Previous studies have highlighted that AAI is found to have a positive effect on VAPB (Chu, 2011; Nur Hidayah et al., 2016). Thus, this study proposes the following hypothesis:

H3: *Users AAG positively affects their VAPB.*

AAB refers to an individual's affective reaction towards a particular brand after the exposure of advertising of that brand to that person (Tan et al., 2013). AAB captures how an individual responds, favourably or unfavourably, towards the advertised brand. It implies that if a user perceives a high value of a brand advertised on Instagram, Facebook, Twitter, or on another social-media platform, they might form a favourable attitude towards the advertised brand. Users' AAB stimulates them to become involved in VAPB (i.e., passing on and sharing and receiving feedback about information related to the advertised brand. It has been found that users actively become involved in connecting and interacting with a brand's activities through their own activities such as VAPB when they attach their positive affection towards the advertised brand (Chu, 2011; Nur Hidayah et al., 2016; Tan et al., 2013). Thus, this study proposes the following hypothesis:

H4: *Users' AAB (in this study, MAC) positively affects their VAPB.*

Users' favourable or unfavourable attitudes towards social media are positively related to their use of the information available on social-networking sites, which in turn could enhance their tendency to engage in viral message (Chu, 2011; Nur Hidayah et al., 2016; Tan et al., 2013). Advertising messages that are relevant and useful affect consumer attitudes positively (Aydin, Gulerarslan, Karacor & Dogan, 2013). For example, such messages can stimulate users to buy a product or drive them to engage in positive VAPB. Therefore, AAM is important in influencing users' VAPB. In fact, it is the most important construct of consumer attitudes affecting viral-advertising behaviour (Aydin et al., 2013). Thus, this study proposes the following hypothesis:

H5: *Users' AAM (in this study, MAC's message) positively affects their VAPB.*

3 METHODOLOGY

This research adopts a survey approach, employing self-administered questionnaires completed online. This research attempts to determine users' VAPB towards MAC cosmetics from the online-consumer perspective. The questionnaire developed for this study was based on the following:

- Four attitudinal factors of consumer attitudes that are widely employed in research of VAPB were utilized: six items each for AI and AAI (Chu, 2011; Nur Hidayah et al., 2016); five items of AAIG (Chu, 2011; Nur Hidayah et al., 2016); and four items of ATAB (Nur Hidayah et al., 2016; Tan et al., 2013).
- A new construct of consumer attitude (AAM) was added to measure consumer attitudes in relation to VAPB: five items were developed to measure AAM (Mahmood, 2016; Soyoen et al., 2014).
- Five items were constructed to measure VAPB (the dependent variable) (Chu, 2011; Ketellar, 2012; Nur Hidayah et al., 2016).

Multiple-choice questions were constructed to gather information about the use of Instagram 'followers' and 'non-followers of MAC's Instagram account, and their use of MAC cosmetics. This was to ensure that only users of Instagram who have accessed and/or use MAC cosmetics were able to participate in this study. Moreover, an advertisement of MAC cosmetics was attached to the survey to allow respondents to refer to the advertisement when completing the questionnaire. The questionnaire had 31 questions. Given that the total population (i.e., users of Instagram) that has accessed and/or used MAC cosmetics is unidentified, 310 sets of questionnaires are necessary for a 5% margin error (Hair, Black, Babin, Anderson & Tatham, 2006). Thus, 310 questionnaires were emailed to the prospective respondents, 280 of which were fully completed and usable for further analysis.

Assessment of instrument validity was conducted using exploratory factor analysis (EFA) to ensure the questionnaire was valid. The results indicated the following:

- For the first round of EFA, the results of Bartlett's test of sphericity were (chi-square 10815.553, Sig. 0.000), and 0.942 for the Kaiser–Meyer–Olkin (KMO) test, indicating the adequacy of sample to proceed with EFA. The total variance explained of EFA was 78.913%, with five component matrix of consumer attitude. All factor loadings for AI, AAI, AAG, ATAB and AAM were >0.50, except for one item of AI. This item was removed in the second round of EFA.
- In the second round of EFA, the results for Bartlett's test of sphericity were (chi-square 10940.991, Sig. 0.000), and 0.940 for the KMO test. The total variance explained of EFA was 84.450%. All items of consumer attitude were well loaded for AI, AAI, AAG, ATAB and AAM, with all factor loadings >0.50.
- For VAPB, the KMO was 0.898, Bartlett's test of sphericity was (chi-square 2260.411, Sig. 0.000) and the total variance explained of EFA was 90.719%. All the five items were well loaded in one component (i.e., VAPB), with a factor loading of >0.50.

For reliability analysis, the results of the Cronbach's alpha for all variables used in the study were >0.7. Specifically, the alpha values are as follows 0.930 (AI); 0.931 (AAI); 0.791 (AAG); 0976 (ATAB); 0.954 (AAM). The alpha value for VAPB was 0.974.

4 RESULTS

As presented in Table 1, the multiple regression results indicated that AI (β 0.129, t 2.659, Sig. 0.008); AAI (β 0.251, t 4.668, Sig. 0.000); AAG (β 0.257, t 4.970, Sig. 0.000); ATAB (β 0.138, t 2.710, Sig. 0.007); and AAM (β 0.197, t 4.351, Sig. 0.000) of consumer attitude positively and significantly affect users' VAPB. Therefore, H1, H2, H3, H4, and H5 are supported. AAG of users' attitude AAG (β 0.257, t 4.970, Sig. 0.000) exerted the strongest effect on VAPB.

Table 1. Regression results for AI, AAI, AAG, ATAB, AAM and VAPB.

Model	Unstandardized coeff. B	Std. error	Standard-ized coeff. β	t	Sig.	Collinearity statistics Tolerance	VIF
1	(Constant) .289	.165		1.754	.048		
	AI .133	.050	.129	2.659	.008	.539	1.854
	AAI .240	.052	.251	4.668	.000	.442	2.264
	AAG .252	.051	.257	4.970	.000	.477	2.098
	ATAB .134	.049	.138	2.710	.007	.491	2.038
	AAM .190	.044	.197	4.351	.000	.622	1.608
	F stat	33.083	R^2			0.624	

a. Dependent variable: VAPB

5 DISCUSSION

The findings indicate that the five constructs of consumer attitude: AI (β 0.129, t 2.659, Sig. 0.008); AAI (β 0.251, t 4.668, Sig. 0.000); AAG (β 0.257, t 4.970, Sig. 0.000); ATAB (β 0.138, t 2.710, Sig. 0.007); and AAM (β 0.197, t 4.351, Sig. 0.000) positively and significantly affect VAPB. The findings demonstrate that the attitudes of Instagram users affect their VAPB for MAC cosmetics. This finding is in line with Chin (2011) and Nur Hidayah et al. (2016). The positive effects of ATAB (β 0.138, t 2.710, Sig. 0.007) and AAM (β 0.197, t 4.351, Sig. 0.000) on VAPB indicate that users of MAC cosmetics tend to viral the brand and its advertising messages on Instagram. Viral advertising on social media is a marketing strategy that uses emails, social networking, and blogs to achieve specific marketing goals, and developing favourable AI, AAI, AAG, AAB and AAM of consumers is vital. Some studies have found that the effect of viral advertising on achieving marketing goals is limited (Brkic, 2012). However, the findings of this study contradict such conclusions, with the results demonstrating the importance of understanding consumer attitudes in designing effective viral-advertising strategies on social media. An appropriate and well-designed viral-advertising strategy focusing on developing favourable consumer attitudes promotes faster sharing of advertising messages among users of social media and more effective viral-marketing campaigns or VAPB (Rollins, Anitsal & Anistal, 2014).

This study extends the current literature on VAPB (AI, AAI, AAG and ATAB) by adding AAM as a new predictor of VAPB. The R^2 of .0624 demonstrates that the variation in VAPB is explained by the underlying factors of consumer attitude. The total variance explained of 84.45% (EFA) of consumer attitude is very high. Thus, a proposed model of VAPB from the perspective of consumer attitude provides a better understanding of online consumers' VAPB. The inclusion of AAM in the conceptualization of VAPB provides stronger effects of consumer attitude on VAPB than are found in studies of VAPB that do not consider AAM (Chu, 2011; Nur Hidayah et al., 2016; Tan et al., 2013). This demonstrates that the content of advertising messages posted on social media must be able to have a positive effect on users to enable a favourable attitude to be formed in the users towards the messages, and in turn drive them to become involved in viral advertising. Consumers tend to disseminate and share the advertising messages when they feel the message is useful and beneficial to them and other users. If consumers are considering buying a product, they might become involved in VAPB to receive feedback and relevant information about the product from other users before committing to a purchase decision. E-marketers need to create a positive effect through viral-marketing messages to boost the promotion of brands, products, and services (Haryani, Motwani & Sabharwal, 2015), which then leads to a purchase (Gholamzadeh & Jakobsson, 2011). Further, if users trust the advertising messages and the advertisers, they are willing to pass on these messages, thus engaging in positive VAPB (Aydin et al., 2013; Mahmood, 2016; Soyoen et al., 2014). This positive association of AAM with VAPB contradicts Kelly et al. (2010), who found consumers have confidence in neither the messages nor the media who post the messages. Consumers' confidence in advertising messages depends a great deal on the extent to which the content of the messages touches or affects their positive feelings towards the message. Thus, for e-marketers, designing a useful and meaningful advertising messages is crucial to ensuring users of social media react favourably to the message, attach positive feelings towards the message, develop a favourable attitude towards the advertisement, and finally, engage in positive VAPB.

6 CONCLUSION

The principal concern of this study is that the background of the Instagram users participating in the survey varied, as did the extent to which these users have accessed and/or consumed MAC cosmetics. It was observed that some of the participating users use Instagram actively, and some do not treat Instagram as part of their daily routine. Among users of Instagram, some of them consume MAC regularly, and some do not. Therefore, future research should address these user differences when examining the VAPB of online consumers because considering differences in usage patterns and rates, and exposure to a brand might lead to different results.

AAM is a new variable added to measure the effect of consumers attitudes on VAPB. The findings of this study demonstrate that AAM influences users VAPB on Instagram for MAC cosmetics, which supports the findings of Aydin et al. (2013). The inclusion of AAM resulted in a better prediction of online users' VAPB from the consumer-attitude perspective. It is suggested that future research further validate the effect of AAM on VAPB on various social-media platforms, using different products and groups of users.

REFERENCES

Aydin, D., Gulerarslan, A., Karacor, S. & Dogan, T. (2013). Value of sharing: Viral advertisement. *International Journal of Social, Human Science and Engineering, 7*(5).

Brkic, A. (2012). Uticaj viralnog marketinga na medijsku produkiju-Amereicka iskustva. *Kultura, 135,* 271–281.

Chu, S-C. (2011). Viral advertising in social media: Participation in Facebook groups and responses among college-aged users. *Journal of Interactive Advertising, 12*(1), 30–43.

Duffy, G. (2016). *How cosmetic brands drive huge engagement on Instagram.* Retrieved from http://www.newswhip.com/2016/05/cosmetics-brands-drive-engagement-instagram

Fishbein, M. & Ajzen, I. (1975). *Belief, attitude, intention and behavior: An introduction to theory and research.*

Gardner, J., Sohn, K., Seo, J. & Weaver, J. (2013). A sensitivity analysis of an epidemiological model of viral marketing: When viral marketing efforts fall flat. *Journal of Marketing Development and Competitiveness, 7*(4), 25–49. Retrieved from http://www.na-businesspress.com/JMDC/SeoJY_Web7_4_.pdf

Gholamzadeh, C. & Jakobsson, K. (2011). A quantitative study about how viral marketing affects the consumer buying act. A Dissertation in Business with Emphasis in Marketing 15hp.

Hair, J. F., Black, W. C., Babin, B. J., Anderson, R. E. & Tatham, R. L. (2006). *Multivariate data analysis* (6th ed.). Upper Saddle River, NJ: Pearson Hall.

Haryani, S., Motwani, B. & Sabharwal, S. (2015). Factors affecting the consumers attitude towards internet induced viral marketing techniques. *Arabian Journal of Business and Management Review, 5*(4), 1–4.

Hoffman, H. (2016). How the makeup and cosmetic industry is ruling social media. Retrieved from https://www.likeable.com/blog/2016/6/how-the-makeup-and-cosmetic-industry-is-ruling-social-media

Johnston, K., Chen, M. M. & Hauman, M. (2013). Use, perception and attitude of university students towards Facebook and Twitter. *The Electronic Journal Information Systems Evaluation, 16*(3), 201–211.

Ketelar, P. (2012). Why do people pass on viral advertising on social network sites? Investigating the effects of social and attitudinal factors. *Journal of Advertising Research,* 1–34.

Mahmood, J. A. (2016). The trust of viral advertising messages and its impact on attitude and behaviour intentions of consumers. *International Journal of Marketing Studies*, *8*(50), 136–145.

Salleh, N. M., Ariff, M. S. Zakuan, N., Sulaiman, Z. & Saman, M. Z. M. (2016). *IOP Conference Series: Materials Science and Engineering*, *131*, 1–10.

Paquette, H. (2013). *Social media as a marketing tool: A literature review*. Major Papers by Master of Science Students, Paper 2, University of Rhode Island. Retrieved from http://digitalcommons.uri.edu/tmd_major_papers/2

Rollins, B., Anitsal, I. & Anistal, M. (2014). Viral marketing techniques and implementation. *Entrepreneurial Executive*, *19*, 1–17. Retrieved from https://www.questia.com/library/journal/1G1-397579775/viral-marketing-techniques-and-implementation

Schiffman, L. & Kanuk, L. (2014). *Consumer behaviour* (11th ed.). New Jersey: Prentice Hall International.

Soyoen, C., Huh, J. & Faber, R. (2014). The influence of sender trust and advertiser trust on multistage effects of viral advertising. *Journal of Advertising*, *43*(1), 100–114. Retrieved from http://www.tandfonline.com/doi/abs/10.1080/00913367.2013.811707

Tomse, D. & Snoj, B. (2014). Marketing communication on social network solution in the times of crisis. *Marketing*, *45*(2).

Tan, W. J., Kwek, C. L. & Li, Z. (2013). The antecedents of effectiveness interactive advertising in social media. *International Business Research*, *6*(3), 2013.

Inclusive Development of Society – Lumban Gaol (eds)
© 2020 Taylor & Francis Group, London, ISBN 978-1-138-33476-2

Knowledge management in agile teams of flexible projects of the enterprise

S.N. Apenko & M.A. Romanenko
Dostoevsky Omsk State University, Omsk, Russia

ABSTRACT: This article presents the results of research on the grant of the Russian Foundation for basic research on the topic "Methodology of evaluation and formation of green (sustainable) project management in the regions of Russia (on the example of the Omsk region)." The purpose was to study the features of knowledge management in flexible projects and to offer a specific technology of knowledge management, which would allow achieving success of a flexible project in social indicators of sustainable development, research issues and how they are logically filed, and what is presented. The actual problem is, on the one hand, the lack of development in the practice of knowledge management in flexible sustainable projects using agile technologies, and on the other hand, the high need to adapt to the conditions of flexibility of knowledge management technology in projects. Our research question is as follows: "What should the technology of knowledge management look like taking into account the modern practice of managing flexible sustainable projects?" The answer to this question was obtained by means of a study using the method of expert assessments at 26 enterprises of Russia. The result was the technology of knowledge management developed by us in agile teams of flexible projects of the enterprise. A feature and novelty of the proposed technology is the repeated cycle of information source refinement, information diagnostics and its transformation into knowledge, and systematization and evaluation of the knowledge quality of the project. Our dedicated cycle takes place in each iteration of the project work according to the agile concept of project management. The relevance of this technology and the willingness to use it are confirmed by a survey of 54 experts at 26 enterprises in one of the regions of Russia.

1 INTRODUCTION

Currently, a project form of organizational activity is developing intensively all over the world and project management acquires the status of professional work with a certain set of specialized tools. These tools include knowledge management in the form of a set of technologies and methods of diagnosis, accumulation, analysis, storage, and use of useful information for current and future projects. Knowledge management is especially important for projects implemented with a flexible methodology. The relevance of knowledge management for flexible projects stems from the fundamental principles of the flexibility concept, associated with the emphasis on people as knowledge holders, their interaction, their experience of teamwork in the project, learning lessons from the project stages, and taking into account the needs of all stakeholders of the project.

Additional relevance of our research is attached to the fact that the technology of knowledge management makes the project sustainable and allows an increase of its success and achievement of social indicators within the concept of sustainable development.

However, as world practice shows, only 6% of projects are recognized as fully successful. Our research at Russian enterprises, which will be discussed in this work, showed that only 23% of enterprises use a diverse set of professional methods and techniques of project management, and 19% of enterprises are trying to implement in their projects a knowledge management system. The situation is slightly better in companies in the IT sector. Here, 36% of organizations use some methods of knowledge management, but these indicators are low. In general, most enterprises use only some methods of project management and fragments of knowledge management. In addition, almost all the experts of the organizations studied acknowledged the lack of readiness of flexible projects teams' staff to participate in knowledge management. Project managers say that the lack of effective knowledge management practices in projects is one of the significant factors affecting the success of projects. Leading companies in project management recognize that building knowledge management technology significantly increases the chances for success, especially in flexible projects. Therefore, our research is related to the urgent problem of the need to organize knowledge management in companies for the growth of the percentage of successful flexible projects.

2 THEORETICAL BASIS

The issues of flexible project management technologies have been in the focus of researchers' attention in recent years. With regard to project management, flexible methodologies in the management of it projects and software development projects have been developed. For the works that laid the foundations of flexible project management or popularize them, we can refer to the publications of Martin et al. (2004), Cohn (2011), Rassmuson (2012), and Wolfson (2015). These works highlighted the project flexibility parameters, such as a high degree of project management system adaptation to changing environmental factors; the adjustment of the project parameters (timing, quality, cost, and other) under evolving product customers of the project; the implementation of the project in the form of short iterations of work, after which there is a process of testing and approval of the manufactured product by the customer; and systematically ascertaining the project implementation quality. Projects that meet these and other parameters are referred to as flexible.

As a result, flexible project management requires flexible teams. Some comments about human resources of flexible projects can be found in publications. But at the moment there is no reflection on the essence of a flexible team. Therefore, we present our author's interpretation of agile teams, which we propose to understand the totality of the performers and managers of the flexible working of the project in terms of multitasking, the increased dynamism of environmental factors, the need to concentrate many of project roles and project functions into one, creating a product of the project with short iterations and systematic discussion of project work progress, and producing the product with all stakeholders.

Knowledge management is recognized by practice and scientific direction of organizational management. The concept of knowledge management emerged in the 1990s. It considers knowledge management as a global structuring of information and knowledge in the organization. For example, in 1994 T. Davenport proposed considering knowledge management as a process of formation, dissemination, and effective use of knowledge (Davenport, 1994). In 1998, B. Duhon proposed formulating the definition of knowledge management as "...a discipline that promotes an integrated approach to identifying, capturing, evaluating, retrieving, and sharing all of an enterprise's information assets. These assets may include databases, documents, policies, procedures, and previously un-captured expertise and experience in individual workers" (Duhon, 1998).

A. I. Vlasov and S. L. Lytkin in the work "A brief practical guide for the developer of information systems based on Oracle DBMS" gave the definition of "knowledge" as "a combination of experience, values, contextual information, expert assessments, which sets the general framework for the evaluation and incorporation of new experience and information" (Vlasov & Lytkin, 2000).

Our research is devoted to knowledge management in projects as a separate functional area of project management. Many authors point to the importance of knowledge management in projects, for example, Razu, (2011), Larson (2013), Meredith and Mantel (2014), Polkovnikov (2015) and others. Most often, knowledge management in projects is understood as the accumulation and subsequent use of knowledge on projects, programs, and project portfolios in the form of sets of knowledge, standards, project documentation, expert opinions, and so on.

Knowledge in project management is understood to be different in content components. For example, a team of researchers, including Laura Diaz Anadon, Kira Matus, and Suerie Moon, include knowledge in technology. They say that 'technology is seen as knowing how to accomplish certain human goals in a certain and reproducible way." Therefore, the knowledge management technology we propose further can be considered as an element of corporate knowledge in itself (Diaz Anadon et al., 2012).

Our research focuses on sustainable projects, that is, contributing to the maintenance of sustainable development policies. Issues related to the transition of projects to sustainable development have been actively discussed recently. For example, scientists say that technological innovations that are implemented in projects do not always provide sustainability. The urgent task therefore is to find solutions on how to make technological innovations more sustainable in projects. In this way, scientists see improved functioning of the "global innovation system," in particular, its fairness to "meet the needs of the poorest, most vulnerable or marginalized segments of the population in the present and future generations."

Scientists Tim Stock, Michael Obenaus, Amara Slaymaker, and Günther Seliger develop the idea of a more effective policy implementation of sustainability technological innovation. They write that "targeted development of new sustainable innovations is one of the key activities to ensure sustainable industrial growth." In response to practice requests, scientists propose a model for the development of sustainable innovation, which focuses on the generation of ideas at an early stage of the innovation process, focusing on indicators of sustainability and innovation. The ideas of these authors are in good agreement with flexible projects, as these projects are able to take into account the sustainability indicators at the early stages of generating ideas for an innovative product. For our research, the principle of iterating over the creation of a new product in the project is important, which we will take as a basis for the development of knowledge management technology. So, scientists Tim Stock, Michael Obenaus, Amara Slaymaker, and Gunter Seliger say that the process steps in their proposed model of sustainable innovation development, "should not be considered as a rigid sequence, but rather should be considered as an iterative process to the idea of solution. This concept is based on the assumption that a human-generated process can be ideally obtained in such an iterative way, as it follows the natural behavior of a human being in solving problems. The iterative approach also provides a possible reorientation of the solution search process and allows ex post to integrate new relevant aspects" (Stock et al., 2017).

Our research is also interested in developments on the relationship between the human situation in the enterprise and the requirements of sustainable development. We cite Lisa Larsson and Johan Larsson's study as an example. In their work, they say that from the standpoint of sustainability "human development is something much more than the growth or decline of national income. It is about creating an environment in which people can reach their full potential and lead a productive, creative life in accordance with their needs and interests" (Larsson & Larsson, 2018).

Flexible teams, which we will talk about further, create such conditions for the disclosure of the potential of team members. The aforementioned authors propose two approaches for analysis: the first concerns the sustainability of the labor activities associated with the technological choice; the second discusses the development of labor activity with the purpose of introducing an enabling technology for sustainable development. Our research is consistent with the first approach, as knowledge management is a technology that contributes to the social dimension of sustainable development.

To clarify the object of our study, it is important to divide the projects into different types. For example, the division into two types of projects proposed by Lisa Larsson and Johan Larsson is of interest. They say that "innovation can arise as a result of solving exploitative problems in business projects, and as a result of research projects in specific organizations, followed by implementation in inter-organizational business projects" (Larsson & Larsson,

2018). That is, business projects and research projects are highlighted. Our research will be more focused on research projects, as the need for flexible teams is higher in those areas.

In addition, Lisa Larsson and Johan Larsson conducted research on case studies and showed that the success of innovative projects was influences by such factors as the flexibility of the process, the commitment of senior management, roles of key individuals, internal and external cooperation, and customer orientation (Larsson, L., Larsson, J., 2018). For our study it is important to recognize the flexibility of the processes that support cooperation as success factors in innovative projects. The knowledge management technology offered by us further supports the flexibility and collaboration of stakeholders in the project.

Therefore, there are studies that provide answers to questions about what knowledge management in projects is and why knowledge management is necessary in projects. The answer to the question on how to manage knowledge in agile projects is presented in fragmented ways. Therefore, it remains unclear what to do exactly in managing knowledge in flexible projects. There are also studies on the sustainability of innovative projects, in which different aspects of sustainability are studied. But the flexibility of the team and its use of knowledge management technology to ensure social sustainability have not yet been addressed by scientists.

3 FORMULATION OF THE RESEARCH QUESTION

Our research question was as follows: "What should the technology of knowledge management look like taking into account the modern practice of flexible sustainable project management?" We hypothesize that the technology of knowledge management in flexible projects should be different from the technology in conventional projects and nonproject activities. Understanding the technology allows more efficient implementation of knowledge management. And the introduction of technology will contribute to the achievement of social indicators of sustainability. Therefore, the purpose of our research was to develop a technology of knowledge management taking into account the generalization of best practices in this field and expert assessments. Knowledge management should cover the whole enterprise, but we put in the spotlight a flexible project team and knowledge management in this team focused on sustainability indicators.

3.1 *Research method*

The methods of data collection in this study were analysis of documents on the project activities of enterprises and expert assessment of the state of knowledge management in flexible projects. The study was carried out at 26 enterprises of Omsk, one of the regions of Russia typical from the point of view of project management development. The selection of enterprises was based on the following criteria:

- Availability of project activities at the enterprise
- Availability of at least some elements of knowledge management at the enterprise
- Use of flexible project management technologies at the enterprise
- The need to cover the study of different enterprises: by size, industry, field of activity

The following were investigated:

1. 38% were large enterprises, 35% were medium enterprises, and 27% were small enterprises.
2. 15% were petrochemical enterprises, 27% were machine-building enterprises, 31% were enterprises in the IT field, 15% were in the construction industry, and 12% were in the trade industry.
3. 100% of the enterprises implement projects, have a project management system, implement sustainable development policies, and to varying degrees are engaged in knowledge management.

The sample consisted of 54 experts of these enterprises, who are experienced project managers. Their experience in project activities was not less than five years, including experience in managing flexible projects for at least two years. We emphasize that the enterprises participating in the study are leaders in flexible project management, and they demonstrate the best experience that allows us to develop recommendations for those who want to improve their knowledge management in flexible projects.

3.2 Results of the study and the rationale for their novelty

When listing facts use either the style tag List signs or the style tag List numbers.

To substantiate the technology of knowledge management, we asked experts: "What is the difference between knowledge management in flexible projects and knowledge management in conventional projects?"

The differences are as follows:

1. Knowledge is updated faster, is more diverse, and can often be contradictory because of the increased dynamics of the factors of the project environment. This feature was confirmed by 89% of experts.
2. The knowledge management cycle is shorter and repetitive owing to multiple iterations of product development and a return to the periods of diagnosis and knowledge accumulation. This feature is indicated by 70% of experts.
3. Knowledge holders are not only project documents or project databases but also the members of the project team and various stakeholders, whose knowledge often takes a hidden form. This feature was confirmed by 81% of experts.

We also asked whether knowledge management was a technology that could contribute to sustainable development. The majority of experts (85%) gave an affirmative answer.

In particular, according to experts, knowledge management technology will achieve the following sustainability indicators:

- Development of cooperation and partnership of participants and stakeholders of flexible projects (96% of experts pointed to this figure)
- Equal access to training of different participants of project flexible teams (89% of experts named this indicator)
- Availability of relevant information to all stakeholders (59% of experts noted the indicator)

Taking into account the peculiarities of knowledge in flexible projects, as well as the practice of advanced enterprises, we have compiled a knowledge management technology that includes a list of sequential actions at different stages of flexible sustainable project management, as well as a description of the subjects of these actions (Figure 1).

The peculiarity and novelty of this technology support the iterative cycle of further sources of information, diagnosis of information and its transformation into knowledge, and systematization and evaluation of the knowledge quality of the sustainable project. Our dedicated

Figure 1.   Knowledge management technology in Agile teams of flexible projects of the enterprise.

cycle takes place during each iteration of the project work according to the Agile concept of project management.

Different diagnostic, knowledge storage, and transfer tools can be used for flexible sustainable projects. Moreover, flexible teams should use the same flexible methods of working with knowledge. For example, enterprises use "cloud" software for project and program management quite successfully in remote communication. This tool was named as useful by 87% of the experts we interviewed. Project teams can structure their documents and build remote project communication, as well as capture all their knowledge using a browser or mobile device.

However, even more adequate to the specifics of the flexible team is a set of methods of direct team interaction. The peculiarity of the information about the flexible sustainable project is that its significant part is owned by individual members of the project team. And the information in the heads of the team members is often not structured, not identified, and has not taken the form of knowledge. Therefore, special tools for the diagnosis of information and transformation into knowledge are needed. Such tools, for example, are scram meetings and sessions on retrospective analysis. These practices were named as actively used in the enterprises by 57% of experts.

Often descriptions of these methods show examples of visualization of knowledge information, for example, Kanban boards (visual boards). These visualization techniques help to track current knowledge and transform it into team knowledge. But it is also possible to combine methods of group development of knowledge on the project and information technologies of their storage and distribution.

We give examples of the benefits obtained in the course of cooperation within a flexible team, showing how the results of group work are converted into knowledge in "cloud" software products:

- Project questions, answers, and brainstorming are integrated in one place as part of the team's natural workflow.
- During work with a flexible team, documents are created and processed through the cloud.
- All materials related to project meetings and project progress analysis during implementation and after project completion are stored in the cloud and available for search.

Another specific feature of flexible teams is their constant interaction with many different stakeholders, for example, with the customers of the product, contractors, and potential consumers of the product. For example, contractors who have important roles in a team become the most valuable because they have very important information/knowledge. It is necessary to identify and preserve this knowledge until the project is completed and the contractors leave the project. To avoid knowledge leakage, it is recommended to use simple project management platforms with such social tools as news feed and discussions. When these technologies are integrated into the workflow, knowledge is extracted and captured in the cloud. In other words, the contractor is engaged in the management of knowledge, not knowing about it.

4 SUMMARY

Knowledge management in projects is often considers as a very complex area of activity; as a consequence, many companies are afraid to turn to knowledge management tools, losing possible benefits. We argue that the time has come when knowledge management, especially in Agile and sustainable projects, becomes a necessity. This requires the development of knowledge management technologies, the involvement of suitable methods and techniques for specific conditions, and the training of a flexible team. Our research is devoted to solving these urgent problems. The introduction of the proposed technology will allow achievement of social indicators of sustainable growth. Although we offer knowledge management technology, it should be said that effective knowledge management cannot be achieved by blindly following any standards for the implementation of processes, techniques, and tools. Moreover, each organization and project team in particular should identify and discuss the processes and methods used and develop and implement improvements that may be unique within the framework of ongoing project work.

REFERENCES

Cohn, M. 2011. *Scrum: Agile software development. Succeeding with Agile: Software Development Using Scrum.* Addison-Wesley Signature Series. Moscow: Williams.

Davenport, T. H. 1994. *Managing Knowledge.* Cambridge, MA: Harvard Business School Press.

Diaz Anadon, L., Matus, K., & Moon, S. 2012. *Innovation and Access to Technologies for Sustainable Development.* Available at: https://www.hks.harvard.edu/centers/mrcbg/programs/sustsci/activities/program-initiatives/innovation/projects/innovation-and-access-to-technologies-for-sustainable-development

Duhon, B. 1998. It's all in our heads. *Inform,* 12(8):8–13.

Larson, E. 2013 *Project Management: The Managerial Process.* rev. ed. Erik W. Larson, Clifford F. Gray. M. New York: McGraw-Hill Education.

Larsson, L., & Larsson, J. 2018. *Sustainable Development in Project-Based Industries: Supporting the Realization of Explorative Innovation.* Available at: www.mdpi.com/2071-1050/10/3/683/pdf

Martin, R. C., Newkirk, J. V., & Koss, R. S. 2004. The rapid development of programs: Principles, examples, practice. In *Agile Software Development. Principles, Patterns, and Practices.* Upper Saddle River, NJ: Pearson.

Meredith, J., & Mantel, S., Jr. 2014. *Project Management,* 8th ed. Hoboken, NJ: John Wiley & Sons.

Polkovnikov, A. V. 2015 *Project Management. Full MBA course.* Moscow: ZAO "Olimp-Biznes."

Rasmusson, G. 2012. *Flexible Management of IT-Projects: A Guide for Real Samurai.*

Razu, M. 2011. *Project Management. The Basics of Project Management.* Moscow: KnoRus.

Stock, T., Obenaus, M., Slaymaker, A., & Seliger, G. 2017. *A Model for the Development of Sustainable Innovations for the Early Phase of the Innovation Process.* Procedia Manufacturing 8:215–222. 14th Global Conference on Sustainable Manufacturing, GCSM October 3–5, 2016, Stellenbosch, South Africa.

Vlasov, A. I., & Lytkin, S. L. 2000. *A Brief Practical Guide for the Developer of Information Systems Based on Oracle DBMS.* Moscow: Mashinostroenie.

Wolfson, B. I. 2015. *Agile Project Management and Products.*

Inclusive Development of Society – Lumban Gaol (eds)
© 2020 Taylor & Francis Group, London, ISBN 978-1-138-33476-2

The evolution of the Internet of Things is a new technological breakthrough in the global economy

L.A. Kargina, S.L. Lebedeva, O.V. Mednikova, I.I. Sokolova & M.A. Britvin
Russian University of Transport (MIIT), Moscow, Russia

ABSTRACT: Another inevitable technological revolution is brewing. It is much simpler and at the same time potentially more significant than the one of a single device. This is a data revolution that can end many of the shortcomings, troubles, dangers and unsafe aspects of modern life.

1 ECOSYSTEM – THE BASIS OF THE INTERNET OF THINGS

In its simplest form the industrial Internet of Things (IoT) is a computerization of a full complex of jobs at a particular enterprise. In this case all production facilities (the equipment, work places, etc.) are added to uniform the network. Due to this approach the ideal working environment in which machines understand their own environment is created and special online-protocols are used for communication, independently solving problems of production efficiency increase or, for example, preventing various hazardous situations. As a result, significant increase in overall performance of all enterprise participants is provided.

In the current research the following issues were considered: the basic concepts of the Internet of things as methodologies to use the extensive computer network with a large number of sensors to increase the production efficiency. The reasons and prerequisites of the described approach emergence have been considered.

The object of research was the network of the Internet of things built on the basis of heterogeneous network technologies. The subject of research was the application of communication models and algorithms in networks of the Internet of things. The purpose of the study was to identify the key scenarios of the Internet of things use in marketing.

The Internet of things generates the spread of digital networks within the physical space - the network blood of the smart city. Smart cities unite in themselves not only the networks of municipal services, such as electricity and water, but also all participants in city life, including citizens, the state and business. There is a wide range of models for the implementation of new opportunities in different parts of the world.

According to industry analysts, today there are about 10-20 billion objects connected to the Internet. This ecosystem of connected objects forms the basis of the Internet of things.

Today, the number of connected objects is negligible compared to how many will be connected in just a few years. According to various estimates, the range of connected facilities by 2030 will be from 60 to 80 billion and include everything.

According to the RAND Europe survey, by 2020, the top estimate of the annual global economic potential of the Internet of things in various sectors will range from 1.4 trillion USD (1.09 trillion euros) to 14.4 trillion USD (11.2 trillion euros) and will be equivalent to the current GDP of the European Union. In fact, by then the Internet of things will no longer be an isolated segment of IT, but rather, it will become a driving force of the world economy. In five years, a rare industry will not be changed by the Internet of things. Even today, few industries receive nothing from the use of Internet objects of things in their processes or products. At the same time, there are several advanced sectors where the Internet of things has become irreplaceable.[1]

The world of the Internet of things will lead to the economy complexity. The rules that industries and governments have adopted to promote growth and competition will not be effective in the long run. The Internet of things will affect every country and economy on the planet, even the developing countries, which historically have been deprived of the benefits of technological progress.

It is safe to say that the world industry is on the eve of another revolution. The growth in the number of connected objects, known as the Internet of things, can eclipse such technological wonders of the past as a printing press, a steam engine and electricity.

Unlike previous industrial revolutions, this one can be predicted. Each industry and each individual company will benefit and prosper by introducing objects of the Internet of things into their business models and, as a result, will open new and better ways of doing business.

There will be new numerous industries, and old ones will disappear. But the phenomenon of the Internet of things is unique as it allows companies to prepare, adapt and prosper in this new economic era.

The world is rapidly moving towards a new era, the distinguishing feature of which is the emergence of technologies of the Internet of things. Familiar devices, such as refrigerators or cars, learned to connect to the network and exchange data without human intervention.

The hardware, software, communication infrastructure, as well as the connected devices involved in the data exchange process, are combined into a technological ecosystem called the Internet of Things (IoT).

The pace of distribution of connected devices and new data streams in production is impressive. It expects further acceleration of implementation rates, especially as infrastructure costs (for example, sensors, computing power, data warehouses) decrease, and procedures for analyzing data received from devices and processed by cloud IoT-platforms are simplified. Applications based on the platform, in turn, contribute to the emergence of completely new types of analytics, which can be obtained in real time or as close to it as possible. According to some estimates, in the next decade, Internet of things can bring producers about $ 4 trillion by increasing revenues and cutting costs.

Thanks to the introduction of data collection, storage and analysis technologies (for example, sensors, controllers, analytical software and telemetry, large data sets and cloud computing), manufacturers can predict the need for maintenance of the equipment prior to the occurrence of malfunctions and thus prevent malfunctions in production processes. The appearance and quality of individual parts can be controlled by using optical sensors or cameras. After that, a barcode or RFID (Radio Frequency Identification) tag is attached to the part that allows to track electronically the entire supply chain for this part, including even its use by the customer. Thus, the connections are formed between the machines, as well as between the machine and the person.

Manufacturers are beginning to use the technology of nternet of things and data analysis in order to maintain their competitiveness. According to the Global Data & Analytics Survey conducted by PwC, over the past two years, 57% of plant managers around the world have changed their approach to making serious decisions as a result of using large amounts of data or analyzing data. These changes affected the use of data modeling tools, training managers in interpreting data analysis results, as well as involving specialists in data analysis.

Obviously, some manufacturers have advanced further than others in the way of mastering the Internet of things.

---

1 www.aig.ru/content/dam/aig/emea/russia/documents

All manufacturers can roughly be divided into three groups: pioneers (those who were among the first to use the data), outside observers (those who do nothing or practically nothing in this direction) and newcomers (those who just started or plan to start using these technologies).

It can be assumed that in a few years practically all manufacturing enterprises will widely use technologies and systems based on the Internet of Things. The study showed the following:

- 35% of manufacturers collect data from smart sensors and use them to optimize production and operational processes;
- 17% of producers plan to do this in the next three years;
- 24% claimed they have such plans, but the exact terms have not been determined yet;
- according to 34% of manufacturers, the introduction of the results of using the Internet of things in operational activities is very important;
- 38% of manufacturers build in their products sensors, through which the consumer can collect data and monitor the operation of the device.[2]

2 METHODOLOGY

Internet of things - the concept of interaction between devices that can collect, store, process and send data to the Internet, directly to others.

Devices can be very different: from the camera sending images to the sensor on the wind turbine, collecting data on the parameters of the environment and turbine performance.

Sensors collect a lot of data, including pressure, humidity, optical performance, speed of moving parts, sound and so on. For industrial enterprises Internet of Things becomes a whole ecosystem, when thanks to the software coupled with cloud resources and analysis tools, data streams are converted into analytical information or forecasts, and this information can be accessed through a simple and convenient interface (for example, in the form of information panels, mobile or WEB-applications), which allows users to monitor and manage, and in some cases also automate the management of equipment or systems.

For the incredible real-time monitoring capabilities, the truth that not all manufacturers have the resources to scale up and manage data across the organization maybe hidden. Table 1 presents data from market research of the Internet of things in Russia, conducted by analytical center TAdviser together with Rostec State corporation.[3]

Figure 1. Chart of the enterprises' need to use Internet of things.

Table 1. Enterprises' need for the use of IoT.

Name of the enterprise	Necessity %
energy industry	33
municipal services	32
automobile industry	31
industrial production	25
health care	20
technology	17
financial services	13

2 /https:/www.pwc.ru/ru/publications/iot-for-indusrty.html
3 https://www.pwc.ru/ru/publications/iot-for-indusrty.html

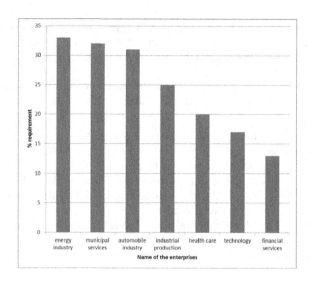

Figure 1. Enterprises' need for the IoT,%
Source: 6th annual PwC Digital JQ study

If we talk about big international corporations, they invest heavily in data collection and processing systems in attempt to increase productivity and improve product quality, and sometimes even be able to track the performance of the product even after it leaves the workshops.

Back in 2012 the American corporation General Electric (GE) announced that its investment in the implementation of Internet industry in the production and service processes will amount to $ 1.5 billion[4]. As part of this program, the company, for example, installed more than 10,000 sensors at its plant in Schenectady, New York. These sensors are connected to the internal Ethernet LAN and used to track the consumption of materials, control the sintering temperature used in the accumulators of high-tech ceramics, as well as atmospheric pressure. Via the factory Wi-Fi network, data are sent directly to the tablet devices of the workshops.

In GE wind turbines, there are about 20,000 sensors that produce 400 measurements per second. The incoming data are analyzed almost in real time, which allows optimizing the turbines operation. At the same time, the collected data are stored and used for predictive analytics in order to improve the efficiency of maintenance operations and replacement work.

EAM systems are also used to solve less complex problems, such as detecting energy losses.

Receiving data on real-time energy consumption, it is possible to adjust the schedule of operation of energy-intensive equipment at the plant in such a way as to avoid the costs associated with the peak load and get a more accurate picture of the share of electricity costs in the total production cost.

Sensors are used not only to reduce costs. With their help, manufacturers plan to ensure the quality of their products and even its safety. Consider, for example, the pharmaceutical industry: optical sensors are used here to measure optical properties in the context of product quality control. Packaging is checked for compliance with the established norm of light transmission. This is a big step forward in comparison with the previously used verification method by statistical sampling. Optical sensors are also used in difficult conditions, in particular in mines, rigs, welding machines or steelmaking. In the oil and gas industry, sensors are used to measure certain indicators, for example pressure, and they can transmit information to shut-off devices by closing or opening them, in order to prevent leakage.

---

4 https://www.vedomosti.ru/

Looking to the future, the Internet of things promises to unite the financial activity of the consumer with other aspects of his life. One example is the connection of the user's health monitor to its financial portfolio. As Deloitte noted, the health problems recorded by the monitor can give a signal to the user's bank to automatically balance his portfolio in order to minimize financial risks.

Indeed, there is no area where the technology of the Internet of things has not been applied or will not be used in the future. In healthcare at the patient level, gadgets with Internet-enabled things will allow doctors to obtain health data that otherwise would remain unknown. Annual medical examinations can become unnecessary, because physicians will already have enough individual patient data that will enable them to understand whether the examination will be justified. In addition, doctors will be able to detect diseases that are asymptomatic before they lead to more serious problems. Physicians will be able to use this data not only to better understand the health of a particular patient, but also to create detailed data sets for subgroups of patients, with the aim of treating and preventing the most ancient diseases of mankind.

Using the IoT technology promises to sharply reduce the number of production injuries and deaths. According to the International Labor Organization, 2.3 million people die from accidents and illnesses in the world each year. According to the European Commission, every year more than three million workers are victims of serious industrial accidents, of which 4,000 die. Internet technology of things will help to protect employees, especially those who work alone in hazardous areas, for example, at construction sites. Thus, a built-in sensor equipped with diagnostic technology can be attached to the worker's clothing when the worker is at risk or performs dangerous actions. Sensors can also monitor dangerous environmental conditions, such as extreme temperatures and the presence of toxic substances. In addition, behavioral data collected by sensors attached to clothing will help health and safety managers understand when the likelihood that an accident may occur to an employee is greatest. It is worth noting that in the technology of the Internet of things there is a forecast element, in many respects still theoretical, but it is one of the most exciting functions.

3 RESULTS

As more manufacturers master the technologies of the Internet of Things to establish smart production, the technologies are being applied throughout the entire value chain in the organization. Enterprises are even considering the possibility of creating new service-oriented business models that involve remote monitoring of assets in the course of their operation and the provision of maintenance services. To understand how actively solutions based on the Internet of things are used in the ecosystem of manufacturers, we asked them which areas of their activity they switched to the technologies under consideration, including smart sensors, machine-to-machine interaction and human-machine interaction, large arrays of data and so on. Among those manufacturers who use these solutions, one in three uses technology on the basis of data only at the production site, and one in four only in production and storage. According to the survey results, about 20% of manufacturers use technologies based on data throughout the value chain: on the production site, in the warehouse, in the extended supply chain and in the process of customer work.

For some manufacturers, building inter-machine data flows and implementing analysis tools can be a challenge.

4 CONCLUSION

The Internet of Things continues to build up a global information network in the length of a whole generation. Here we can distinguish two overlapping directions: the creation of smart connected products and the collection of data to improve the results of business activities. Different sectors and companies concentrate their efforts in one direction or another. The first direction includes well-known consumer devices - from smart watches and thermoregulators to robotic home helpers and even connected cars. The second direction includes the Industrial Internet of Things, through which manufacturers and other market participants collect data from equipment and other sources and analyze them in order to optimize their processes, predict and prevent problems and, ultimately, create better ecosystems for new products and services.

At the same time, the Internet of things can bring significant difficulties to all sectors of the economy and for all industries. Despite the fact that it solves the problems that pursue business for decades, if not centuries, it also creates entirely new dilemmas, both procedural and ethical. Concern about the use of personal data, cybersecurity, property and responsibility for product quality will quickly become as relevant as the Internet of things. Companies should start implementing Internet technologies of things if they hope to survive in the long term. And they also need to implement strategies that will take into account the many risks associated with using the Internet of things.

REFERENCES

1. Karen Tillman, a blog of Cisco Systems and a counter of connections to the «Internet of things» in real time (Cisco System's Internet of Everything Connections Counter). The data is downloaded from the site http://blogs.cisco.com/news/cisco-connections-counter.
2. Article ARC Advisory Group «Manufacturers can get 3.88 trillion US dollars thanks to the «Internet of Things» (IoT) Could Enable $ 3.88 Trillion in Potential Value to Manufacturers, 2014.
3. PwC International Data Collection and Analysis (PwC Global Data & Analytics Survey), 2014.
4. Joint research and analysis of PwC and Zpryme «Disruptive Manufacturing Innovations Survey», 2014.
5. Article «Internet for Production» (An Internet for manufacturing), magazine MIT Technology Review, January 28, 2013
6. The article «GE makes machines and then uses sensors to listen to them» (G.E. Makes the Machine, and Then Uses Sensors to Listen to It), Quentin Hardy (Quentin Hardy), The New York Times, June 19, 2013.
7. The article «Internet of Things» can become a source of innovation, you just need to understand how sensors work» (The Internet of Things can Drive Innovation – If you understand sensors), Altera website, http://www.altera.com/technology/system-design/articles/2014/internet-things-drive-innov ation.html.
8. Practical examples from the Opto22 website, http://www.opto22.com/site/documents/doc_drilldown. aspx?aid=4038. 2014g.
9. Ersue, M .; Romascanu, D .; Schoenwaelder, J .; Sehgal, A. 2015. «Management of Networks with Constrained Devices: Use Cases». IETF Internet Draft.
10. Management of networks with constrained devices: Use cases. M Ersue, D Romascanu, J Schoenwaelder, A Sehgal - 2015 - rfc-editor.org
11. https://www.pwc.ru/en/communications/assets/the-internet-of-things/PwC_Internet-of-Things_Rus. pdf. «Prospects for the development of the Internet of things in Russia»
12. https://www.vedomosti.ru/
13. www.pwc.ru/ru/publications/iot-for-indusrty.html
14. fastsalttimes.com/sections/obzor/1875.html
15. aig.ru/content/dam/aig/emea/russia/ documents/business/iotbroshure.pdf

Inclusive Development of Society – Lumban Gaol (eds)
© 2020 Taylor & Francis Group, London, ISBN 978-1-138-33476-2

Main trends in tourism development in Russia and abroad

E.V. Mikhalkina & A.V. Gozalova
Southern Federal University, Rostov-on-Don, Russian Federation

A.I. Makarova
Russian State University of Tourism and Service, Russian Federation

ABSTRACT: This article views the industry of recreation as the basis of the economy of many developed and developing countries. This topic is relevant today because tourism is one of the fastest-growing industries in terms of both national and global economies. The rapid development and significant volumes of foreign exchange earnings actively influence various sectors of the economy and contributes to the development of tourist and recreational potential of the examined territories. The findings of this research in the field of tourism show that it has become not just a socioeconomic phenomenon but also a critical factor in the development of territories, regions and countries. Functional features of types and forms of tourism, such as mass character, profitability, creativity and innovation promote development of territories and increase their competitiveness. The main object of this article is the identification of key trends in tourism development on both national and international levels.

1   INTRODUCTION

Tourism as one of the fastest-growing industries has become a centre of interest for many contemporary economists. It is particularly important to analyze some obvious trends of tourism development in Russia and abroad. The relevance of the chosen topic can be explained by the timeframe of the research because it only takes into consideration theoretical and statistical data from the last few years. This fact explains the 'modernity' of the research highlighted in the title.

The research aims to determine key trends of the modern recreation industry in Russia and abroad which serve as the object of the present study. The research attempts to answer the following questions:

a) What are the main trends of the modern recreation industry in Russia and abroad?
b) What kind of impact does the tourist industry have on economic development: positive or negative?
c) What are the key figures in Russian tourism development of the last two years (e.g. internal tourist flow, number of Russian and international tourists)?

2   METHODOLOGY

The present paper reflects a theoretical and applied study of modern trends of the recreation industry in Russia and abroad. The study uses statistical analyses based on data retrieved from recent electronic resources, such as official documents published on governmental websites and the most recent articles. This method allowed the authors to access quantitative data and to analyze them along with the corresponding literature. There was also a detailed analysis of current trends on the tourism market and forecasts concerning the tourism business.

Due to the chosen methodology, the research results suggest that the tourist industry may have both positive and negative impacts on economic development. They also show that tourism cannot be the only branch of the economy to be focussed upon.

3 MAIN TRENDS OF THE MODERN RECREATION INDUSTRY IN 2017–2018

3.1 *Rapid growth of tourism revenue*

From an economic point of view, tourism is one of the leading and most rapidly developing sectors of the world economy. Due to such rapid development, it was declared as the economic phenomenon of the century. Profits obtained from tourism in terms of the total volume of exported goods and services has taken second place globally after exports of oil and oil products. This is the first key trend of the modern recreation industry associated with the rapid growth of tourism on the basis of the population's income thanks to the attractiveness of mass tourism. Thus, according to the World Tourism Organization (WTO) (2017), tourism's contribution to the world economy is equivalent to 11–12% of global gross national product. The benefit obtained from tourism is 6–10% of gross domestic product in the countries of the EU and other industrialized countries (Kalimullin & Tishkov 2017). According to WTO forecasts, the tourism sector will develop rapidly in the future. The number of travelling people will reach 1.6 billion people per year by 2020. Tourist arrivals will thus increase by 2.4 times in comparison with 2000. Benefits gained from tourism, according to WTO forecasts, will reach 1,550 million in 2018, which is 3.3 times more than in 2000, and by 2020 the income is projected to increase up to 2,000 million dollars (Pyatkova 2017). With an annual growth of 8%, the number of tourist arrivals in China will reach 137.1 million dollars by 2020. The second most popular tourist destination will be the USA (102.4 million), followed by France (93.3 million), Spain (71.0 million) and Hong Kong (59.3 million) (Shol`c 2004).

Today, the tourist industry includes not only Europe, which is a traditional tourist centre, but also America, Africa and Asia. The rapid development of outbound tourism is expected according to WTO forecasts. Germany, Japan, the USA, China and the UK will become the largest countries/suppliers of tourist flows. The relative economic backwardness of Eastern Europe is a real obstacle for involving the population of these countries in international tourism. Their share of international tourism accounts for approximately 7% of world exports (Ligidov et al. 2012).

The positive impact of tourism on the economy of the state occurs only when it develops comprehensively. The economic effectiveness of tourism suggests that tourism needs to develop in conjunction with other sectors of socioeconomic complex of the country.

3.2 *Development of related sectors of economy*

The second trend of the modern recreation industry is the development of related sectors of the economy.

Tourism stimulates the development of other sectors of the economy, primarily building, trade, agriculture, consumer goods production, communication, etc. Businesspeople are attracted to the following conditions: low start-up investments, growing demand for tourism services, high level of profitability and minimum term of return on investment.

In a number of foreign countries, tourism is one of the highest priority sectors, the contribution to the gross national income of which is 15–35% (Hungary, the Czech Republic, Austria, Sweden, Italy, France, Spain, Portugal) (Bucenko & Kulakova 2017).

Due to its geographical location, Russia is not a country of mass arrival of tourists for the purpose of traditional summer beach holidays and cannot become one in the foreseeable future. However, the cultural, historical and natural potential of the country is significant. Inbound tourism will become the most profitable component of the tourism sector with the help of a well-built marketing strategy of Russia, focused on the key guides of tourist markets.

There are currently more than 15,000 tourist-based organizations in Russia (Grigoriev 2017). The development of tourism promotes budgetary recharge, industry de-monopolization, development of other economic sectors such as trade, transport, communications and production of consumer goods, as well as developing citizens' constitutional rights.

Despite the fact that tourism is developing rapidly in Russia, its impact on the economy is insignificant. It is adequate in terms of its contribution to the state, with the development of this industry being mainly constrained by the lack of real investment, low level of hotel service, insufficient number of hotels and lack of qualified staff. According to the most optimistic estimates, the Russian tourism industry employs only one employee out of 300, which is 30 times lower than the same global figure (Saburov 2016, Ignatiev 2015).

Underdeveloped tourism infrastructure, poor service quality and a stable myth about Russia being a high-risk zone have led to the country currently accounting for less than 1% of global tourist flow.

According to forecasts, tourism could have a significantly positive impact on the economy of the country and its major cities in the next 10–15 years. For example, by 2020, the number of foreign tourists arriving in Moscow could reach 10.6 million people, while foreign exchange earnings could amount to 2.4 billion dollars (Il'in 2017). Thus, tourism as a profitable sector of the economy can become an important item of gross national income for Russia under appropriate conditions.

3.3 *Development of tourism infrastructure*

The third trend is related to the development of tourism infrastructure.

The successful development of tourism in Russia requires an inflow of investment, both Russian and foreign primarily in the development of tourist infrastructure for the formation of a network of tourist class hotels and, in particular, small hotels and motels which should be located on federal-aid highways and could provide comfortable accommodation at low prices.

We should also note the importance of integrated development of tourism infrastructure. It includes not only accommodation facilities, but also tourist attractions, transport, including car hire, roads, parking, airports, catering, souvenirs, means of communication, etc. Investing in the development of the above-mentioned tourist industry objects should be carried out with the involvement of both Federal and regional budgets and extra-budgetary resources. For example, building, reconstruction and maintenance of airports, federal-aid highways, the restoration of architectural monuments to turn them into tourist attractions and general arrangements for preservation and development of historical centres should be financed by the state. Regional and local authorities are responsible for the development of local infrastructure and improvement of tourist centres located on the territories of these authorities.

3.4 *Monoculture of tourism*

Along with the positive effects of tourism, there is also the 'other side of the coin' - the so-called monoculture of tourism. There is a constant struggle for land, resources and financial capital. The industry is pressing agriculture and other resources of inhabitants' income. Tourism workers are paid higher wages, which affects agriculture because of the outflow of labour. As a result, agricultural production is declining, while consumption is increasing due to the arrival of tourists. The traditional way of life and the natural landscape in the centres of mass tourism are also being disturbed or completely destroyed. The monoculture of tourism destroys its beginning itself. Diversity is the basis of economic stability. When one industry is experiencing acute economic distress, the other is thriving, thus reducing the possibility of a crisis. Therefore, instead of promoting diversification, tourism sometimes replaces the agricultural sector. For many reasons, tourism cannot be allowed to become the replacement industry. First of all, tourism depends on seasons. Sometimes seasonal fluctuations in demand can be reduced, but they can never be completely avoided. If tourism provides the main income of the region, the low season can lead to serious employment problems. Secondly, the demand for tourism and travel depends largely on the tastes and income of tourists. Thus, full dependence of the region on a single industry sector is highly objectionable. Tourism generates certain problems with social and additional costs for the maintenance of the environment, which the host region and its inhabitants must solve. If they stop the rapid development of tourism, it will entail an economic crisis. If they do not constrain the development of the industry, it can lead to the impoverishment of cultural and natural resources of the country because of excessive use, and they will become unusable and worthless. In this case, as a rule, it is very difficult to make a decision.

Sometimes tourism increases the inflation rates of the region where it develops. Tourists invest their money, which is earned in another region or country, in the economy of the tourist region. As this fact increases the income of the region, it may cause inflationary pressure. Consumer goods such as food, clothing, housing and transport increase in price. Lands in tourist regions are very often overcharged without any reason (price increases can reach 20,000%). The price that foreigners pay for their stay in a tourist region can suddenly reduce the demand for housing of inhabitants. Inhabitants are displaced from the housing market in areas with developed tourism.

We can therefore conclude that tourism is not the key to solving all economic problems at both national and global levels even if it has considerable potential as a tool for economic development. Governments of different countries should make efforts to optimize, but not to maximize profits from tourists, taking into account all the negative aspects of the industry.

Regarding Russia, tourism was relatively successful at the end of 2017. The flow of Russian tourists abroad increased by 30%, but the situation was opposite for domestic destinations. Difficult situations and negative events could not be avoided in 2017–2018. The bankruptcy of the Vim-Avia airline spoiled 38,000 Russian tourists' holidays, who had to return from vacation on board this airline. This year, there was also the bankruptcy of Ted travel, while 755 Russian citizens were abroad. They were returned to the country due to efforts of the Federal Agency for Tourism, Tour Assistance and market participants. Losses to tour operators, according to their data, amounted more than two milliard Rubles (Bezchastnaya 2017).

According to domestic tourism analyses for the year, the sale of air tickets for domestic destinations decreased by 7%. However, not everyone agrees with statements that Russians have begun to travel less in Russia. The head of the Federal Agency for Tourism of the Russian Federation Oleg Safonov (2017) believes that the domestic tourist flow stayed at a record level of 2016, and it increased by 5–10% in a number of areas. According to the Department of Sports and Tourism of Moscow, 19 million tourists, including 4.5 million foreigners, visited the city in 2017. According to statistics, 63% of Russians decided to vacation in Russia. The main location of choice is Moscow (22%), followed by St. Petersburg (10.5%) and Krasnodar (5%). Sochi took 4th place (4.3%): more than 500,000 people spent their New Year holidays there, with 358,000 of them staying in ski resorts. But the Crimea took only sixth place (3.2%). The top 10 domestic destinations also included Simferopol, Mineralnye Vody, Kaliningrad, Novosibirsk, Rostov-on-don and Yekaterinburg (Figure 1). However, according to Frolova et al., Russians generally treat foreign resorts as a better choice due to their ecological and climatic peculiarities, while social groups which they belong to do not play any role (Frolova, E.V. et al. 2017). A new cruise line was opened between Sochi and the Crimea in 2017. Regular cruise communication on the ring route of Sochi-Novorossiysk-Yalta-Sevastopol began on June 11, 2017. Only during summer, approximately 4,000 passengers made cruises with the ship «Prince Vladimir». Domestic tourist flow is growing rapidly: in 2014, the number of tourists traveling in Russia increased by 30%, in 2015 — by 18%, in 2016 — by 10% and, according to experts, 2017 was a year of some stabilization (Figure 2). Some factors promoting domestic tourism include the rapidly developing hotel market and recent achievements in transport machine building (Lepeshkin et al. 2016).

The flow of foreign tourists to Russia also reached a maximum this year for the eight-year period. According to the border service of the Federal Security Service of Russia, the tourist flow of foreigners increased by more than 16% compared to the same period

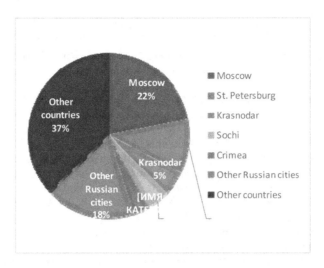

Figure 1. Tourist destinations chosen by Russians in 2017.

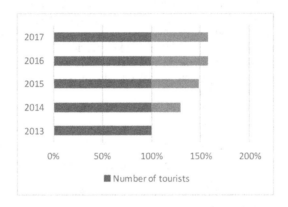

Figure 2. Growth of domestic tourist flow.

last year. The leaders in terms of entry were the Chinese: their tourist flow amounted to approximately 957,4000 only on the visa-free group regime for 2017, which is 24% more than for the same period last year. The Germans were in second place (368,000) and U.S. visitors third (207,000). The top 10 also included South Korea, Israel, Italy, France, Spain, Iran and Japan. According to Interfax-Moscow, more than 21 million guests visited the Russian capital in 2017. National Geographic magazine noted Moscow to be among the seven cities recommended for travel in 2017. The Wall Street Journal included Moscow in the top 5 most interesting destinations for winter holidays. These ratings are not particularly surprising because Russia is unique, both in terms of its size and in the variety of travel types (Glagolev 2017).

In terms of outbound tourism, in 2017, Russians made about 31 million trips abroad. The main destination was Turkey, which was not on the market for approximately one year. Tourist flow to Turkish resorts amounted to more than 4 million tourists, and several million of them were travellers who rested in Russian resorts the previous year. Guests brought the city 117 million rubles, 57 million being sent to the regional budget. The number of trips by Russians on this route were more than 3.9 million, which amounts to eight times more than in 2016. China is not located so far in the leader board (more than 1 million +25%) (Glagolev 2017).

4 INBOUND AND OUTBOUND TOURISM IN RUSSIA

Both types of tourism are influenced by factors such as geographical location of the country, its environmental and cultural peculiarities and economic status on the international market. These are general characteristics of any tourist destination.

Outbound tourism is often discussed with regard to the social status and economic situation of a tourist. Obviously, a lack of financial resources is the main reason preventing Russian people from travelling abroad. In addition, they are often more attracted by domestic tourist destinations promoted by mass media. Despite this fact, Russians generally have more positive attitudes towards foreign resorts, being impressed by the environmental and cultural peculiarities of the tourist destination, as mentioned previously. TurStat statistical data show that the top 5 countries for outbound tourism in 2018 were Abkhazia, Finland, Kazakhstan, Thailand and Ukraine. Except for accessibility of transport to the neighbouring countries, there are also economic factors influencing the choice of Russian tourists.

Some concrete reasons preventing foreigners from travelling to Russia include insufficient development of international flights, visa restrictions, safety issues, expensive tour packages, low-quality service, weak infrastructure, foreign policy tension and, lack of information about tourism in Russia (Ermakova 2017). There are also some positive factors promoting incoming tourism. They include a slightly simplified visa regime, introduction of the so-called 'tourism police', the development of infrastructure both in Moscow and in Russian regions, and creation of national 'Visit Russia' tourist centres in different countries. The latter provide foreigners with some image information about Russia. According to TurStat, the most popular destinations for incoming tourism in 2018 are Ukraine, Kazakhstan, China, Finland and Azerbaijan. Thus, the popularity of these destinations is influenced by their geographical location. However, last year saw a larger number of Ukrainian people visiting Russia, due to the political situation in the former country.

5 CONCLUSION

Summarizing the 2017–2018 tourist year, it can be stated that Russia has significant potential to become another unique centre of tourism on the world stage. The famous sights of Moscow, St. Petersburg, Yekaterinburg, Sochi, as well as the cities of the Urals, of the Far Eastern regions and the Volga coast attract increasing numbers of foreign tourists. The development of various types of active tourism such as skiing, hiking, water-skiing, sailing, horsing, cycling and extreme tourism has significant opportunity in the vastness of our country. However, if the state continues to adopt the experience of countries with developed tourism, adapting it and implementing its own innovative solutions, it will allow approaching the share of the tourism industry in GDP to the value of the leading European countries (which amounts to 10%). The state should make efforts to optimize income from tourism, taking into account the costs that its development can provoke. The tourism industry should develop in parallel with other sectors of the economy.

Tourism is developing thanks to major international sports events, such as the preparation and holding of the World Cup 2018, which took place in 11 Russian cities, and where matches were held, stadiums were built and the infrastructure of cities was improved. According to the Federal Tourism Agency, the championship was attended by approximately one million foreign fans and people accompanying them. They brought 1.5–2 million dollars. The World Cup became the starting point of the tourism industry and attracting foreign tourists. According to the President of the Federal Agency for Tourism Sergey Shpilko (2017) «it will be a reserve for many years, and for several years then it will have an impact on the development of the market».

REFERENCES

Bezchastnaya, M. 2017. Collapse of the Vim-Avia: blow to the authorities, tourism and aviation. [Electronic resource] Available at: https://svpressa.ru/economy/article/182067/ (Accessed on July 10, 2018).

Butsenko, I. N. & Kulakova, D. S. 2017. *International travel market: dynamics of the development and basic participants*: 3. Simferopol: Institute of economics and management 'V. I. Vernadsky Crimean Federal University'.

Ermakova, S. 2017. Inbound tourism in Russia: The concept, problems and prospects. [Electronic resource] Available at: http://fb.ru/article/320693/vyezdnoy-turizm-v-rf-ponyatie-problemyi-perspekti vyi (Accessed on August 12, 2018).

Frolova, E.V. et al. 2017. Tendencies and prospects of tourism industry in Russia: Sociological analysis of stereotypes among population during tourism trips. *European Research Studies Journal XX* (2B): 308–320.

Glagolev, A. B. 2017. Tourist market Russia 2017. The results and prospects. [Electronic resource] Available at: http://sci-article.ru/stat.php?i=1522656267/ (Accessed on July 7, 2018).

Grigoriev, L. 2017. Recovery of demand for tourism services in Russia. *Bulletin on current trends of Russian economy* (31): 7. Analytical Center for the Government of the Russian Federation. [Electronic resource] Available at: http://ac.gov.ru/files/publication/a/15422.pdf (Accessed on July 12, 2018).

Ignatiev, A.A. 2015. Development of the Tourism Industry in the North of Russia Based on the Environment. *Mediterranean Journal of Social Sciences* 6 (5): 387–401. Rome: MCSER Publishing, S4Features. [Electronic resource] Available at: http://citeseerx.ist.psu.edu/viewdoc/download?doi=10.1.1.830.2501&rep=rep1&type=pdf

Il'in, V. N. 2017. *The impact of tourism industry on the structure of the economy*: 2. The Russian Presidential Academy of National Economy and Public Administration (Altai branch).

Kalimullin, D.M. & Tishkov, A.A. 2017. Current state and prospects of the tourism industry in Russia 2017. *Scientific forum: Economics and management: proceedings of IX International scientific and practical conference* 7 (9): 57–64. Moscow.

Lepeshkin, V.A. et al. 2016. Inbound tourism within the context of Russian tourist market transformation. *Education for Entrepreneurship - E4E: International Journal of Education for Entrepreneurship* 6 (1): 121–135.

Ligidov, R. M. et al. 2012. Innovative development tourist-recreational complex of the region. *Modern problems of science and education* (4). [Electronic resource] Available at: http://science-education.ru/ru/article/view?id=6819 (Accessed on July 10, 2018).

Pyatkova, S.G. 2017. Organization of tourist activities. *International journal of experimental education*: 65-67. Surgut: Surgut State Pedagogical University.

Saburov, O. B. 2016. *Innovation processes in tourism*: 3. Omsk: Omsk State Institute of Service.

Shol`ts, K.A. 2004. Tourism: Forecast until 2020. [Electronic resource] Available at: https://p.dw.com/p/5JLi (Accessed on July 12, 2018).

Analytical agency TurStat. Inbound tourism in Russia 2018 [Electronic resource] Available at: http://tur stat.com/inboundtravelrussia3month2018 (Accessed on August 12, 2018).

Analytical agency TurStat. Outbound tourism in Russia 2018 [Electronic resource] Available at: http://turstat.com/outboundtravelrussia3month2018 (Accessed on August 12, 2018).

Federal Agency for Tourism. Interview of the head of the Federal Agency for Tourism Oleg Safonov. [Electronic resource] Available at: https://www.russiatourism.ru/content/9/section/85/detail/14546/ (Accessed on July 25, 2018).

The system of statistical registration of inbound tourism: barriers of industry development. Bulletin on competitive situation in Russia (17): 4–5. Analytical Center for the Government of the Russian Federation. [Electronic resource] Available at: http://ac.gov.ru/files/publication/a/12533.pdf (Accessed on July 13, 2018).

The World Cup-2018 will be the best presentation of Russia as a tourist destination. [Electronic resource] Available at: http://fineworld.info/rst-chm-2018-stanet-luchshej-prezentaciej-rossii-kak-turistiches kogo-napravleniya/ (Accessed on July 25, 2018).

Inclusive Development of Society – Lumban Gaol (eds)
© 2020 Taylor & Francis Group, London, ISBN 978-1-138-33476-2

Potential of ecotourism in city branding: A case study of Kabupaten Pinrang, South Sulawesi, Indonesia

Eli Jamilah Mihardja
Universitas Bakrie Jakarta, Indonesia

ABSTRACT: The explanation and identification of the special characteristics of a particular area is a criterion for such area to have a strong brand. The development of ecotourism has been observed to be a potential strategy of city branding in Pinrang. From the results of the surveys and in-depth interviews conducted, there are opportunities and challenges in the development of local ecotourism as a strategy for city branding in Kabupaten Pinrang (Pinrang Regency).

1 INTRODUCTION

The Indonesian Ministry of Tourism website (2015) reveals that branding of a city requires careful thought. It must begin with the exploration of potential regions in Indonesia that "qualify" for city branding specifications in the world. Indonesian government stressed that the Pemda (Pemerintah Daerah) or local government must also build a brand for the region, certainly in accordance with the potential and positioning of targeted area. The implementation of city branding strategy comes with various benefits, which include awareness, reputation and good perception as a tourist destination. In addition, the city branding concept can encourage the investment climate, as well as increase tourist visits to the place where it has been implemented.

In this case, Pinrang Regency is located in the South Sulawesi Province area of the north. This area has a lot of potential such as its natural wealth, especially from the agricultural sector and landscapes which have the potentials of becoming a place that gains tourist attractions.

A legendary brand that is capable of staying for decades or even hundreds of years does not just appear. They take steps that are planned, clear, and different from those of their competitors. Likewise, to have a strong brand, an area must have special characteristics that can be explained and identified. For example, the physical appearances of the city, the experience of people in the area, as well as the type of people that live in the area are of utmost importance. The development of ecotourism in this region requires building on these characteristics. Pinrang has considerable potential for ecotourism such as mountain tourism and beach tourism. This is because of its waterfalls that are being visited by many people despite the fact that it has not yet been officially opened. These waterfalls include Kalijodo, Karawa and Latta Pitu waterfalls. Pinrang also has beaches, dams, hot springs, and islands in the middle of the lake.

This research explores whether the ecotourism potential of Pinrang could be developed as its city branding. Simple survey, indepth interviews and documentary study as well as observation were used as data collection methods. Then, theme analysis of qualitative data was implemented. The results were displayed in a descriptive narrative manner so as to have a portrait of the data.

2 LITERATURE REVIEW

City branding is an integral part of city marketing (Karavitz, 2004). When implemented, city branding can serve as way by which the economy of a particular place can be developed (Allen, 2007) and can also help in building national identity (Hall, 2004). As a result of this, the role of local government is very significant (Deas and Ward, 2000; Guler 2017). Branding also shows the development of a regional agenda (Jensen and Richardson, 2005).

The use of the potentials provided by ecotourism in city branding is based on studies carried out by Latupapua (2007) and Tafalas (2010). Latupapua (2007) aims to determine the potentials of nature such as beach attractions, caves, natural springs, culture, flora, and fauna in tourism and also in knowing tourist perceptions, community perceptions and their roles about these natural habitats. The study also aims to know the suitability of tourism product components in the determination of the direction of development in Tual Southeast Maluku Regency. The Southeast Maluku Regency was found to have potentials for ecotourism and the perception of visitors and the public was positive to the fact that ecotourism aids the development of the community. Object development and ecotourism attractiveness in Southeast Maluku was carried out by developing potential objects and tourist attractions, structuring spatial planning, increasing the active role of the community, increasing human resources, structuring management, institutions and more intensive promotion and marketing.

Thus, ecotourism ensures sustainability and conservation of nature. Ecotourism activities increase income, expenditure and production assets (Tafalas, 2010). Therefore, development of ecotourism should be used as a tool in city branding.

Studies show that ecotourism experience affects tourist loyalty (Anahid and Ardabili, 2018). Likewise, a study was carried out on Hong Kong that uses the concept of green revolution and sustainable energy as a form of city branding (Chan and Marafa, 2014). According to them, the green city branding has raised the interest of people in visiting the city (Chan and Marafa, 2017).

3 PUBLIC PERCEPTIONS REGARDING THE UTILIZATION OF ECOTOURISM AS CITY BRANDING

Public knowledge about the place of ecotourism in Pinrang is quite adequate and, generally, most people knows about it through the use of social media. The data from the surveys carried out in March and July 2018 shows that the community also likes the place of ecotourism in Pinrang Regency. However, according to the respondents, there is no proper management of the tourist attraction areas. Elements of security and comfort are poorly maintained. Therefore, the improvement of management related to the comfort and safety of tourist attractions is highly needed according to the respondents.

Respondents generally do not know about government programs and policies on tourist attractions in Pinrang. In addition, management and marketing of tourist attractions in Pinrang is still considered to be very minimal. As a result of this, respondents generally doubt and do not think that tourism in Pinrang can be used as a city identity.

The tourist destinations in Pinrang are meant for the enjoyment of culinary and natural scenery. This can be a potential input for development. From the total respondents in the two surveys carried out (a total of 115 people), as many as 63% said they enjoyed visiting a tourist attraction in Pinrang. This data shows that the visit satisfaction is relatively still around half of all respondents. This means that the potential for tourism in Pinrang still need improvement for better management and wider promotion.

It was also discovered that social media has a big role to play in the promotion of tourism in Pinrang. In addition, the public preferences towards Pinrang are on culinary tourism and enjoying the natural scenery.

As a result of this, the combination of culinary tourism and natural scenery is recommended as a form of ecotourism in Pinrang. At the same time, management and marketing must be made broader and better.

4 DEVELOPMENT OF LOCAL ECOTOURISM POTENTIAL AS AN EFFORT TO ESTABLISH CITY BRANDING IN PINRANG REGENCY

It is important to point out that despite the opportunities and positives of the use of ecotourism in city branding; there are other precarious challenges that come with it.

The opportunities of ecotourism in Pinrang Regency are mainly in the regional assets in the form of various types of natural tourist attractions which are scattered in several districts as showed in the Table 1 below:

Table 1. Ecoourism Spot in Pinrang.

No	Tourism Spot	District
1	Lue River and Rajang Balla hot spring, Lemo Susu hot spring, Karawa Waterfall, Kali Jodoh waterfall, Batu Pandan spring, Balaloang Permai spring, Paniki caves, and Kajuanging Beach and Kanipang Beach	Lembang
2	Batu Lappa	Batulappa
3	Bukit Tirasa, Lamoro WAterfall, Pasandorang spring, Kappe Beach, and Maroneng Beach	Duampanua
4	Bulu Paleteang, Sulili hot spring	Paleteang
5	Batu Moppangnge	Patampanua
6	Beach Ammani, and Ujung Tape Beach	Mattiro Sompe
7	Wakka Beach	Cempa
8	Wiring Tasi Beach, Ujung Lero Beach, Ujung Labuang Beach, Sinar Bahari Sabbang Paru Beach, Bonging Beach, Ponging Desa Lotang Salo, Marabombang Port Beach and Kamarrang Island	Suppa
9	Wae Tuwoe Beach	Larinsang
10	Artificial Lake of Bakaru Hydroelectric Power Plant	Lembang
11	Benteng Dam in Benteng Village and floating restaurants	Patampanua
12	Points curing fish, traditional boat building yard, coconut plantation and fishing port	Suppa

Source: RJPMD KabupatenPinrang 2014-2019

In the 2014-2019 Regional Medium-Term and Medium-Term Development Plan (*Rencana Jangka Panjang dan Menengah Daerah*—RJPMD), it was stated that tourism activities are one of the sectors that need to be boosted through the development of tourist attractions, given that the tourism sector has the capabilities of developing a broad range of related sectors through multiplier effect. However, in Rencana Induk Pariwisata Daerah Kabupaten Pinrang (RIPDA) or Regional Tourism Master Plan Pinrang, it was also stated that tourism in Pinrang Regency supports the efforts of its flagship sector, namely agriculture (including paddy field and fisheries) which makes the district the 'ricebowl of the region'.

The tourism development plans in Pinrang Regency include cultural tourism, natural tourism and built-place tourism. But, the development of tourism in Pinrang Regency has not been able to fully contribute to the regional income because the patronage of the tourist centers are majorly by domestic tourists while patronage by foreign tourists is still very low. *Pinrang dalam Angka* (2014) notes that tourist visits to Pinrang in 2013 were 19,272.

From the results of the survey, it was found that respondents complained about the safety and comfort of tourist attractions and that the management was not at the optimal level. While addressing this issue, it was explained in RJPMD that tourism had not been a priority. In Kabupaten Pinrang Regional Tourism Plan (2017) it was reported that this sector is still facing some fundamental problems, such as:

1. Lack of community participation in the development of tourism, creativity, innovation and competitiveness of tourism spot. The quality of human resources and tourism business actors is not yet optimal;
2. Integration and synergy between tourism actors in tourism development is still low.
3. Limited tourist objects as well as problems with ownership status/land certificate.

The same thing happens with natural tourism. Natural tourism is divided into special interest tourism and recreation-based tourism. Eco-edu is placed as a second prioritity after tourism development of integrated marine-based family recreation.

In-depth interviews conducted as well as the data gotten revealed that the development of natural tourism is constrained by access to infrastructure and land ownership. Observations of these tourism spot showed that there are no adequate roads to reach these ecotourism sites. The management of these tourist spots was also discovered to be poor.

On the other hand, the tourism industry players stated that they manage the tourist sites independently and are looking forward to investment and assistance from the government.

Adapun cara-cara kami cukup tradisional, mulai dari pembagian lokasi, penataannya, sangat tradisional sekali. Tidak ada cuma ah. kami sendiri bahwa apabila di salah satu. pada umumnya di suatu daerah, apabila kita ingin dikunjungi cukup menata tempat dengan pertama-tama keamanan, yang kedua kebersihan, sebagaimanasupaya kiranya tamu merasa nyaman dan aman, seperti itu. (As for the ways, we are quite traditional, ranging from the distribution, location, arrangement, we are very traditional in all. There is no just ah ... we ourselves that if in one ... in general in an area, if we want to visit, just arrange the place, firstly, with security, secondly cleanliness, as much as possible so that guests feel comfortable and safe.)

As a result of the limitation of infrastructure, special interest tourism can be targeted towards limited markets (niche) through the intensification of promotion of special interest products like trekking, history, culture, sailing and caving. These promotions can be carried out on special portals on the website (specialized online portals).

Regarding digital sources, Dinas Informatika dan Komunikasi (Infokom) atau the Information and Communication Service has made provisions for devices to be used.

Infokom menyediakan aplikasi di android 'KemanaPinrang' itu, di situ, publik juga ada tempat-tempat destinasi pariwisata. Di situ lengkap dengan foto dan diarahkan, diarahkan dan diberi cara ke sana. Bisa diarahkan ke sana melalui apps, terus di aplikasi itu bisa juga chat, bisa juga apa keluhannya masyarakat ketika kalo menuju ke sana. Tapi kami di diskominfo memang hanya sebatas menyediakan aplikasi. Promosi terkait aplikasi tersebut tentu dari divisi terkait. (Infokom provides applications on the Android 'KemanaPinrang', with provisions for tourism destinations for the public. It is complete with photos and directions to these destinations. The application also gives room for chatting and also makes it possible to know the complaints of the community as regards the destination. At Infokom, we only make provisions for the applications. The promotions of this is carried out by the relevant division.)

5 CONCLUSION

The requirements for the development of ecotourism as citybranding in Pinrang include natural potential, community support, the use of social media, and government policy planning. However, the implementation of planning in the field must be realized to the fullest. Thus, the development of local ecotourism potential as an effort to establish city branding in Pinrang Regency still requires a process that requires special attention.

ACKNOWLEDGEMENT

This study was funded by the Ministry of Research and Universities through the Higher Education Primary Research Scheme in 2018.

REFERENCES

Allen, G. (2007). Place branding: New tools for economic development. *Design Management Review*, *18*(2), 60–68.

Anahid, M., & Ardabili, F. S. (2018). The Relationship of Tourists' Loyalty with Individuals' Normative Beliefs and Tourism Experience of Ecotourism. *Marketing and Branding Research*, *5*(1), 1.

Chan, C. S., & Marafa, L. M. (2014). Developing a Sustainable and Green City Brand for Hong Kong: Assessment of Current Brand and Park Resources. *International Journal of Tourism Sciences*, *14*(1), 93–117.

Chan, C. S., & Marafa, L. M. (2017). How a green city brand determines the willingness to stay in a city: the case of Hong Kong. *Journal of Travel & Tourism Marketing*, *34*(6), 719–731.

Deas, I., & Ward, K. G. (2000). From the 'new localism'to the 'new regionalism'? The implications of regional development agencies for city–regional relations. *Political geography*, *19*(3), 273–292.

Eli Jamilah, Suharyanti, Mirana Hanathasia (2017). "Potensi Ekowisata sebagai Upaya City Branding: suatu Study Pendahuluan di Kabupaten Pinrang."Prosiding the 1ˢᵗ*Communication, Culture, and Tourism Conference* 2017.Buku 1.Aspikomdan FISIP Universitas Riau.

Guler, E. G. (2017). The Role of Local Governments in City Branding. In *Global Place Branding Campaigns across Cities, Regions, and Nations* (pp. 251–269). IGI Global.

Hall, D. (2004). Branding and national identity: the case of Central and Eastern Europe. *Destination branding*, 87–105.

Jensen, O. B., & Richardson, T. (2005, May). Branding the contemporary city-urban branding as regional growth agenda. In *Plenary paper for Regional Studies Association Conference 'Regional Growth Agendas'. Aalborg, 28th to 31st May*.

Kavaratzis, M. (2004). From city marketing to city branding: Towards a theoretical framework for developing city brands. *Place branding*, *1*(1), 58–73.

Latupapua, Y. T. (2007). *Studi potensi kawasan dan pengembangan ekowisata di Tual Kabupaten Maluku Tenggara* (Doctoral dissertation, Universitas Gadjah Mada).

Laporan Akhir Penyusunan Rencana Induk Pembangunan Kepariwisataan Daerah (RIPPDA) Kabupaten Pinrang (2017).

Rahayu, R. K., Damayanti, R., & Ulfah, I. F. Tourism Branding In Asean Countries. In *International Conference of ASEAN Golden Anniversary 2017* (pp. 165–173). Brawijaya University.

Rencana Pembangunan Jangka Menengah Daerah Kabupaten Pinrang Tahun 2014-2019.

Tafalas, M. (2010). Dampak Pengembangan Ekowisata Terhadap Kehidupan Sosial dan Ekonomi Masyarakat Lokal (Studi Kasus Ekowisata Bahari Pulau Mansuar Kabupaten Raja Ampat).

Yananda, M. R., & Salamah, U. (2014). *Branding Tempat: Membangun Kota, Kabupaten, dan Provinsi Berbasis Identitas*.

Walean, R. H., & Mandagi, D. W. (2017). Factors Influencing Perceived Attractiveness of the Ecotourism City. *International Business Management*, *11*(6), 1374–1378.

Inclusive Development of Society – Lumban Gaol (eds)
© 2020 Taylor & Francis Group, London, ISBN 978-1-138-33476-2

Gnoseological bases of interaction of employees in network companies of infrastructural type

I. Ayham Hussein Ahmar
Postgraduate student, Maykop State Technological University, Maikop, Russia

S.K. Kuizheva
Department of Mathematics and Systems Analysis, Maikop State Technological University, Maikop, Russia

L.G. Matveeva & O.A. Chernova
Faculty of Economics, Southern Federal University, Rostov-on-Don, Russia

A.G. Gaboyan
Postgraduate student, Southern Federal University, Rostov-on-Don, Russia

ABSTRACT: Improving the efficiency and quality of interaction between employees of healthcare infrastructure companies is essential for improving the quality of service and achieving a multiplier effect. This effect is manifested, on the one hand, in a more effective use of qualifications by employees in medical institutions, and on the other, in improving the overall quality of public health. In this article, the challenge is approached from a gnoseological position, which provides a more accurate understanding of the essence of subject–object relationships between employees in health care infrastructure companies (for example, pharmacy chains) and the consumers of their products.

Keywords: communicative interactions, competence of the employees, gnoseology, infrastructure companies, pharmacy chains

1 INTRODUCTION

The modern phase of developing competitive business models is characterized by the expansion of networks of spatial interactions based on the consolidation of various fields, including pharmaceuticals. A network model for the development of infrastructure companies has found widespread use in the sale of medicines. Network models in healthcare services have resulted in widespread sales distribution chains of medical products. The share of retail chains in the global pharmaceutical market is estimated at 70–80%, while in some countries, pharmacies occupy up to 50% of retail trade. Most networks are subsidiaries of large distribution companies. According to AlphaRM the top 20 pharmaceutical corporations whose products are represented in the retail segment already account for 53% of the market (Baranova, 2012). Many pharmacy chains in complex financial situations that affect most of the global pharmaceutical market are united in associations. This is confirmed by the example of the Medical and Pharmaceutical Association of Russian pharmacy chains, in which the number of retail outlets increased from 952 to 2,232 units in 2017 alone (Russian Pharmaceutical Market, 2018). Furthermore, despite the redundancy of pharmacies noted by experts, this sector of the economy dynamically contributes to development.

The change in the number of sales points of the top pharmacy chains in 2018 compared to 2017 is shown in Figure 1.

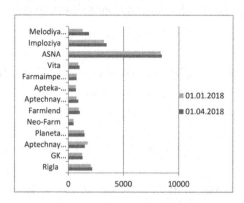

Figure 1.   Sales points of the top pharmacy chains in 2017 and 2018.

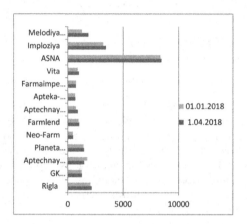

Figure 2.   Turnover growth of pharmacy chains in 2018 compared to 2017, (%).

Figure 2 shows the increase in trade turnover of pharmacy chains in 2018 compared to 2017.

The current situation in this area substantiates the study of the specifics of labor interactions of employees of network companies on a new methodological basis, in order to identify opportunities for quality improvement. As such, the clarification of the gnoseological foundations of communications between pharmaceutical networks and consumers of their services, in the example of Russia, will form a theoretical and methodological platform for the effective management of these interactions in developing countries, in the Republic of Lebanon in particular. This is due, first, to the more extensive experience in successfully operating pharmacy chains in Russia compared to developing countries; second, to the formation of an effective management system of labor interactions; and third, to the presence of a common "macroeconomic background" in which the network pharmaceutical companies operate.

2   RESEARCH METHODOLOGY

The methodological foundation of any research subject area is formed on the basis of systematization and generalization of knowledge about the subject studied and analysis and understanding of the relationship between the subject and the object of cognition, including the structure of the cognitive processes and forms and methods used. Hence, it all applies to the gnoseological problems and determines the objectives of this article: to clarify the understanding of the correlation of content and criteria for the effectiveness of communicative interactions of employees in infrastructure companies and the factors and conditions of their professional activities.

Considering the peculiarity of the interactions of employees of companies of the infrastructural type, it is necessary to note the prevalence of network forms of business organization. This is due to the fact that in the face of increasing competition in both local and global markets, the prospects for the development of companies are largely related to the possibility of combining the potential of various participants in the form of different configurations of networks. For instance, today trading networks occupy more than 95% of all retail markets. A network model for the development of infrastructure companies has found widespread use in the sale of medicines. Advantages of pharmacy chains are manifested in the possibility of increasing working capital, reducing the cost of purchases, reducing advertising costs, etc. Moreover, another important feature of pharmacy chains, which affects communicative interactions among the employees, is the expansion of organizational and functional boundaries and the emergence of dynamically developing global markets (Kosenkova, 2012). The need to exchange various types of information is directly associated with the communicative risks, which can be reduced only on the basis of knowledge of the partner's characteristics, the correct choice of a set of tools, and suitable technologies.

The features of the interaction of employees of network infrastructure companies and consumers of their services include

- The availability of competencies that allow the effective use of various information technologies while interacting
- The importance of correlating the speech characteristics of workers with an understanding of what category (by level of medical knowledge) of subjects they interact with
- Maintaining effective communication with the external environment, forming consumer target groups

In addition, the emergence of new medicines and medical technologies leads to an increase in the quality of healthcare institutions and population needs in modern medicines and treatments. This encourages pharmacies to focus on the individual needs of their customers.

In this context, one should speak about the strategic diagnostics of the potential interaction of employees of the network company and consumers of its services, which "allows for the formation of a model for the effective development of these processes based on timely recognition and overcoming of limiting factors" (Matveeva et al., 2015). From a methodological perspective, while analyzing the content and the essence of communicative interactions among the employees of infrastructural companies (between the branches of networks), as well as with consumers of their services, we may conclude that the client-oriented approach must be applied in the practice of managing communication processes. Not only does such an approach allow for the achievement of most interaction goals and balance of interests of participants in the process, but it also neutralizes the negative factor in a timely manner. The process is possible thanks to first, a clear hierarchy of service priorities, taking into account both the economic and social effects of interactions; second, the fullest possible account of the consumers and preferences of medicines offered by the branches of the pharmacy network; and third, stimulating the employees of the network company to increase their professional competencies. Among the latter, a special role is played by communication and information competence, as well as mobility, creativity, openness, and readiness for innovation. These circumstances lead to the conclusion that it is necessary to consolidate the client-oriented model of behavior of employees in infrastructure companies within the framework of corporate culture.

3 METHODS AND ANALYSIS OF RESULTS

For pharmacy networks, communicative and informative interactions of employees are fundamental. At the same time, the most important is the fact that the quality of such interactions largely depends on both communicative factors and social factors, taking into account the specificity of the goods being sold—medicines. As some social psychologists point out, communication objects are not autonomous and exist in a particular social environment. Therefore, in some societies (groups of the population) for effective communication, it is necessary to take into account and understand the established social values and national traditions. For example, awareness of the "prestige" of involvement in the process of communication interaction can lead to a change in the habitual model of an employee's behavior.

That is exactly why the study of what requirements are imposed on the competence characteristics of the employees in infrastructure companies acquires special significance in the gnoseological sense. It should be emphasized that with regard to the various social groups with which an employee of an infrastructure company interacts, he must exhibit various professional competencies. Variations in communications can be manifested in network workers even in the process of differentiated use of medical terms, names of medicines, etc.

Thus, based on the aforementioned information, it can be concluded that the gnoseological aspect of considering the interactions of employees in infrastructure companies among themselves, as well as with consumers of their services, allows us to reflect the cumulative role of diverse, complex interacting factors that determine the effectiveness of the communication process.

The whole set of factors can be grouped as follows:

- Competence factors, which determine the level of medical knowledge of the interacting parties, which in turn depend on the social status, scope of activity, motivation for communication, etc.
- Social; associated with social norms of behave our adopted in society, the overall level of socio-economic development of the region, the social role of interacting groups of the population, etc.
- Info-communicational (interaction of employees of network companies using information technology), which directly determine the communicative abilities of the employee, including speech etiquette, the ability to effectively use modern information technologies, etc.
- Cognitive: associated with the ability of the employee to adequately interpret the characteristics of the partner, his social values, the level of medical knowledge, language competence.

These groups of factors are determined empirically based on the analysis of pharmacy chain companies.

Furthermore, it is vital to note that effective communication interactions in a network infrastructure company are provided not only by the competencies of the individual employee, but also by the ability to form a unified system of participants, balanced by interests and needs. Specifically, we would like to clarify that these participants include not only employees of the grid company, but also consumers of the services provided by this company. In other words, the communication system is formed and viewed both within the network and in the external environment. This is the most important factor for obtaining joint synergies from communication interactions, that is, a factor in the formation of an effective system of synergistic relations in such a company, as the interactions are focused on accounting and achieving a balance of interests of both employees of the network and consumers of its services.

At the same time, the form of the network organization of business determines not only the specifics of the aforementioned socio-psychological relations, but also the specific features of the applied model of managing its activities. As a result, the specifics and sizes of the synergy effects produced by different types: operational, synergy of sales, financial and managerial synergy, and the effect of sharing knowledge and experience in a particular area of the company's activities. The source of operational synergy is the impact of a number of factors that provide savings in costs, including the effect of the scale of the company's operations through joint investments in economical technologies; saving on the scale of the sphere of activity, as integration ensures the rational use of the production and human resources potential of the network; reduction of innovation costs through the sharing of knowledge and experience among employees from various branches of the chain; and the introduction of new products (e.g., the expansion of the range of medical products and equipment) and ways of providing services (delivery of goods to consumers). In addition, "the effect of operational synergy is achieved by increasing the size of the market niche of the integrated structure resulting in an increase in the market power of the corporation and increase in operating income" (Avdonina, 2012). Synergy of sales is achieved through access to new markets (suppliers of medical drugs and medical equipment), channels for the sale of pharmaceutical products, state procurement, etc. This situation is especially typical for large network companies that have a distribution network and a recognizable brand that use their advantages to increase sales. In addition, this kind of synergy can be achieved when the same delivery channels are used for several types of goods, and the sales process is managed from a single center or the same warehouses are used. As such, it could be clearly illustrated in the example of the Lebanese network company in the field of pharmaceuticals and healthcare Germanos Health Care Supplies (Figure 3), which is organized by an "umbrella" network.

The company has a focal point and three storage facilities from which goods are delivered. The company is engaged in marketing, promotion, and distribution of an exclusive wide range of products, including generic products, over-the-counter products, supplements, medical devices, and other health care products. The activities of this company are based on registered and certified procedures in accordance with the Lebanese GSDP guidelines (good storage and distribution practices for medical products). This company serves more than 1600 pharmacies throughout Lebanon with the support of a qualified team of medical representatives to increase interest in products and increase sales.

The source of financial synergy of this network company is the mutually beneficial production links of the center with its branches, which leads to an increase in the operating assets of the central organization and an increase in development opportunities for its branches. Integration of organizations contributes to the growth of their creditworthiness, because at integration, aggregate profit and cash flows become more stable and predictable. Reducing risks by dispersing assets and minimizing losses due to negative environmental impacts is also a source of financial synergy. The managerial synergy of this company is a consequence of improving management effectiveness: the benefits of integration arise, as a more capable and efficient management of the central organization of the network can more rationally manage

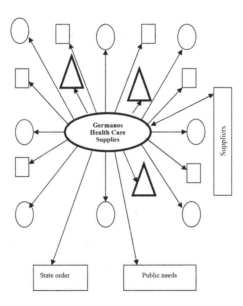

Figure 3. Scheme of production and information communications of employees of the pharmacy network company Germanos Health Care Supplies. Circles, pharmacies; squares, hospitals; triangles, pharmacy warehouses.

the resources that the branches have. The effect of knowledge sharing is possible because of the formation of new knowledge in the medical field, their archiving, use and interchange, not only between employees of the network, but also with customers.

Knowledge of the following areas is of high value:

- Innovate experience and scientific achievements in medicine generally, new medical drugs and medical equipment, new technologies for storage and distribution of medical products
- Technological knowledge within the participants of the technological chain of medical products
- Knowledge in the field of management in grid companies
- Marketing of new medical products and services, identification of needs of consumer expectations

Thus, in the framework of a networked pharmaceutical company financial, labor, production, and other relations that the center has with the branches (as well as with the customers) are in place to obtain positive effects. At the same time, the relevant staffing of the innovative development of the grid company is of great importance, which means not only the appropriate personnel management, from its recruitment to maintenance of constant development, but also covers the whole process of managing the company's operation, the continuous development of employees in all areas of activity. Strategic management of the network company's personnel consists in coordinating the interests, efforts, and resources of all its divisions, which allows receiving positive synergetic effects.

4 CONCLUSIONS

The following conclusions were drawn from the conducted research: the communicative function of employees, shown in labor interactions, can ensure the development of a company through the expansion of mutually beneficial cooperation and partnerships and regular accumulation of information about consumers and the market (positive synergic effects). Similarly, ineffective communication (negative synergistic effects) may restrict the development of a company. This conclusion is important for further research, to pursue methods and technologies that improve the efficiency of interaction between employees of network companies and their contractors, and consequently, the quality of health care services. The methodology proposed by the authors, based on the gnoseological approach, can be used to identify opportunities of growth in communication quality and other problems areas; as a result, the efficiency of management decisions in network infrastructure companies may be improved on the basis of practical implementation of the developed recommendations.

ACKNOWLEDGMENTS

The article was prepared within the RFBR grant 18-010-00623 "The intellectual modeling of organizational –economic mechanism of water supply management in regional water management complexes."

REFERENCES

Avdonina, S. G. 2012. Synergetic effect of cluster formations and parameters of its estimation. *Regional Economy and Management*, 1.

Barona, O. Russian pharmaceutical market traditionally demonstrates abnormal trends. *Pharmaceutical Bulletin* No. 32. Available at: https://pharmvestnik.ru/publs/lenta/v-rossii/bez-sjurprizov-prnt-17-m9-903.html#.W0x2zqC3pSk

Kosenkova, E. L. 2012. *Organization, Evaluation and Management of the Personnel Potential of the Network Company: Methods, Tools, Information Technologies*. Taganrog: Publishing House of Southern Federal University.

Matveeva, L. G., Mikhalkina, E. V., Chernova, O. A., & Nikitaeva, A. Yu. 2015. The strategic context of interaction between financial and real sectors of the Russian economy. *Journal of Applied Economic Sciences*, 7(37):1085–1092.

Russian Pharmaceutical Market. 2018. Analytical review. Data of DSM Group retail audit. URL: Available at: http://dsm.ru/docs/analytics/march_2018_pharmacy_analysis.pdf

Inclusive Development of Society – Lumban Gaol (eds)
© *2020 Taylor & Francis Group, London, ISBN 978-1-138-33476-2*

The methodical tools of formation of coal mining enterprises' strategies

O.G. Andryushchenko
Novocherkassk Engineering-Land Reclamation Institute of Don State Agrarian University, Novocherkassk, Russia

A.Y. Nikitaeva
Southern Federal University, Rostov-on-Don, Russia

M.A. Komissarova & M.M. Afanasyev
Platov South-Russian State Polytechnic University (NPI), Novocherkassk, Russia

ABSTRACT: The article discusses the need to work out the issues of strategic management of coal mining companies, provides an improved economic and mathematical tool for constructing a system of indicators for assessing the effectiveness of management of the coal-mining enterprise. The development of methods for aggregating indicators in assessing the effectiveness of enterprise management is provided in the paper. The proposed method of aggregation allows to identify the most effective strategy for the development of the coal mining enterprise. To assess the effectiveness of alternative strategies for the development of coal mining enterprises of the Rostov region a set of indicators built at the stage of modeling of system dynamics was considered. Using the method of aggregation of indicators, a priority strategic alternative to the development of these enterprises was chosen.

1 INTRODUCTION

The focus on achieving future goals in a rapidly evolving external environment based on the results of theoretical and practical developments in the field of strategic planning and management can be considered the fundamental basis of the modern concept of strategic management (Agajanova, 2013). In the domestic economy the use of the basics of strategic management is quite difficult due to a variety of different reasons: the peculiarities of the Russian mentality, the lack of the necessary organizational culture, weak strategic thinking, lack of knowledge of the basics of strategic marketing, etc.

Today, the conceptual setting of strategic management is the idea of the enterprise as an open system, the efficiency of which is determined by both the interaction of its individual structural parts and the completeness of the impact of external forces on it. Such orientation of strategic management predetermines its internal content (Tikhonov, 2013) and allows to include the following areas of governance:

– development of long-term objectives affecting technical, technological and economic aspects of activities and assessment of them from the perspective of social and environmental perceptions;
– structural policy: creation of new structures, modernization of old ones, enterprise restructuring aimed at increasing efficiency, liquidation of unprofitable and inefficient structures;
– development of measures aimed at preventing bankruptcy;
– development of measures for the implementation of the adopted guidelines, which cover the objectives of the enterprise and its structural policy.

All the tasks of strategic management are closely interrelated and are implemented as a complex. This requires staff to be flexible, adaptive, provide an entrepreneurial type of behavior, enthusiasm, an ability to set goals and organize business in accordance with them. (Astafieva, 2013). Therefore, strategic management with no routine procedures and specific instructions but only general recommendations is a symbiosis of intuition and art of achieving goals.

The development of enterprises of the economic industrial sector, including enterprises of the fuel and energy complex (FEC), can be achieved through the implementation of various, sometimes mutually exclusive strategic measures aimed at the foreground satisfaction of the economic interests of certain territorial subjects (Nikitaeva & Andryushenko, 2012). In this regard, it is necessary to take into account not only macroeconomic factors and operating conditions of industrial enterprises but also the main target-oriented strategic documents.

The prepared draft of the Energy strategy of Russia for the period up to 2035 (ES-2035) from this point of view is not just a prolongation of the previous strategy, but it also forms new strategic guidelines for the development of the energy sector in the framework of the transition of the Russian economy into an innovative development path, stated in the Concept of long-term socio-economic development of the Russian Federation.

2   METHOD

The reform of the fuel and energy complex of the country in the mid-1990s was based on the energy strategy which actually retained the inertia of the previously pursued policy of foreground gasification of the national economy. The strategy of "gas pause" focusing on the Directive distribution of labor and financial resources did not correspond to the ideology of market reforms, which ultimately led to negative results. Over the years the problem of substitution of other heat and energy carriers with coal is becoming more and more urgent. The entire domestic economic complex should focus on meeting the demand for energy and the economy as a whole by increasing the share of Russian coal in the fuel and energy balance of the country, and coal should become the main fuel resource in both the European part of the country and the areas of its traditional use.

Despite the very difficult market situation Russia remains one of the world leaders in coal production. A third of the world's coal resources (173 billion tons) and a fifth of the proven reserves are concentrated in its depths. The energy coal reserves amount to about 80%. Industrial reserves of existing enterprises are almost 19 billion tons, including coking coals (about 4 billion tons).

Strategic management of coal mining enterprises is the implementation of a comprehensive long-term action program (strategy) in an ever-changing environment, taking into account the specifics of the enterprises of the industry. Thus, the strategy allows you to focus on the main problems and discard the secondary ones, coordinate the company's actions to implement the mission and main goals, provide the company's strong competitive advantage in the global and domestic markets. With the help of the strategy the strategic potential of the company is created, which, in general, is the basis of the company's practical activities and the process of its adaptation to the external environment (Yershov & Kobylko, 2015).

The development of the modern economy today is closely connected with the use of high technologies and innovative products, the natural transition from the sale of raw materials and material products to the sale of services and technologies (Nikitaeva & Andryushchenko, 2014). If it was not for it, it would be impossible to speak about any positive shifts and rates of qualitative growth in any sphere of national economy (Babikovaet al, 2012).

Structural policy should be supplemented by scientific, technical and technological policy aimed at updating the material base, ensuring energy and resource saving, creating conditions for the implementation of advanced technologies in the energy sector. In developing the strategy of its activities, the Ministry of energy of the Russian Federation as a subject of budget planning set a number of main goals and identified appropriate development objectives (Figure 1). The first group of goals in the figure assumes a comprehensive increase in the efficiency of fuel and energy resources; the second group - the development of the competitiveness

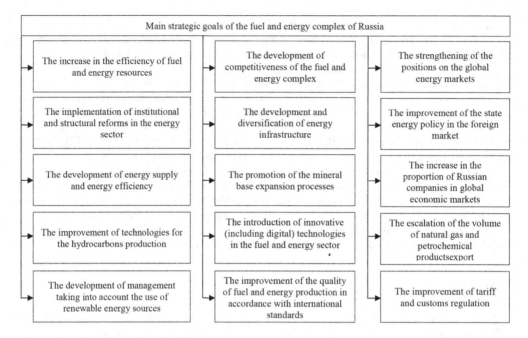

	Main strategic goals of the fuel and energy complex of Russia	
The increase in the efficiency of fuel and energy resources	The development of competitiveness of the fuel and energy complex	The strengthening of the positions on the global energy markets
The implementation of institutional and structural reforms in the energy sector	The development and diversification of energy infrastructure	The improvement of the state energy policy in the foreign market
The development of energy supply and energy efficiency	The promotion of the mineral base expansion processes	The increase in the proportion of Russian companies in global economic markets
The improvement of technologies for the hydrocarbons production	The introduction of innovative (including digital) technologies in the fuel and energy sector	The escalation of the volume of natural gas and petrochemical productsexport
The development of management taking into account the use of renewable energy sources	The improvement of the quality of fuel and energy production in accordance with international standards	The improvement of tariff and customs regulation

Figure 1. The main goals and objectives of the strategic development of the Russian fuel and energy complex.

of the Russian fuel and energy complex. The third group of strategic goals is aimed at strengthening Russia's position in the foreign market.

The achievement of these goals and objectives requires a comprehensive approach to their implementation, taking into account the requirements of the state policy at the federal, regional and municipal levels. Moreover, the fuel and energy complex is an important sector of the Russian economy for other sectors too. Our state has a powerful resource potential in the field of fuel and energy, which acts as the basic competitive advantages of the Russian Federation in the global energy market, and, consequently, is its national treasure. However, there is a whole set of problems in activity of fuel and energy complex nowadays:

– extreme deterioration of basic production assets;
– low level of reproduction of resource base;
– weak use of innovative technologies;

The task of energy sector transformation into a dynamic, cost-effective and corresponding to the world standards complex is particularly relevant in the current crisis conditions.

Figure 2 presents extraction and processing of the main types of minerals in Russia.

As follows from the figure, there is a positive trend during the study period in coal mining, while the volume of natural gas production is slightly reduced. This means that the coal complex of Russia continues to develop actively. The Russian economy has always been characterized as the one of resource and raw material, so the development of the fuel and energy complex is a priori the most important component characterizing the well-being and prosperity of the national economy. The importance of the coal industry in the domestic fuel and energy sector also remains unchanged, as the transition to "coal" energy and other alternative fuels remains relevant.

Today 25 domestic regions are engaged in coal mining, 16 coal basins are operating. According to the magazine "Coal" and Rosstat 385 million tons was extracted in 2016, the coal production has increased from 198 to 285 million tons over the past 5 years. The volume of coal production in Russia is shown on Figure 3.

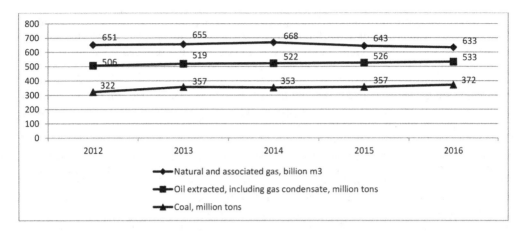

Figure 2. Extraction and processing of the main types of minerals.

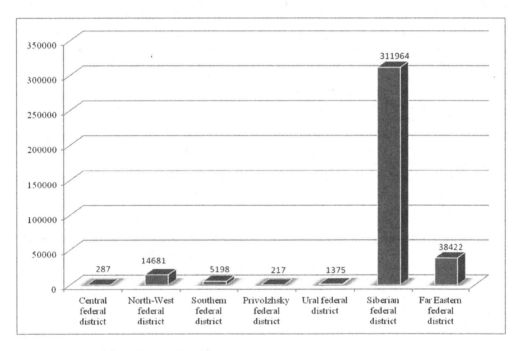

Figure 3. Coal mining, (thousand tons).

According to the draft of the long-term programme of development of the coal industry for the period up to 2035, signed by the President of the Russian Federation V. V. Putin, the results of investment in the industry are expected to be approximately 2.75 trillion rubles. At the same time, among the most important priorities in the state policy in the field of fuel and energy should remain: the strengthening of the importance of the coal component in the power industry, the increase in the export potential, the reproduction of basic assets, the increase in the efficiency of coal transportation, the creation of a favorable investment climate, the formation of reflected mechanisms to support the enterprises of the industry on the basis of business integration, etc. Despite the presence of a rich raw material base, the practice of developing not only the most favorable reserves has taken root in Russia: more than a quarter of the mines operating in recent years developed reserves that did not allow the use of modern means of coal mining complex mechanization. Similar fields in other major coal producing countries are not developed and are excluded from conventional reserves. That is the reason why the technical and economic indicators at existing coal mining enterprises in Russia are significantly inferior to foreign ones. In this regard, the new standards for classification of coal seams are being developed according to their certified capacity, ash content, consistency of the elements of the occurrence, the degree of stability of the lateral rocks, the presence of tectonic zones, increased fracturing of host rocks, etc. There is toughening of the requirements to exploration work and initial data, laying the basis for the evaluation of counting of coal reserves in relation to their increased importance in improving the technical and economic indicators of the construction and operation of mines and quarries for developing and planned coal deposits. The requirements to the study of tectonics, hydrogeology, gas content and structure of coal seams and host rocks are increasing. It is necessary to reconsider the standards for the reference to the conditioned reserves of layers confined to the zones of tectonic disturbances, lines of splitting and wedging of coal seams, to the contours of their distribution in zones with low production capacity and increased values of ash content of coal.

The study identified a list of the above reasons existing at the macro and micro levels, currently weighed down by political instability in the world and the strengthening of economic sanctions by the West. For enterprises of the coal industry external and internal factors can be classified as follows (Table 1). In addition to the above, the weak investment activity is due to the lack of an effective strategic plan for the development of the coal mining industry and companies in its structure; the company owners' and other investment project participants' lack of a well-thought-out and effective program of action formulating strategies for the financing and development of coal enterprises.

The strategy that is being formed should take into account the industry characteristics of the coal industry and the specifics of each individual coal mining region, while relying on the existing powerful resource and knowledge-intensive potential of our state. In addition, it is necessary to take into account the experience of Western countries in the conduct of resource conservation and subsoil use policy.

A number of innovative projects have been proposed by Russian scientists recently that contribute to the successful development of coal mining industries, but due to the reasons mentioned above, many of them remain only a kind of policy documents, the practical implementation of which is not yet possible.

The authors formed a possible classification of coal mining companies' development strategies in the fuel and energy complex structure. The results of the studies show that in case of refusal of state support for coal mining enterprises, within the next five years, almost all of them can cease to exist. Therefore, coal mining enterprises need to develop a set of measures and strategies for their further existence. We consider it possible to classify the development strategies of these companies in the complex, ever-changing environment as follows:

Table 1. The analysis of external and internal factors affecting low investment activity.

The factors of the external environment	The factors of the internal environment
– Long investmentlag – poor awareness of the amount of coal reserves in the reservoir, the reservoir power, etc. – limited lifetime of coal enterprises due to the magnitude of the reserves (the lifetime of a mine is 40 years, on average) – regular legislative transformations – lack of a clear policy in the system of taxation and pricing – a high degree of risk of investments	– high capital intensity – adverse natural conditions – the need for regular reconstruction of enterprises – deteriorating of conditions of reserves – low production efficiency as work moves deep into the earth's interior – severe conditions and high hazard operations

1. Strategies of passive existence. They ensure the functioning of the enterprise in the most favorable conditions to its activities; the existing positive factors in the developed environment.
2. Adaptation strategies. They are aimed at the active use of existing environmental factors, existing technologies, possible adaptation and the search for the most favorable ways to move to new market positions. This should be reflected in the company's ability to rapidly restructure internal structures and to adopt more complex forms of strategic behavior.
3. Strategies of active growth, i.e. strategies of the company's influence on the external environment. They ensure sustainable production growth through the use of technological and organizational innovations and contribute to the direct influence of the enterprise on environmental factors (Komissarova, 2014).

One of the tools to implement the strategy in a visual form is a balanced scorecard (Balanced Score Card, BSC). In the Russian literature it is possible to find various versions of the translation of the term balanced score card (Kaplan Robert S. &Norton David P., 2009).

The BSC attaches great importance to an integrated set of criteria linking the financial component with such indicators as the customer base, internal business processes, and personnel work in the company (Shevchenko I. K., 2007).

In the classical view, the structure of the BSC is represented by 4 structural elements, we considered it necessary for the convenience of the coal mining company to add the fifth element, taking into account their specificity (Figure 4). The implementation of the objectives of this element will provide an environmental component, which plays an important role for the extractive industries, and meet the interests of society at the municipal, regional and state levels. In addition, the indicators in this element will take into account the beneficial for the company difference between penalties, the amount of which can significantly increase in accordance with international standards, and the cost of waste disposal of coal mining.

In terms of the use of modern methods of multivariate statistical analysis, it is advisable to consistently use the methods of aggregation of indicators in assessing the results of management of the coal mining enterprise in order to reduce the dimension of the problem and build an integral criterion of efficiency. To do this, it is necessary to perform the approbation of this method on the example of one of the coal mining companies of the Rostov region.

In many practically important problems of multicriteria choice the preference ratio > is considered to be invariant with respect to the linear positive transformation. A sign of invariance of the relation > is the presence of the properties of additivity and homogeneity (Storozhenko, 2016). In other words, for any pair of vectors $b', b'' \in \Re^m$ connected by a relation $b' > b''$, both the relation $(b' + c) > (b'' + c)$ for any vector $c \in \Re^m$ and the relation $xb' > xb''$ for any positive number X must be executed.

In this context it is necessary to consider a set of indicators built on the basis of a modified balanced scorecard for each of the selected criteria subgroups (Table 2).

As three alternative scenarios Coal company strategies, the classification of which is described above, will act as three alternative scenarios. These are: alternative A1-passive existence strategy, alternative A2-active growth strategy, alternative A3-adaptation strategy. The

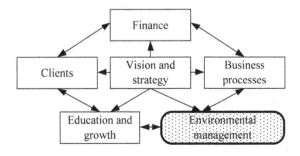

Figure 4.   Components of the BSC structure (for coal mining companies).

Table 2. Ranking indicators of the strategies of the Rostov region mining enterprises.

№ subitem	Name of the indicator	Designation	Value of the indicator for the corresponding alternative		
			A_1	A_2	A_3
I. Financial component					
1.	Current liquidity ratio	k_{11}	0,4	0,43	0,43
2.	Ratio of own funds availability	k_{12}	0,12	0,05	0,02
3.	Solvency restore ratio	k_{13}	0,19	0,22	0,28
4.	Financial dependency ratio	k_{14}	0,24	0,31	0,42
5.	Cover ratio	k_{15}	0,45	0,4	0,43
6.	Product profitability	k_{16}	0,15	0,26	0,12
II. Client component					
7.	Market segment proportion	k_{21}	0,30	0,24	0,36
8.	Growth in coal consumption	k_{22}	0,46	0,38	0,26
9.	Delivery time	k_{23}	0,20	0,25	0,40
10.	Regular customers proportion	k_{24}	0,85	0,90	0,85
11.	Product quality ratio	k_{25}	0,71	0,90	0,7
12.	Unit price of the product compared to competitors	k_{26}	0.9	0.9	1
13.	Level of capacity utilization	k_{31}	0.9	0.86	0.73
14.	Level of labor mechanization	k_{32}	0.7	0.75	0.75
15.	Novelty ratio	k_{33}	0.2	0.25	0.25
16.	Utilization rate of machinery working time fund	k_{34}	0.5	0.65	0.6
17.	Rate of norm renewal	k_{35}	0.55	0.6	0.6
18.	Unit cost	k_{36}	646	712	755
IV. Training and development					
19.	Number of specialists with higher education	k_{41}	0.45	0.52	0.39
20.	Coefficient of staff constancy	k_{42}	0.65	0.59	0.54
21.	Staff turnover factor	k_{43}	0.42	0.37	0.51
22.	Coefficient of working conditions	k_{44}	0.8	0.8	0.7
23.	Coefficient of labor discipline	k_{45}	0.85	0.9	0.9
24.	Cost of staff training	k_{46}	0.017	0.02	0.015
V. Environmental management					
25.	Certified stocks proportion	k_{51}	0.87	0.92	0.96
26.	Coefficient of environmental load	k_{52}	0.46	0.67	0.54
27.	Resource sharing	k_{53}	0.27	0.43	0.42
28.	Total natural potential	k_{54}	0.92	0.88	0.79
29.	Level of natural resources use	k_{55}	0.62	0.66	0.64

presented set of indicators for alternative strategies was calculated by the expert group for the Obukhovskaya OJSC mine in the conditions of different scenarios of the company development using the elements of system dynamics (Ivanchenko et al, 2008).

Based on the formal model of the general problem of decision-making, the purpose of the multi-objective optimization problem is to allocate a set of effective (weakly effective) elements out of X. The relationships $\Re_1$, $\Re_2$, in general, are not linear, i.e., there are incomparable in $\Re_1$, $\Re_2$ elements of the X set (Larichev, 2004).

One of the reasonable methods for solving multicriteria optimization problems is the method ELECTRE I. ELECTRE I Method is based on the use of indices pairwise comparison of alternatives called the index of agreement and disagreement. Then the hypothesis of alternative Xi superiority over alternative Xj is put forward. The set I consisting of N criteria is divided into three subsets:

K+ is a subset of criteria by which X_i is preferable to X_j;

K= is a subset of criteria by which X_i is equal to X_j;
K^- is a subset of the criteria by which X_j is preferable to X_i.

The index of agreement based on the hypothesis of superiority of A_i over A_j is formulated. The index of agreement will be determined based on the values of the criteria weight coefficients. In accordance with the provisions of the ELECTRE I method the index of agreement is given by the ratio of the sum of the weight coefficients of the criteria of subsets I^+ and $I^=$ to the total weight coefficients:

$$\delta_{X_iX_j} = \frac{\sum\limits_{i \in I^+, I^=} \pi_i}{\sum\limits_{i=1}^{N} \pi_i} \tag{1}$$

According to the hypothesis of A_i dominance over A_j the value of the disagreement index Δ_{xy} is determined by the antagonistic criterion – the criterion according to which A_j is maximally superior to A_i.

The criteria are normalized to the maximum scale:

$$\Delta_{x_ix_j} = \max_{i \in I^-} \frac{l^i_{\alpha x} - l^i_{x_i}}{L_i} \tag{2}$$

where $l^i_{A_j}, l^i_{A_i}$ - values of the i-th criterion for alternatives A_i and A_j;
L_i is the length of the scale of the i-th criterion.

When implementing the ELECTRE I algorithm the binary superiority ratio is determined by the levels of agreement and disagreement. In the case of $\delta_{x_ix_j} \leq X_1$ and $\Delta_{x_ix_j} \leq \gamma_1$, where x_1, γ_1 are given levels of agreement and disagreement, X is the dominant alternative to Y.

If it is not possible to compare alternatives at the corresponding level values, such alternatives should be considered incomparable, which is of particular interest, since such contradictory alternatives require closer study.

3 RESULTS

The values of the mentioned coefficients allow (by gradually lowering the value of the coefficient of agreement and increasing the coefficient of disagreement) to narrow the space of alternatives reducing it eventually to the only alternative. Thus, the set values of the levels allow us to identify a subset of non-dominant alternatives (generally equivalent or incomparable). Successive changes in the level values allow the analyst to narrow the subset to generate different solutions (including the only "best" alternative).

To assess the effectiveness of options for the implementation of strategies for the development of enterprises of the Rostov region it is necessary to identify the core of non-dominant alternatives from the set $X = \{X_i\}$. The application of the algorithm for determination of the Pareto set can be attributed to the alternatives given in Table 2 to the sought core A'.

Table 3. Matrix of the agreement indices $\delta_{X'_iX'_j}$.

$\delta_{A'_iA'_j}$	A'_1	A'_2	A'_3
A'_1	0.24	0.87	0.62
A'_2	0.13	0.37	0.44
A'_3	0.38	0.56	0.61

Table 4. Matrix of the disagreement indices $\Delta_{X_i X_j}$.

$\Delta_{X_i X_j}$	X'_1	X'_2	X'_3
X'_1	0.18	0.34	0.41
X'_2	0.34	0.71	0.27
X'_3	0.41	0.73	0.01

For the set of alternatives related to core A' build the matrix of indices of agreement $\delta_{X'_i X'_j}$ (Table 3).

Similarly the disagreement indices are defined $\Delta_{X_i X_j}$ (Table 4).

In accordance with the provisions of the ELECTRE I method, the levels of agreement and disagreement are significant criteria for the attribution of alternatives. Thus, based on the hypothesis of the superiority of alternatives with confidence intervals $\chi_i = 0.5, \gamma_i = 0.5$ (adopted heuristically), the outcome is that among the compared alternatives the alternative X_2 (active growth strategy) has an advantage. This alternative assumes: the streamlining of sales aimed at supporting regular customers and increasing their proportion in the structure of consumers; the choice of the policy of formation of highly professional staff, which will reduce the turnover rate; the increase in the labor discipline rate provided by the system of motivation of employees.

4 DISCUSSION

The analysis of theoretical and methodological approaches to management shows that in the absence of a single theoretical basis a universal structure of the strategic management process has not yet been developed, and the set of strategic alternatives put forward mainly by western scientists and economists has not yet justified itself in the conditions of the domestic economy. Therefore, it is necessary to search for alternative directions of strategic development of Russian industrial companies, which will allow them to embark on the path of further sustainable development. Within the framework of the considered problem the enterprises of the coal mining sector were in the center of attention.

The specifics of coal mining companies require the development of strategic groups that will ensure the interests of both the owners of these companies and the interests of the economy and social sphere of the region and the country as a whole. The main function of strategies adapted to the conditions of the domestic economy and the specifics of the coal industry should be for them to ensure the development of enterprises in an unstable environment. An important role is played by the environmental factor, as it addresses the issues of rational subsoil use and minimization of environmental damage caused by the activities of such enterprises. We believe that in order to successfully adapt the financial and economic activities of enterprises to the requirements of a rapidly changing market and the new needs of consumers, special attention should be paid to methods that allow to not only quantitatively but also qualitatively assess the objective internal capabilities in order to successfully confront external influences and ensure effective strategic management. Therefore, according to the authors, the use of a balanced scorecard as an element of the formation of a new management concept will complement the instrumental apparatus of strategic management of industrial enterprises at the stages of analysis, control, and efficiency of decision-making and implementation of management decisions.

5 CONCLUSION

In modern conditions a certain integral criterion or criterion operator, which allows to unambiguously rank a sample of comparable strategies of enterprise management (i.e. the enterprises themselves basically) and their performance enterprise, should be understood as an assessment of the effectiveness of strategic management. At the same time, one of the basic relations, which allows to set this criterion set mathematically, is the preference relation. This mathematical tool was used in the present study in order to justify the choice of the method of aggregation of indicators from the standpoint of modern multi-criteria decision theory, since the use of methods of convolution of indicators in most works: additive, multiplicative, etc., is often used unfounded, without verification.

To assess the effectiveness of options for the implementation of strategies for the development of coal mining enterprises of the Rostov region, a set of indicators built at the stage of modeling system dynamics was considered. Using the method of aggregation of indicators from among the selected alternatives of development strategies of these enterprises on the basis of appropriate calculations, preference was given to one of the available strategic alternatives.

REFERENCES

Astafieva L.I. 2013. Management of strategic change programs as management innovation. *Management in Russia and abroad.* 5: 59–62.

Azhakhanova D.S. 2013. Modern approaches to strategic management. *Modern trends in Economics and management: a new view.* 19: 16–18.

Babikova A.V., Bogomolova I.S., Borovskaya M.A. et al. 2012. *Innovative mechanisms of strategic management of development of social and economic systems. Ed. By* Borovskaya M.A., Shevchenko I.K. Taganrog: Publishing house TIT SFedU. 112–127. (12.4 p.sh/1.0p.sh.).

Energy strategy of Russia for the period up to 2035 approved by the decree of The government of the Russian Federation № from June 9, 2017 № 1209 - p.: static.government.ru

Ivanchenko A.N., Zajcev R.G., Kuz'menko P.G., Mal'cev I.V. 2008. Intellektual'nyj graficheskij redactor modelej sistemnoj dinamiki dlja sredy Sphinx SD Tools [Intelligent graphic editor of the system dynamic model for Sphinx SD Tools system]. *Izvestij avuzov. Severo-Kavkazskij region. Tehnicheskie nauki.* [University News. North-Caucasian Region. Technical SciencesSeries]. 3.

Kaplan Robert S., Norton David P. 2005. *Strategic maps: Transformation of intangible assets into tangible results.* Moscow: CJSC "Olimp-Biznes": 436.

Komissarova, M.A. 2014. *Strategicheskoe upravlenie predprijatijami ugledobyvajushhej promyshlennosti.* Strategic management of coal mining enterprises: dissertation... Dr. of Economy: 08.00.05 (Unpublished doctoral dissertation). Rostov on Don: 269.

Larichev O.I. 2004. *Theory and methods of decision-making.* Moscow: Publishing house EKSMO: 284.

Market analysis: coal mining in Russia. Retrieved from http://moneymakerfactory.ru/biznes-plan/analiz-ryinka-uglya-rossii/

Nikitaeva A.Yu., Andryushchenko O.G. 2012. Modern management tools in the mechanism of industrial development of the regions of the South of Russia. *Economics and entrepreneurship.* 3 (26): 34–38.

Nikitaeva A.Yu., Andryushchenko O.G. 2014. The Role of state institutions in ensuring innovative development of industry at the regional level. *Regional economy. The South of Russia.* 1 (3): 20–27.

Regions of Russia. 2016.

Shevchenko I.K. 2007. Balanced scorecard as a decision-making tool. News of TSURE. *Thematic issue "System analysis in Economics and management".* 17: 0.6 p. sh.

Storozhenko V.V. 2016. Modern approach to strategic management of industrial enterprises, using modular modeling. *Economics and management in the XXI century: development trends.* 26: 185–188.

Territorial organization of fuel and energy complex. Retrieved from http://uchebnik-besplatno.com/uchebnik-geopolitika/territorialnaya-organizatsiya-toplivno.html.

The role of the fuel and energy complex in the Russian economy. Retrieved from http://tgk-12.ru/structur a_funkc_energ/rol_top_en_kompl.php.

Theory and practice of application of mathematical and instrumental methods in economics, business and education. 2015. Monograph. Ed. By Kryukova S.V. Rostov-on-don.

Tikhonov A.A. 2013. Strategic management of enterprise development in the context of the evolution of the theory of strategic management. *Prospectsofscience.* 7: 98–103.

Website of the Federal state statistics service: http://www.gks.ru/(accessed 19.04.2018).

Yershov D.M., Kobilka A. 2015. The choice of a comprehensive enterprise strategy taking into account the compatibility of strategic decisions. *Economics and mathematical methods.* 1: 97–108.

Inclusive Development of Society – Lumban Gaol (eds)
© *2020 Taylor & Francis Group, London, ISBN 978-1-138-33476-2*

Energy efficiency as a key factor for achieving energy security

A.Y. Nikitaeva, E.V. Maslukova, E.P. Murat, L. Molapisi & D.V. Podgajnov
Faculty of Economics, Southern Federal University, Rostov-on-Don, Russia

ABSTRACT: The challenge of energy security and energy efficiency is one of the key priority areas of research, as the level of a country's energy resources largely determines the stability and dynamics of its socio-economic development. On the basis of the DEA-BBC I model, estimates of the levels of energy efficiency in the regions of Southern Russia were obtained. The regions with suboptimal use of resources, for the relatively more efficient subjects of the Russian Federation were identified, showing that the significant potential of energy efficiency in the regions of Southern Russia can be realized by promoting innovation and bridging the technological gap between the regions, in the interests of a balanced and sustainable development of the mezzo-economic systems under consideration.

1 INTRODUCTION

zAchieving a sufficient level of energy security is among the key development priorities of most nations given that the level of energy security largely determines the dynamics and stability of socio-economic development. As such, the methods and means of ensuring energy security largely depend on the conceptual platform that underlines this term. This substantiates the study of the modern-day content and factors of energy security in the context of global economic transformations.

Numerous scholars concur with the modern day concept that energy security is multidimensional and that there still has not been a single encompassing definition that combines all the priorities and distinguishing factors.

According to the European Commission, energy security is the uninterrupted physical availability of energy products on the market, at a price that is affordable for all consumers (private and industrial) (EC-European Commission, 2001).

The World Bank connects energy security with ensuring that countries can sustainably produce and use energy at reasonable costs in order to facilitate economic growth and reduce poverty while directly improving the quality of life by broadening access to modern energy services. The key pillars of energy security are energy efficiency, diversification of energy resources, and resilience to volatility (World Bank Report, 2005).

One of the most influential contemporary conceptualizations of energy security is the 4A's, proposed by the Asia Pacific Energy Research Center. The center defines energy security as the ability of an economy to guarantee the availability of energy resource supply in a sustainable and timely manner, with the energy price at a level that will not adversely affect the economic performance (APERC, 2007). The 4A's concept states that energy security can be achieved by ensuring the following:

1. Availability of energy resources
2. Accessibility the energy resources
3. Affordability of the energy resources
4. Acceptability of the energy resources.

In addition, the three fundamental elements of energy security are physical security, economic energy security and environmental sustainability.

A. Cherp and J. Jewell look at energy security beyond the 4As and define energy security as 'low vulnerability of vital energy systems' (Cherp & Jewell, 2014). They further suggest that the 4A's should be accompanied by three important questions: security for whom; security for which values; for what threats. This line of questioning for defining the term is in alignment with the definition of security by D. Baldwin, 'security is the low probability of damage to acquired values' (Baldwin, 1997). In this respect, vital energy systems refer to energy resources, infrastructure and use linked together by energy flows that support critical social functions.

The current state of global affairs requires a new and innovative approach to energy security, broadening its scope beyond conventional boundaries. The traditional understanding of energy security does not fully correspond with current global trends that have been observed in the last decade, which are associated with new industrial revolutions and technological modernization. The global energy trend is characterized by increasing energy demand, the transformation of the energy sector and changing geopolitical dynamics. Under these conditions, one of the key factors for achieving energy security is improving energy efficiency. This ensures that adequate volumes of energy resources (while maintaining all quality, price, and risk prerequisites) are secured not by increasing mining, extraction, reserves or imports, but by the more efficient use of energy at all stages of energy generation, transmission, distribution, and consumption.

2 RESEARCH METHODOLOGY

In the Russian regulatory framework, energy efficiency is understood as "the characteristics reflecting the relation of the useful effect of the use of energy resources, to the cost of energy resources, produced in order to obtain such an effect, with regards to products, technological processes, legal entities and individual entrepreneurs" (Federal Law №261-FZ of November 23, 2009).

Sustainable energy security requires appropriate and sufficient investments that take into consideration economic development and environmental factors. Short-term energy security demands a robust energy system that rapidly responds to abrupt interruptions in energy supply, market instability or political influence by putting in place countermeasures to support system balance. The most significant benefit of energy efficiency is the reduction of imports due to decreased energy demand. Decreasing demand reduces the pressures on an energy system and subsequently minimizes supply disruptions.

According to the International Energy Agency (International Energy Agency), member countries made a $50 billion saving in additional spending on energy imports between 2000 and 2016. In countries where energy imports considerably affect the trade balance, energy efficiency can significantly improve the state of national accounts by cutting down on expensive supply infrastructure and storage facilities. Particularly in the electricity sector, reduced demand can delay the construction of new power plants especially when considerations are being made to transition from conventional sources to renewable energy technologies.

The 2010 European Union Security of Gas Supply Regulation introduced the N-1 standard, which requires countries to be able to sustain gas supply during infrastructure interruptions. N-1 ensures that during a supply disruption, a country is able to run on the available infrastructure to meet peak demand. In Germany and the UK for example, energy efficiency advancements produced a 30% saving in 2015, which was equal to total European imports from Russia, and therefore ensuring the availability of energy resources (International Energy Agency 2017).

In 2016, a global 12% saving was made due to energy efficiency advancements since the year 2000 (International Energy Agency, 2017). Emerging countries have economically benefited from the limited increase in energy consumption related to economic growth. Developing countries also have the opportunity to grow their economies with energy efficiency measures, effectively managing demand from industries and households in a cost-effective manner, as they are able to implement energy efficiency measures right from the beginning of building construction, industrial machinery, and infrastructure development. Furthermore, developing countries can benefit from reducing the energy-intensity of their economies and therefore minimizing the high cost of industrialization. Therefore, energy efficiency contributes to energy security by ensuring the affordability of energy resources.

Understanding the socio-economic and technological factors that influence energy demand is essential for appropriately regulating energy systems. Improving the efficiency of downstream energy use is also as important as converting upstream generation and transmission systems. This requires authorities to effectively coordinate public and private resources to achieve long-term energy security objectives. The efficiency used in converting primary energy and the intensity of electricity service delivery determines the total demand of primary resources required, therefore, optimum savings can be obtained by thoroughly capitalizing on energy efficiency improvements throughout the energy supply chain.

Energy efficiency enhancements that scale down the volumes of energy needed to deliver a particular service or product can be essential for reducing the negative effects of high-energy production. The infrastructure and technical challenges often are experienced during the development of new energy systems become more controllable when energy loses are minimized from generation to consumption. Efficiently modernizing energy systems throughout the supply chain is highly effective as the investment yields high cost-savings in the future.

Concurrently, special attention should be paid to the industrial sector as it accounts for a 38% share of the world's total final consumption by sector, therefore presenting a considerable opportunity for efficiency advancements (International Energy Agency, 2018). The industrial sector is made up of two major groups: energy-intensive industries and light industries. Light industries refer to the manufacturing of finished goods, which include textiles, food, and metal processing. Energy-intensive industries consisting of a broad range of fields, including chemicals, cement, iron and steel and pulp and paper, represent over 50% of the industrial sector energy consumption in many countries (IEA, 2018).

The amounts and types of materials and goods produced, and the level of energy efficiency of the production infrastructure determine a country's industrial energy consumption. In the last decade, the growth rate of energy-intensive industries has declined in most developed economies. However, in developing countries, the production of cement, aluminum, and iron and steel have considerably increased over the same period of time. The principal drivers for energy efficiency improvement include reducing production cost, ensuring resilience to price fluctuations, complying with environmental regulations and increasing consumer demand for environmentally friendly services and products. Energy service providers, including power utilities, also benefit substantially from energy efficiency improvements as this results in reduced costs of energy generation and supply, greater system reliability and stable market prices. In summary, and in accordance with the 4A's concept, energy efficiency ensures the accessibility, affordability and acceptability of energy resources.

The key beneficial effects manifested as a result of improving energy efficiency and having a direct positive impact on energy security (in an economic, social and environmental context) are presented in Figure 1.

Improving energy security and ensuring sustainable development in emerging and developing countries can be achieved by increasing the role of energy efficiency in the economy:

- Economic growth: Energy efficiency improves industrial productivity and reduces fuel import costs (Shahbaz, Zakaria, Shahzad, & Mahalik, 2018; Zhang, 2011; Tursoy, & Resatoglu, 2016).
- Energy Access: Energy efficiency enables the adequate supply of power to more users through existing infrastructure.
- Equity: Energy efficiency increases the affordability of energy services for underprivileged households by reducing the per-unit cost of electricity.
- Environmental pollution: Energy efficiency reduces the requirement for energy production resulting in minimized greenhouse gas (GHG) emissions associated with energy production while supporting economic growth.

Improving energy efficiency is closely related to not only energy sector development, but also the modernization of the industrial sector (as the dominating energy consumer) as a whole. Numerous scientific studies around the world demonstrate a direct positive correlation between innovation and industrial energy efficiency, which fully fits into the conceptual framework of the third (Rifkin, 2011) and fourth (Schwab, 2017) industrial revolutions.

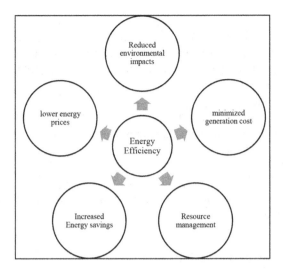

Figure 1.   The positive effects of energy efficiency in the context of energy security.

According to Gerstlberger W., Prast Knudsen M., Dachs B., Schroter M. (Gerstlberger, Knudsen, Dachs, & Schröter, 2016), key factors that determine the adaptation of energy efficiency technologies in European manufacturing firms are product and process innovations. On the one hand, the introduction and use of energy efficient technologies leads to a direct reduction in energy consumption for production and other activities, on the other hand, this in itself is an innovation that generates a number of indirect effects associated with, for example, the dissemination and exchange of knowledge on production technologies.

Kazi Sohag, Rawshan Ara Begum, Sharifah Mastura Syed Abdullah, Mokhtar Jaafar (Sohag, Begum, Abdullah, & Jaafar, 2015) note the crucial role of technological innovations in improving energy efficiency, as their implementation allows not only to produce a given level of commodities at a lower volume of energy consumption but also to provide opportunities for the conversion to the end use of renewable energies. Chao Feng, Miao Wang (Feng, & Wang, 2017) suggest that technological progress is identified as the main factor of energy efficiency, based on the analysis of panel data on the industrial sector of the provinces of China from 2000 to 2014.

Technological innovation unlocks new opportunities, advancing energy efficiency as digitalization is bridging the gap between renewable energy and energy efficiency to provide clean and cost-effective energy services.

Thus, a fairly clear and logical chain is built, according to which a neo-industrial development, with a strong innovative vector and an increased level of technological development, makes it possible to increase the level of energy efficiency. Therefore, improved energy efficiency positively impacts and enhances the level of energy security.

As such, the development of specific measures of improving energy efficiency and consequently, energy security involves the preliminary formation of the joint assessment of the level and factors of energy efficiency. In this paper, such an assessment is carried out using the data envelopment analysis tool (DEA). Data envelopment is an effective service management and benchmarking technique originally developed by Chames, Cooper, and Rhodes to evaluate nonprofit public sector organizations. The DEA has been proven to identify means of improving services where other techniques have failed (Sherman, & Zhu, 2006). The data envelopment analysis tool examines service units taking into account the use of resources and the provision of services, and locates the most efficient divisions, departments, branches or staff as well as the inefficient areas where actual efficiency improvements may be made. This is realized by making a comparison between the mix and quantity of services provided and the resources used by each unit compared with those of other units.

For calculations, the DEA – BBC I model is used, as proposed by A.Y. Nikitaeva, E. B. Maslyukova, and D. V. Podgajnov. This model has been expanded to include electricity generation per capita data (kWh per capita). Therefore, in accordance with the validated relation between technological progress, innovation, and energy consumption, the following structure of energy efficiency factors was used in the calculations:

– Total depreciation of fixed assets (%);
– Investment in fixed capital per capita, RUB;
– The use of special software tools in organizations (%);
– Level of innovation in organizations (%);
– Electricity production per capita (kWh per capita).

As a resulting criterion of energy efficiency, the selected indicators of energy intensity GRP (kg of equivalent/10,000 rubles) and electricity consumption per individual in industrial production (kWh per capita).

3 METHODS AND ANALYSIS OF RESULTS

For analysis, an aggregated annual indicator for the southern regions of Russia was used (administrative subjects of the Russian Federation that are part of the Southern and Northern Caucasian Federal districts). The following estimates of the energy efficiency in the regions of Southern Russia were obtained as results of the calculations (Table 1).

For each region, the model generates values of efficiency θ in the range of 0 (completely inefficient) to 1 (fully efficient). These estimates are comparative characteristics of the effectiveness of a single set of indicators and factors of energy efficiency.

It is important to take into account the fact that this refers to the relative level of energy efficiency in the macro-regional system under consideration. The obtained estimates allow for the determination of both the relative level of energy efficiency and the level of expression of the factors that have a positive or negative impact on the efficiency of energy use.

The testing of the proposed approach also allowed for identifying the regions in which the use of resources is not optimal relative to the more efficient subjects of the Russian Federation.

The results show that promoting innovation and closing the technological gap between the regions in the interest if achieving a balanced sustainable development of the mezzo-economic systems under consideration can realize the significant potential of energy efficiency in the Southern regions of Russia.

Table 1. Energy efficiency indicators of the Southern regions of Russia.

Region	Score
Republic of Adygea	0,850558
Republic of Kalmykia	1
Republic Of Crimea	1
Krasnodar region	1
Astrakhan region	1
Volgograd region	0,795878
Rostov region	1
City of Sevastopol	0,899668
Republic of Dagestan	1
Republic of Ingushetia	1
Kabardino-Balkar Republic	1
The Karachay-Cherkess Republic	1
Republic Of North Ossetia - Alania	0,682465
Republic of Chechnya	0,763025
Stavropol territory	0,999979

Significant attention should be paid to information technologies in the production, distribution, and consumption of energy resources by enhancing the integration of IT solutions into production and technological chains. In this context, the interpenetration of energy efficiency criteria, innovation in energy sector development strategies, industrial development strategies and the substantive agreement of the relevant strategic documents are required.

4 FUTURE RESEARCH

The high level of differentiation in the development of the regional economies of Southern Russia makes it practical for further research of sources, tools and technologies that determine the effectiveness of the use of internal resources to ensure sustainable economic growth and stability of the national economic system. The methodology proposed for the assessment of energy efficiency levels in the Southern regions can be used to identify problematic areas of development. As a result, appropriate steps may be taken to improve the effectiveness of solutions and formulate concrete recommendations for modernizing the energy sector in order to achieve an adequate level of energy security.

REFERENCES

Asia Pacific Energy Research Centre 2007. A quest for energy security in the 21st century. Tokyo.
Baldwin, D. A. 1997. The concept of security. *Review of international studies, 23*(1), 5–26.
Cherp, A., & Jewell, J. 2014. The concept of energy security: Beyond the four As. *Energy Policy, 75*, 415–421.
EC-European Commission. 2001. Green Paper–Towards a European strategy for the security of energy supply. *European Commission DG Energy and Transport, COM (2000)*, 769.
Federal Law No. 261-FZ of November 23, 2009 (as amended on July 29, 2017) «Energy saving and on improving energy efficiency and on making changes in separate legislative acts of the Russian Federation". URL: http://www.consultant.ru/document/cons_doc_LAW_93978/
Feng, C., & Wang, M. 2017. Analysis of energy efficiency and energy savings potential in China's provincial industrial sectors. *Journal of Cleaner Production, 164*, 1531–1541.
Gerstlberger, W., Knudsen, M. P., Dachs, B., & Schröter, M. 2016. Closing the energy-efficiency technology gap in European firms? Innovation and adoption of energy efficiency technologies. *Journal of Engineering and Technology Management, 40*, 87–100.
Nikitaeva, A.Y., Maslyukova, E.V., & Podgajnov, D.V. 2018. Model-analytical instrumentation of estimation of energy efficiency factors in the regions of the South of Russia. *Regional economy. South of Russia, 4 (22)*.
Rifkin, J. 2011. The third industrial revolution: how lateral power is transforming energy, the economy, and the world. *Macmillan.*
Schwab, K. 2017. *The fourth industrial revolution.* Crown Business. *New York: Crown Publishing Group.*
Sherman, H. D., & Zhu, J. (2006). *Service productivity management: improving service performance using Data Envelopment Analysis (DEA).* Springer science & business media.
Shahbaz, M., Zakaria, M., Shahzad, S. J. H., & Mahalik, M. K. 2018. The energy consumption and economic growth nexus in top ten energy-consuming countries: Fresh evidence from using the quantile-on-quantile approach. *Energy Economics, 71*, 282–301.
Sohag, K., Begum, R. A., Abdullah, S. M. S., & Jaafar, M. 2015. Dynamics of energy use, technological innovation, economic growth, and trade openness in Malaysia. *Energy, 90*, 1497–1507.
Tursoy, T., & Resatoglu, N. G. (2016). Energy consumption, electricity, and GDP causality; the case of Russia, 1990-2011. *Procedia Economics and Finance, 39*, 653–659.
World Bank Report, 2005. Energy Security Issues. The World Bank Group Moscow, Washington, DC December 5, 2005.
Zhang, Y. J. 2011. Interpreting the dynamic nexus between energy consumption and economic growth: Empirical evidence from Russia. *Energy Policy, 39*(5), 2265–2272.
International Energy Agency, 2007. Energy Security and Climate Policy: Assessing interactions.
International Energy Agency, 2017. Energy Efficiency: Market Report Series.
International Energy Agency, 2018. Tracking clean energy progress: Industry.
Wilson, A. B. 2017. New rules of energy security of gas supply. *European Parliamentary Research Service, PE 607.271.*

Inclusive Development of Society – Lumban Gaol (eds)
© 2020 Taylor & Francis Group, London, ISBN 978-1-138-33476-2

Human potential increment in the system of innovative economy sustainable development management

E.I. Lazareva, T.Y. Anopchenko & A.D. Murzin
Southern Federal University, Rostov-on-Don, Russia

ABSTRACT: The relevance of the presented research is defined by the qualitative human resources increasing role in modern processes of the economy sustainable innovation-oriented development. The research is aimed at proposing ways to improve the human capital management system in the context of the economic development innovative orientation. Achieving the goal of the research required the solution of such problem tasks as systematization of conceptually new approaches to strategic management of economic systems; structural representation of the human capital characteristics system affecting the level of economic trajectories stability; justification and empirical verification of human capital increment strategy in the economic sustainable development management system. Novation of authors' approach, first of all, is in justification and verification of the concept of four-sector human potential system as a resource of innovative economic growth as well as in a comprehensive study of the strategy of the human resource potential increment in order to convert it into incomes, which are the source of innovation-oriented modernization and increase the competitiveness of economic systems. Components (sectors) of the proposed model accordingly characterize the quality of the individual human capital, standard of living, quality of social sphere and environmental quality. As an adequate research tool for substantiating, diagnosing and monitoring the strategy of human resource potential increasing, the authors applied an economic-mathematical model that integrates the interrelated analytical procedures for the human potential integrated assessment, as well as econometric estimates of the potential innovation effect from the strategy aimed at improving the level of its institutional conversion into the factors of innovative growth. The features of the innovations' reproductive functions of human resources are identified. The mechanism of rent on the human capital public expenditure and strategies of the human resources increment for economic systems' sustainable innovation development are substantiated. The empirically verified regularity expressed in the fact that the increase of the human capital qualitative characteristics facilitated the growth of the innovative activity level of the economic subjects and the national economy is also revealed. The spatial heterogeneity of strategies for the human potential increment, revealed on the basis of clustering of the Russian Federation regions in terms of the economic development subjects' innovative activity, makes it possible to strengthen the scientific and instrumental component of the system of managing innovation-sustainable development of the economy in dynamic and contradictory environmental conditions.

Keywords: sustainable development of the innovative economy, management system, human potential increment strategy, rent on the human capital

1 INTRODUCTION

The problem of sustainability of the economic development innovation-oriented trajectory today is one of the key tasks in a number of modern transformations. In a context of these transformations the essence of human resources is expressed in new aspects – it is converted into the integrated innovation-oriented economic growth resource-factor. This conversion is connected with economic systems movement towards innovative «knowledge economy», competition gravity center transference to the science, education, innovative activity sphere. The development of the countries which realize the policy of human capital quality, high technologies increment provides advantages in world socioeconomic evolution, raises competitiveness of national «intellectual» economy [1].

Modern cyclical development of economic systems is influenced by the transition to an information-mobile society, based on "high-hume" technologies and innovations, which gradually initiates the replacement of the competition paradigm by the information and innovation paradigm [2-6]. The competitive advantages of such systems directly depend on the ability of human capital to create innovations, new knowledge, combining production, intellectual and social resources, to make a variety of non-standard decisions, to implement new methods of management of socio-economic processes that ensure sustainability [7].

A significant rent-forming function of human capital determines its important place in the system of innovative-oriented development [1, 8]. Increasing human development quality importance for economic growth generating and competitiveness initiated the mounting interest of economists to the subjective factor (the human capital) role in economic progress. It gradually promoted the human capital parameters (at first – the individual, especially economic; later – the social, public) inclusion into the economic dynamics resource supply research system.

The research is based upon such founders of the innovation-oriented economic development theory as J. Bessant, S. Kuznets, B. Kuzyk, R. Lucas, P. Romer, J. Schumpeter, J. Jakovets etc.

In the conditions of active search for sources of sustainable-innovative evolution in the late twentieth – early XXI century, various theories in the field of innovative economy' development management were formed. The search for new (additional) factors of value added growth led to the activation of theoretical research in the field of resource approach to the sustainable trajectories of innovation-oriented development analysis, which contributed to the human capital – the basic (fundamental) resource of sustainable innovation dynamics gradual inclusion in the management system of such trajectories. «Human qualities» (high professionalism, quality of education, intellectual and innovative potential) and strategies for converting them into labor productivity/organization income increasing are turning into key technologies for sustainable and innovative development of the world economy. Theoretical approaches from the perspective of human capital focus on individual, separated from each other, the sources of development – technology, innovation management. In addition, the problems of innovative orientation of the economic development are considered mainly within one reproduction cycle, without looking into the future and without showing what will happen in the long term [9-11].

The understanding of systemic and interdependent changes in the social and economic systems, interdependence of the vectors of long-term development of these systems is reflected in the studies of recent years, developing an approach from the perspective of long-term cyclical development, the key source of which is knowledge and intellectual resources. In the works Bessant [2], Capello & Kroll [3], Grant [12], Ling & Jaw [13], Kuzyk & Yakovets [14] implicitly there is a tendency to broad interpretation of human capital, leading to a gradual installation in the system of strategic management not only its economic, individual, but also non-economic, social parametric characteristics. In modern conditions, there are completely different points of view on the mechanism of implantation of new, modernized structures and functions of human capital in the system of strategic decision-making. The authors' point of view is the necessity of implantation of the four-sector strategy of human potential increment in the system of strategic management decision-making.

2   MATERIALS AND METHODS

The expansion of factors and sources of innovative economy' positive evolution, the interrelated transformation of the development resource determinants initiated the formation of conceptually new approaches to the process of strategic management, justifying its focus on human resources increment (Figure 1).

Expanding resource concept of human capital increment as a condition for increasing competitiveness

The concept is based on expanding the number of components that ensure the competitiveness of economic systems; a unified approach to the quality of human capital as a system of its properties that express the aggregate of the properties of individuals and take the form of empirically observed and measurable qualitative characteristics that directly determine the mode of population reproduction and the influence of the demographic factor on economic processes. The concept of social capital provides a theoretical basis for the search for new mechanisms for the development of a social state that promotes, through social cohesion, the growth of the country's competitiveness.

The concept of human potential increment as a way to increase the productivity of labor

It places at the forefront of development not the traditional capacity for productive work, but the widening of the possibilities of choice through the growth of well-being (longevity, education and income), which is seen as a way to increase labor productivity and income. Thus, human development is considered both as a goal and criterion of social progress, and as an important resource of positive economic development.

Socio-ecological and economic concept of stable, balanced growth on the basis of improving the level of well-being

Sustainable development of well-being / poverty reduction is treated as a system of organization of society in which the needs of the present generation are met not at the expense of opportunities to meet the needs of future generations, the development paradigm moves towards balanced growth that explicitly takes into account social (distributive) and environmental goals, especially the goal of reducing the number of the poor and minimizing environmental degradation, and it has the same value as the economic efficiency.

The concept of opportunity, freedom of human choice and decision-making with the participation of the poor

The basis of this approach is the concept of A. Sen' human choice, the central idea of which is that the level of well-being should be judged, if possible, by people to lead a life that they consider worthy, and not in terms of income per capita.

The concept of subjective perception of welfare as a factor of entrepreneurial activity

The level of well-being is determined on the basis of subjective assessments of "well-being" or "happiness" - the degree of satisfaction of needs, which can only be estimated by the individual himself. The degree of satisfaction / dissatisfaction with one's life is one of the important factors of "perception" - the motives of a potential entrepreneur-innovator for doing business, which determine the level of entrepreneurial activity in society.

Figure 1. Modern methodological concepts of mechanisms of sustainable innovation development in strategic management.

The resource concept, which, as opposed to the branch organization, serves to disclose the mechanisms of support for the sustainable competitive advantages of economic agents by assigning them economic/innovative rents takes dominant place among the methodological approaches.

The approach is based on the idea of a business organization as a unique multi-component resource potential, the state and development of which directly depend on the management strategy and serve as a key source of business profitability. The competitiveness of economic agents at all levels of strategic management (macro- (national economy), mezo- (region), micro-level (organizations)) is largely determined by the quality of the practice/management strategy of anthroposocial (human and social) capital and investments in its development.

Representatives of the resource concept pay special attention to human capital. There are such qualities of human resources needed to create competitive advantage as the value, rarity, uniqueness and irreplaceability [15]. Competitive advantage is impossible without the institutional conversion of highly valuable human resources in the global competitive markets into innovative factors of production of competitive, «intellectually capacious» goods/services, which generates revenue in the form of innovative rent. Capitalization of the innovative rent transfers it into the source of the innovation oriented production process modernization (Figure 2).

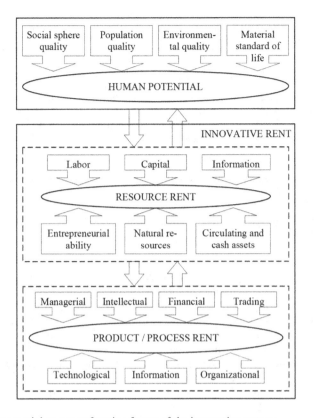

Figure 2. Human potential as a rent-forming factor of the innovative process.

The analysis of functional properties and set of interrelations of human potential components with process of sustainable development of the urbanized territory within the framework
of the basic resource approach and developments in the theory of human capital [16, 17] has
led to its structured representation in the form of four-sector system. The selection of human
potential indicators, carried out on the information base of Russia and the group of countries
with a high HDI [20, 21], revealed a stable correlation between the dynamics of «outcome»
indicators – the characteristics of economic agents innovation activity and the four components of human potential (Table 1) and proved not only theoretically, but also empirically,
that the basic structural elements of the human potential are, in addition to the traditionally
used quality of an individual human capital and welfare level, such its important components
as the quality of the social and environmental spheres. It is important to emphasize that these
four components of human potential as the holistic object determine the main vectors of strategic policy.

Values of the pair correlation coefficients (Table 1) as well as of the calculated elasticity
coefficients reflecting the relationship between the dynamics of the economic entities innovation activity parameters and human potential indicators (average life expectancy, GDP$_{PPP}$
per capita, the Gini Income Index and ecological sustainability index) have shown that the
degree of the human resources potential in innovative growth factor sources conversion
depends to the greatest extent on the social sphere quality – the elasticity coefficient was 1.724
and the correlation coefficient was 0.667 (the negative sign of the coefficient reflects the fact
that the Gini Income Index is a statistical indicator - destimulator). Then, in the order of
decreasing degree of dependence, the quality of the ecological environment is 0.463 and 0.324,
the quality of the population is 0.137 and 0.393 and the material standard of living is 0.057
and 0.442, respectively.

Table 1.   The pair correlation matrix of human potential indicators and the level of plant units' innovative activity.

Indices	1)	2)	3)	4)	5)
1)	1.000	0.442	0.393	-0.671	0.324
2)	0.442	1.000	0.585	-0.466	0.249
3)	0.393	0.585	1.000	-0.346	0.241
4)	-0.671	-0.466	-0.346	1.000	-0.227
5)	0,324	0,249	0,241	-0,227	1.000

Notes:
1) Innovative activity level of plant facilities
2) GDP$_{ppp}$ per capita
3) Average life expectancy
4) Gini coefficient
5) Ecological sustainability index

These results served as the basis for the development of an adequate tool for the human potential increment strategy analytical evaluation in the system of innovative economy' sustainable development management. The advantage of the developed model tools is the possibility of using it to accumulate analytical information on the economic, social, and environmental strategies parameters related to the growth of human potential in order to increase the pace of innovation economic dynamics. The set of instruments integrates the analytical procedures of the human potential complex estimation in the space of the quality of individual human capital, the level of well-being, the quality of social and environmental spheres global coordinates as well as the innovative strategies of strategy forming on the basis of the innovative effect evaluation.

Structural analysis of the model led to a well-founded allocation of its four components that actively influence the trends of sustainable innovation development of economic actors. And the most significant impact on the economic system trajectory is the quality of the social environment, the key indicator of which is the degree of social segmentation of society (Figure 3). This result proves the priority of reproduction of social, socially useful goods - sources of accumulation of social capital, characterized by such important in terms of innovative growth properties, as positive network effects and increasing their marginal utility in the process of use.

It is justified, using the authors' set of tools, that, the indicators of innovative activity of subjects of the modern Russian economy can be increased by about one and a half times due to the implementation of certain strategies of social and economic policy that increase the

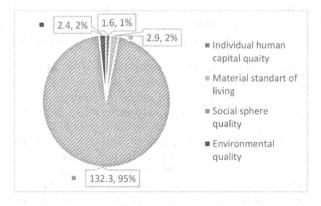

Figure 3.   Innovative effect of human capital qualitative characteristics increment, %.

118

level of conversion of human potential into the factors of innovative economic dynamics. Such strategies should, first of all, include strategies to increase the purchasing power of cash income per capita and reduce poverty; strategies to improve the level/quality of education and reduce morbidity; strategies for the development of social infrastructure, for increasing social and territorial mobility and the level/conditions of employment of the population; strategies for developing small business and increasing the freedom of entrepreneurship; strategies for creating a dynamic information infrastructure and improving access to technology and science, and a number of other strategies.

The initial clusterization of the regions of the Russian Federation in terms of the parameters of innovation activity of economic entities, implemented to assess the effectiveness of existing strategies and determine the priorities for further increment of human potential, identified four different spatial groups in terms of the number of their constituent territories. The first group (cluster A) was formed by subjects of the Russian Federation, characterized by zero innovative activity of economic growth subjects, the second (Cluster B) – regions, the distinguishing feature of which is the weak innovation activity of economic development subjects, the third (cluster C) and the fourth (cluster D) are regions with an average and high level of innovative activity of economic entities, respectively.

The spatial heterogeneity of strategies for the reproduction of human potential is revealed on the basis of the obtained results of clustering of the regions of the Russian Federation in terms of the indicators of innovative activity of economic development entities. In particular, for the territories included in clusters A, B, C and D, the strategies for the development of social infrastructure and the improvement of the level/quality of employment of the population, the strategies for the development of small business and the expansion of freedom of entrepreneurship, strategies for increasing access to scientific achievements and new technologies, development of information infrastructure are in priority, respectively (Table 2).

So the territorial organization of the reproduction of resources (first of all, qualitative human capital) should be the real content of state strategic policy. This policy should be based on a full evaluation of the resource potential and long-term criteria of its dynamics, stimulation of economic growth, creation of a modern public sector of the economy as the main organizer of such reproduction with support, first of all, of public investments. Investments in human capital (education, health, population policy) can directly improve the quality of life. They can also provide a greater interest in investment, since healthier and better educated workers increase the productivity of capital. Thus, in order to sustain growth in the long term, it is necessary to revise priorities in favor of human capital (Figure 4).

Table 2. Ranking of strategies for the human resources increment for sustainable development of various types of Russian territories.

Type of region	Rank			
	1	2	3	4
Type A	1)	2)	3)	4)
Type B	3)	2)	1)	4)
Type C	3)	4)	1)	2)
Type D	4)	3)	1)	2)

Notes:
1) Social infrastructure development
2) Increase of the employment level/conditions
3) Small business development, freedom of entrepreneurship
4) Increase of the access to the science and technologies, informational infrastructure development

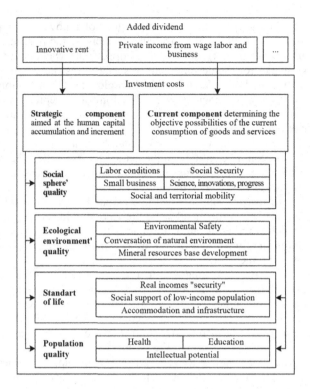

Figure 4. Mechanism of public spending of innovation rent (rent on human capital) in strategic management.

3 DISCUSSIONS

The results of the research made it possible to take a fresh look at the problem of the interconnection of human resources management strategies/practices with the innovative activity of Russian economic entities. The resource concept of strategic management (Barney [15], Prahalad, Hamel [17], Grant [12], Kuzyk, Yakovets [14]) and the theory of human capital (Becker [16]), which consider the company's human resources as a factor of sustainable innovation-oriented development, formed the theoretical basis of the research.

We proceeded from the assumption that the practice of effective human capital increment should include four interrelated directions - the quality of individual human capital, the level of well-being, the quality of social and environmental spheres. The results of the study as a whole confirmed this assumption.

The results of the authors' set of tools approbation make it possible to prove that human potential increment management strategies in all directions have a significant impact on the level of economic entities' innovation activity. So we can recommend these tools, which not only simplify the process of forming a strategic portfolio of sustainable human potential increment system management, but also predict the contribution of various strategies to the innovative effect, for application in the development of long-term socio-economic policies.

As limitations of the study, we can consider the small size of the information base for the empirical verification of model tools, partly related to the small number of the Russian innovation-active companies cohort, as well as to the problems of innovation processes statistical monitoring organizing, which could affect the results of regression analysis To overcome them the application of a more perfect methodology in further empirical research will allow.

4 CONCLUSION

The author's elaborations allow to suggest such identification criteria of the mutual agreement degree of the social and economic strategies of the innovation oriented economical dynamics as: production of the considerable volume of the gross product per capita, achievement of the high level of well-being, social institutions maturity; the place of the human resources policy among the development priorities, the importance of the state redistribution function, allowing to eliminate the considerable incomes differentiation of the revenues in the society and to eliminate poverty and welfare poles.

The results of the research make it possible to change the view on the priorities of strategic management of the economic entities' development trajectories and formulate some recommendations for actors aimed at increasing the level of innovation activity:

- application of the theory of human capital to the strategic management of human resources;
- the development of a strategic policy for human resources management and its integration into a comprehensive development strategy;
- use of the methodology of the national quality management standard for human resources «Investors in People» for a reasoned choice of criteria for evaluating the management of human resources;
- active promotion of long-term employment in various forms, freedom of information exchange, contributing to the achievement of strategic imperatives;
- the development of social infrastructure, aimed at long-term improvement of the quality of social conditions for the human capital increment.

ACKNOWLEDGMENT

This work was supported by the Russian Foundation for Basic Research: projects No. 17-02-00296 and No. 18-010-00594.

REFERENCES

1. Lazareva, E. I. 2009. *National Welfare as an Integrated Resource of Innovation-Oriented Development of the Economy: Theoretical-Methodological Aspect*. Rostov-on-Don: Southern Federal University. (In Russian).
2. Bessant, J. 2013. Innovation in the twenty-first century. In: *Responsible innovation: Managing the responsible emergence of science and innovation in society*: 1–25.
3. Capello, R. & Kroll, H. 2016. From theory to practice in smart specialization strategy: Emerging limits and possible future trajectories. *European Planning Studies* 24(8): 1393–1406.
4. Kuznets, S. S. 1971. *Economic Growth of Nations: Total Output and Production Structure*. Cambridge: Belknap Press.
5. Radjou, N. & Prabhu, J. 2015. *Frugal Innovation: How to do more with less*. The Economist.
6. Schumpeter, J. A. 2017. *Theory of economic development*. Routledge.
7. Lazareva, E. I., Anopchenko, T. Y. & Lozovitskaya, D. 2016. Identification of the city welfare economics strategic management innovative model in the global challenges conditions. In: *SGEM* 2016 *Proceedings*: 835–842.
8. Anopchenko, T. Y., Lazareva, E. I., Bagdasaryan, I., Vasileva, Z. & Almabekova, O. 2015. Human resource innovative management using the tools of econometrics. In: *SGEM* 2015 *Proceedings*: 393–399.
9. Aghion, P. & Bolton, P. 1997. Theory of Trickle-Down Growth and Development. *Review of Economic Studies* 64: 87–110.
10. Lucas, R. 1988. On the mechanism of economic development. *Journal of Monetary Economics* 22: 116–125.
11. Romer, P. M. 1990. Endogenous Technical Change. *Journal of Political Economy* 98(5): 21–37.
12. Grant, R. 2002. *Contemporary Strategy Analysis: Concepts, techniques, applications*. Malden, MA: Blackwell Business.

13. Ling, Y. H. & Jaw, B.S. 2011. Entrepreneurial Leadership, Human Capital Management, and global competitiveness. *Journal of Chinese Human Resources Management* 2(2): 117–135.
14. Kuzyk, B. N. & Yakovets Y. V. 2005. *Russia-2050: Innovative Break-out Strategy*. Moscow. Economics. (In Russian).
15. Barney, J. 2015. Firm resources and sustained competitive advantage. *International business strategy: theory and practice*: 283.
16. Becker, G. 1994. *Human Capital: A Theoretical and Empirical Analysis with Special Reference to Education*. Chicago, IL: The University of Chicago Press.
17. Prahalad, C. K. & Hamel, G. 1994)Strategy as a Field of Study: Why search for a new paradigm? *Strategic Management Journal* 15: 5–16.
18. Lazareva, E.I. 2013. Features of national welfare innovative potential parametric indication information-analytical tools system in the globalization trends' context. In: *CEUR Workshop Proceedings* 9: 339–351.
19. Anopchenko, T. Y., Grinenko, S. V., Edalova, E. S., Zadorozhnyaya, E. K. & Murzin, A.D. 2018. Management of human capital reproduction processes in post-Soviet countries. *Espacios* 39(9): 16.
20. *World Development Indicators*. 2016. Washington, D.C.: World Bank.
21. *Environmental Performance Index* (*EPI*). Yale: Yale Centre for Environmental Law & Policy, University; Ispra, Italy: Joint Research Centre of the European Comission. URL: http://epi.yale.edu

Inclusive Development of Society – Lumban Gaol (eds)
© 2020 Taylor & Francis Group, London, ISBN 978-1-138-33476-2

Industrial complex development initiatives in the fourth industrial revolution

T.O. Tolstykh
National University of Science and Technology "MISIS", Moscow, Russia

L.A. Gamidullaeva
K.G. Razumovsky Moscow State University of Technologies and Management (the First Cossack University), Penza, Russia

E.P. Enina
Voronezh State Technical University, Voronezh, Russia

T.V. Matytsyna
Southern Federal University, Rostov-on-Don, Russia

T.N. Tukhkanen
Don State Technical University, Rostov-on-Don, Russia

ABSTRACT: Global trends raise a great challenge to economics, politics, and society. Among the trends of technological development in the 21st century one may distinguish the following: the transition from "manufacturing" as production by means of human physical forces to "brainfacturing"—intellectual production or production by means of human intelligence; the transition from "high-tech" and "low-tech" in the 20th century to the conception of advanced and disruptive industries in the 21st century; the transition from B2B and B2C to M2M—the conception of IoT ("Internet of Things"). The industrial complex management is incapable of keeping up with all the trends. In these conditions the most important issue is prioritizing. The victory belongs to companies that know how to prioritize and concentrate efforts and understand that no company is capable of coping with all the trends. You have to be open to the world. The article compares foreign and Russian state initiatives regarding the development of high-performance computing systems, cutting-edge materials, and production technologies of the United States, Germany, and Russia. In the United States an example of such initiatives is Advanced Manufacturing Partnership 2.0 (AMP 2.0); in Germany, the conception Industrie 4.0; and in Russia, the National Technological Initiative (NTI). The article characterizes each initiative: its goal, technologies, organizational status, and main participants. It is proved that the leadership of US production is based on system engineering, full life span simulation, and computer engineering. Germany is strong in the development of industrial equipment and control systems thereof but lags behind in software development in comparison with American companies in the lead.

1 INTRODUCTION

Each industrial revolution develops a new mindset. For example, the creation and application of computer-aided design (CAD) systems in development of technical systems and constructions was recognized by the National Science Foundation of the United States as the greatest event capable of significantly increasing productivity, comparable in this regard only with production electrification (Krouse, 1982).

The new industrial revolution at the present time represents an intersection of three branches: research, construction, and design in the form of convergence and synergy, digital platforms, Big Data, intelligent assistants, smart design, and smart production. We are at the threshold of a new era—the era of artificial intelligence, or as some scientists and visionaries say—now is the first day of the creation. It is not, however, just for new technologies, but cardinal changes of the very foundation of our civilization, the way of thinking of all Earth dwellers. The founder of the World Economic Forum Klaus Schwab (2017) called it "the Fourth Industrial Revolution." The third industrial revolution was taking place similarly. Determinacy is inevitable, and society is involved in changes regardless of individuals' desires: decisions on credit granting or employment offers are made by artificial intelligence. The mass media have risen the tide of reports about the success of neural networks and investments in the development thereof, the sums of which match the budgets of million-population cities.

However, if we recall the history of the development of artificial intelligence (AI), the déjà vu is guaranteed: the same surge of interest has already been observed twice—in 1980s unreasonably high expectations of large corporations and the military industrial sector financing research in this field led to the period known as "the winter of artificial intelligence." The cutting off of state support resulted in almost total cessation of research in this field.

AI technologies are actively implemented in the retail and banking sectors, but the introduction into the real sector of the economy is still at the initial stage. Most large enterprises still do not dare to switch from traditional methods of client service or production management to the technologies based on data analysis and fully automatic decision making. However, those few that have made up their minds speak about real efficiency from the implementation.

The fourth industrial revolution is fundamentally changing modern manufacturing as a result of new technological achievements, including digitization and robotization, AI, and the Internet of Things, and new materials and biotechnologies. Owing to these changes manufacturing in the developed countries is again becoming the main source of prosperity and creation of new workplaces. Today, the leading developed and developing countries implement their own initiatives focused on the development of future production, and the leading foreign regions play an active part (Schwab, 1017).

The features of the fourth technological revolution may include the following:

- Pace of development. The development goes nonlinearly at a rather exponential pace. The new technology itself synthesizes newer cutting-edge and efficient technologies.
- Width and depth. Changes apply not only to what we do and how we do it, but also to who we are.
- Systemic action. It provides for integral external and internal transformations of all systems in all countries, companies, industries, and the society in whole.
- Methodological changes. Harmonization and integration of a large amount of various scientific disciplines and inventions increase.

The following three are the key development programs in the Russian Federation:

1. National Technological Initiative (NTI). Initiator: the Agency of Strategic Initiatives (ASI), the Russian Venture Company (RVC). Launched on December 4, 2014. Implementation term: until 2035.
2. Digital Economy of the Russian Federation. Developer: the Ministry of Communications and Mass Media of the Russian Federation (from May 2018, the Ministry of Digital Development, Communications and Mass Media of the Russian Federation). Approved on July 28, 2017. Implementation term: until 2024.
3. The Strategy of Scientific and Technological Development of the Russian Federation. Coordinator: the Ministry of Education and Science of the Russian Federation (from May 2018, the Ministry of Science and Higher Education of the Russian Federation). Approved on December 1, 2016. Implementation term: until 2035.

The goal of the digital economy in the Russian Federation program is to organize systemic development and implementation of digital technologies in all spheres of life: in the economy, in business as a social activity, in the public administration, in the social sphere, and in the municipal economy.

The development of the digital economy is planned to be managed via a "road map" drawn for five directions: regulation, personnel and education, development of research competencies and technological backlogs, informational infrastructure, and information security.

2 RESULTS AND DISCUSSION

Let us consider the state initiatives on the development of high-performance computing (HPC), cutting-edge materials, and manufacturing technologies in the United States (Table 1).

The leadership of US manufacturing is based on systemic engineering, full life span simulation, and computer engineering (Borovkov & Kukushkin, 2018; Borovkov et al., 2018).

Germany has created and is actively implementing Industrie 4.0—a German program of technological development, one of 10 projects of the future, provided for in High-Tech Strategy 2020 Action Plan—the strategy of German industrial competitiveness improvement launched in 2011. The project embraces leading research and industrial organizations of Germany: National Academy of Engineering Sciences (Acatech), Fraunhofer Society, German Artificial Intelligence Research Center (DFKI), Wittenstein AG, Bosch, Festo, SAP, and Trumph.

Table 1. State initiatives on the development of HPC systems, cutting-edge materials, and manufacturing technologies in the United States.

Launching year	Original name	Characteristics
2004	Department of Energy High-End Computing Revitalization Act	Development of HPC systems for the industry
2009	US Manufacturing—Global Leadership through Modeling and Simulation	Manufacturing on the basis of computer simulators
June 2011	Advanced Manufacturing Partnership	Establishment of America Makes—National Additive Manufacturing Innovation Institute
2012	Materials Genome Initiative	The goal is to double the pace and to reduce costs of invention, manufacturing, and mass implementation of cutting-edge materials by means of computer-aided design at the atomic level
September 2013	Advanced Manufacturing Partnership 2.0	Establishment of Manufacturing Innovation Institutions: in digital manufacturing, light materials, next-generation power electronics, cutting-edge composite materials, integral photonics, flexible hybrid electronics, fibers and fabrics, additive technologies, and smart manufacturing in the field of clean power engineering, cutting-edge robotics in manufacturing, bio-growing of fabrics and organisms, biopharmaceutics manufacturing, recycling of various materials and electronic waste, modular intensification of chemical processes
2015	National Strategic Computing Initiative	The goal is to create computing systems of the exa-level intended to integrate hardware and software. The processing power of such systems exceeds by 100 times the existing systems of 10 petaflops; studying the issue of HPC system development beyond Moore's law and upon the exhaustion of the cost efficiency of semiconductor manufacturing technologies

The solution offered by Industrie 4.0 is the formation of enterprise networks that compile cyber-physical systems including hardware, logistic systems, technological equipment, capacity for autonomous information exchange, initiation of operations, and independent control over operations. The tenor of Industrie 4.0 is the transition from embedded systems to cyber-physical ones. The embedded systems are central control units built in various objects to be controlled. The cyber-physical systems are new technologies enabling a uniting of the virtual and the physical world, thus providing interaction of smart objects with each other through use of the Internet/networks and data (Vertakova et al., 2017; Tolstykh et al., 2018a, 2018b).

The basis of Industrie 4.0 is the Internet of Things—the conception of a computing network of physical objects with technologies embedded to provide interaction between the said objects or with the external environment. The goal of Industrie 4.0 is to improve German industrial competitiveness in the conditions during which German is incapable of competing in expenditures with the developing and some developed countries (the United States); Germany is strong in the development of industrial equipment and its control systems, but lags behind American companies in software engineering. Industrie 4.0 will result in a labor productivity leap and a considerable increase of gross domestic product (GDP) growth rates. The Industrie 4.0 project coordinators are the Federal Ministry of Education and Research and the Federal Ministry of the Economy and Technology. The Ministry of Internal Affairs is also involved in the project.

In 2014 the president of the Russian Federation gave a start to the National Technological Initiative (NTI) with its main goal to introduce Russia into shaping of standards of future global markets and to earn Russian companies a significant share in these markets.

A comparison of Industrie 4.0 (Germany) and NTI (Russia) is shown in Table 2. Besides the state initiatives of the United States, Germany, and Russia described in Table 2, the core principles of technological development have been extended in national documents of other countries. Examples include Made in China 2025, Smart Factory in the Netherlands, Usine du Futur in France, High Value Manufacturing Catapult and Future of Making Things in Great Britain, Fabbrica del Futuro in Italy, Made Different in Belgium, etc.

The main tool of Russian technological breakthrough is TechNet.

The following are TechNet's cutting-edge manufacturing technologies:

Table 2. A comparison of Industrie 4.0 (Germany) and NTI (Russia).

Industrie 4.0	NTI
Making Germany a leader in cyber-physical systems development and implementation by 2020 both inside and outside the country	Growing national companies in the markets that do not exist today
Industrial equipment, the Internet of Things, automation, service robotics, smart factories, M2M and H2M, etc.	Cross-cutting technologies, cutting-edge manufacturing technologies (TechNet) + nine markets of the future
One of 10 projects of the future within the High-Tech Strategy 2020 Action Plan	Independent interdepartmental cross-industry initiative with the society and business being its key elements
The Ministry of Education and Research, the Ministry of the Economy and Technology	The Council under the President of Russia on economic modernization and innovative development, the Ministry of Industry and Trade, the Ministry of Science and Higher Education of the Russian Federation, the Ministry of Energy of the Russian Federation, etc.
Large, medium, and small companies, Fraunhofer Society, National Academy of Engineering Sciences (Acatech), Association of IT companies (BITCOM), Association of Mechanical Engineers (VDMA) and Electronics Manufacturers Association (ZVEI), etc.	Agency of Strategic Initiatives (ASI), the Russian Venture Company (RVC), medium and small companies

- Advanced Simulation & Advanced Optimization–Driven Design & Manufacturing: CAD/CAE/FEA/CFD/FSI/MBD/EMA/CAO/HPC/PDM/PLM … MES/ERP/CRM
- Additive and hybrid technologies
- New materials: composite materials, polymers, ceramics, alloys, metal powders, and metal materials
- Smart Big Data at the input and output as a basis for Advanced Predictive Engineering Analysis/Analytics
- ICS, sensing, industrial robotics, the industrial Internet, etc.

At the present time, as a result of changes in the global economy (structural, technological, geopolitical) it is necessary to digitally transform the activities of Russian enterprises regarding their managerial and technological processes and models, as well as to strengthen the interaction of companies between each other and with other members of the industrial sector at meso-, macro-, and mega-levels to ensure companies' international competitiveness. This will enable complete fulfillment of the digital economy's system potential by direct introduction of digital economies into companies' economic mechanisms.

Industrial companies are the foundation of the Russian economy. The implementation of digital technologies in the industrial sector is a strategic priority for the Russian economy. However, the economy still remains based on raw materials and focused on the export of natural resources. The share of the digital economy in GDP is just over 2%, and this value has been stagnating since 2014, while in other countries it is growing. The density of manufacturing robotization at Russian enterprises is more than 20 times lower than the world average (Babkin, 2014, 2017).

Customization, rebalancing of the value of individual engineers' experience and the value of digital models, decentralization of design and manufacturing, new certification requirements —all these changes are becoming the reality of the global competitive industry.

With the beginning of the fourth industrial revolution the world finds itself on the border of major transformations, characterized by the joint effect of new technological achievements, including AI, the Internet of Things, robotization, 3D printing, portable devices, genetic engineering, nanotechnologies, new materials, biotechnologies, etc. Provided there is interaction among them, these technologies may become a driver of accelerated economic growth and an increase of productivity. According to the latest Index of Global Competitiveness by the World Economic Forum covering 138 countries, Russia has consolidated its economic indicators associated with the fourth industrial revolution.

Digital transformation is a major project requiring alteration of many usual procedures, implementation of new methods and work techniques, and organizational changes. The success of the digital transformation is conditioned not just by technologies. To a great extent it implies teamwork, customer relations, liaisons with suppliers, and logistics (Bloching et al., 2015).

Until recently the technological industry was just another economic sector. Today the situation has cardinally changed. Technological companies have not just reached the top of the business mountain. Having elaborated new business models, they are now inventing new managerial standards for all industries, state bodies, and the third sector in the whole world. Technological development has led to a situation in which virtually any business has become technological. Any large company today has a set of objectives in the field of technologies. It is the technologies—with regards to infrastructure, automation, skills, rate of new solution implementation—that more often become important competitive advantages for business in general (Lee et al., 2014; Schweer & Sahl, 2017).

In addition, it is important to create a unified ecosystem of a digital economy in the form of an ecosystem of virtual clones of individual systems (avatars) using methods of industrial analytics (Big Data), which will make it possible to generate an effective system of communications in science, technology, and innovations, thus ensuring improvement of the economy's and population's to innovation by creating the conditions for knowledge-intensive business development. Achieving the goals of the scientific and technological development of the Russian Federation are the guideline of the Strategy of Scientific and Technological Development of the Russian Federation from January 12, 2016, No. 642).

Digital manufacturing is the kernel of the digital economy. It is sometimes referred to as "the real sector" of the economy, i.e., the sphere of manufacturing, but at a new technological level based on computer (supercomputer) technologies. The digital manufacturing is a broad-scale application of software in the whole production cycle.

3 CONCLUSION

The most promising areas that need to be developed in Russia today are digital design and modeling, new materials, additive technologies, industrial Internet, and robotics. This is the aim of the NTI road maps (as of June 2018, five action plans have been developed (NTI road maps in the Energy, AutoNet, Aeronet, Marinet, and Neuronet areas). The experience of organizing digital factories in Russia can be characterized as follows:

- Automotive-1 on the basis of FGUP "NAMI": Development of a set of technological solutions that ensure the integration of advanced production technologies into the production target of FSUE "NAMI"
- Automotive-2 on the basis of UAZ LLC: Development of a complex of technological solutions that ensure the integration of advanced production technologies into the company's production target for the creation of the UAZ Patriot 2020
- Automotive-3 on the basis of OOO VOLGABS: Development of full-scale mathematical models, design studies, and design of structural elements of new generation passenger car buses, modular platform of unmanned passenger and cargo transport, and municipal machinery
- Automotive-4 on the basis of PAO "KAMAZ"
- Tractor-1 on the basis of OAO "KIROVSKY FACTORY" (cabin, gearbox, bridges)
- Helicopter-1 on the basis of AO "Helicopters of Russia": Development of full-scale mathematical models, design studies and design of structural elements of a civil/military high-speed helicopter
- Shipbuilding-1 on the basis of AO "SPMBM Malakhit": Development of full-scale mathematical models, design studies, and design of structural elements of fourth- and fifth-generation nuclear submarines "Yasen"
- Engine-2 on the basis of PAO NPO Saturn: Development of an applied software package for the design and analysis of parts made of polymer composite materials with a 3D woven reinforcing internal structure
- Engine-3 on the basis of PAO "NPO Saturn": Development of multidisciplinary mathematical models of metalworking of spare parts and laser processing of complex surfaces
- Aerospace-1 on the basis of OAO "ORKK": Creation of a distributed center for virtual testing for the rocket and space industry
- Defense-1: Development of full-scale mathematical models, computational research, and design of structural elements of the AIS

Effects from the implementation of Digital Fabric are possible: 10–50% cost reduction, up to a four times reduction of production time reduction, up to two times profit growth, 50–70% increase in the number of new products, 7–15% reduction in the number of equipment units,- and up to four times predictability growth.

ACKNOWLEDGMENTS

The study reported in this article was funded by RFBR according to research project No. 18-010-00204-a.

REFERENCES

Babkin, A. V. (ed.) 2014. *Economics and Industrial Policy: Theory and Instrumentary.*.

Babkin, A. V. (ed.). 2017. *Economics and Management in the Conditions of Nonlinear Dynamics.* St. Petersburg.

Bloching, B., Leutiger, P., Oltmanns, T., et al. 2015. The digital transformation of industry. Roland Berger Strategy Consultants und Bundesverband der Deutschen Industrie, Munich.

Borovkov, A. I., & Kukushkin, K. V. 2018. Center for NTI "New Production Technologies" on the basis of the Institute of Advanced Production Technologies of SPbPU. *Trampoline for Success,* 13:22–27.

Borovkov, A. I., Marusev, V. M., & Ryabov, Yu. A. 2018. New paradigm of digital design and modeling of globally competitive products of the new generation. In *Digital Production: Methods, Ecosystems, Technologies,"* pp. 24–43.

Frolova, I. V., Panfilova, E. A., Matytsyna, T. V., Lebedeva, N. Y., & Likhatskaya, E. A. 2016. Integration of the corporate reporting instruments of situational and matrix modeling. *Social Sciences and Interdisciplinary Behavior: Proceedings of the 4th International Congress on Interdisciplinary Behavior and Social Science, ICIBSOS 2015 4th,* Pp. 317–324.

Frolova, I. V., Pogorelova, T. G., Likhatskaya, E. A., & Matytsyna, T. V. 2017. Harmonization of the tax portfolio of an organization by means of situational matrix modeling. In *Managing Service, Education and Knowledge Management in the Knowledge Economic Era: -Proceedings of the Annual International Conference on Management and Technology in Knowledge,* Service, *Tourism and Hospitality, SERVE 2016 4th,* pp. 69–74.

Karapetyants, I., Kostuhin, Y., Tolstykh, T., Shkarupeta, E., & Krasnikova, A. 2017a. Establishment of research competencies in the context of Russian digitalization. In *Proceedings of the 30th International Business Information Management Association Conference (IBIMA),* November 8–9, 2017, Madrid, Spain, pp. 845–854.

Karapetyants, I., Kostuhin, Y., Tolstykh, T., Shkarupeta, E., & Syshsikova, E. 2017b. Transformation of logistical processes in digital economy. In *Proceedings of the 30th International Business Information Management Association Conference (IBIMA),* November 8–9, 2017, Madrid Spain, pp. 838–844 (0,81/0,2).

Krouse, J. K. 1982. *What Every Engineer Should Know about Computer-Aided Design and Computer-Aided Manufacturing: The CAD/CAM Revolution.* Boca Raton, FL: CRC Press, T. 10.

Lee, J., Kao, H. A., & Yang, S. 2014. Service innovation and smart analytics for industry 4.0 and big data environment. *Procedia Cirp,* 16:3–8.

Schwab, K. 2017. *The Fourth Industrial Revolution.* Moscow: Publishing House "E."

Schweer, D., & Sahl, J. C. 2017. *The Digital Transformation of Industry: The Benefit for Germany,* pp. 23–31. The Drivers of Digital Transformation. Cham, Switzerland: Springer, Cham.

Tolstykh, T. O., Gamidullaeva, L. A., Shkarupta, E. V. 2018a. Key factors of development of industrial enterprises in industry conditions 4.0. Economics in Industry, 1:4–12.

Tolstykh, T. O., Shkarupta, E. V., & Gamidullaeva, L. A. 2018b. Approaches to the design of an innovative ecosystem in the conditions of digitalization of socio-economic systems. In A. V. Babkin (ed.), *The Formation of the Digital Economy and Industry: New Challenges,* pp. 117–135. St. Petersburg.

Tolstykh, T., Shkarupeta, E., Kostuhin, Y., & Zhaglovskaya, A. 2018c. Digital innovative manufacturing basing on formation of an ecosystem of services and resources. In *Proceedings of the 31st International Business Information Management Association Conference (IBIMA),* April 25–26, 2018, Milan, Italy, pp. 4738–4746.

Tolstykh, T., Shkarupeta, E., Kostuhin, Y., & Zhaglovskaya, A. 2018d. Key factors of manufacturing enterprises development in the context of industry 4.0. In *Proceedings of the 31th International Business Information Management Association Conference (IBIMA),* April 25–26, 2018, Milan, Italy, pp. 4747–4757.

Tolstykh, T. O., Shkarupeta, E. V., Shishkin, I. A., Dudareva, O. V., & Golub, N. N. 2017a. Evaluation of the digitalization potential of region's economy. In: E. Popkova (ed.), *The Impact of Information on Modern Humans: HOSMC 2017,* pp. 736–743. Advances in Intelligent Systems and Computing, Vol. 622. Cham, Switzerland: Springer. DOI: 10.1007/978-3-319-75383-6.

Tolstykh, T., Vertakova, Y., Shkarupeta, E., Shishkin I., & Krivyakin, K. 2017b. Assessment of the Impact of higher education development on the social and economic processes in the region. In *Proceedings of the 29th International Business Information Management Association Conference (IBIMA),* May 3–4, 2017, Vienna, Austria, pp. 2180–2191.

Vasin, S. M., & Gamidullaeva, L. A. 2015a. Increasing the efficiency of state institutional aid to small innovative enterprises. *Review of European Studies.*

Vasin, S. M., & Gamidullaeva, L.A. 2015b. Modeling and development of a methodology for assessing the socio-economic processes in the management of business incubators. *Mediterranean Journal of Social Sciences.*

Vertakova, Yu. V., Tolstykh, T. O., Shkarupta, E. V., & Dmitriyev, E. V. 2017. *Transformation of Management Systems Under the Impact of the Digitalization of the Economy.* Kursk.

Inclusive Development of Society – Lumban Gaol (eds)
© 2020 Taylor & Francis Group, London, ISBN 978-1-138-33476-2

Feedback mechanism for digital technology use in enterprise strategies

L. Goncharenko, E. Sharko, S. Sybachin & M. Khachaturyan
Plekhanov Russian University of Economics, Moscow, Russia

Z. Prokopenko
South Federal University, Rostov-on-Don, Russia

ABSTRACT: The subject of this article is the process of strategy formation based on the use of digital tools. Its aim is to develop a mechanism for feedback on the use of digital technologies in enterprise strategies. The authors set out the concept and structure of the media space, highlighting the elements of the macro - and mesospheres of the enterprise. In addition, a study was conducted to identify the profile characteristics that consumers present when choosing a company for further interaction. The methodology used consisted of consumer reviews using Google Forms. At the end of the research, the following results were obtained: before purchase of products future customers should carefully read the information on the enterprise network, analyze the website, and read the reviews and the blogs that influence decision-making. The obtained results can be used in the concept development of the strategy.

1 INTRODUCTION

The government of the Russian Federation has been realizing the digital economy program for 2017–2030 approved by President Vladimir Putin (Order of the Government of the Russian Federation No. 1632, 2017). A main goal of the program is the creation and development of a cyberspace that will facilitate the solution of problems of competitiveness and national safety of the Russian Federation. According to the priority concept, during global informatization (authors' comment: "informatization is policies and processes aimed at building and developing telecommunications infrastructure that combines geographically distributed information resources") managers are obliged to introduce modern digital technologies for realization of their strategies. Today the borders of the physical and virtual markets are particularly erased so the struggle for consumers is getting an aggressive character from marketing specialists.

Digital technologies are designed as much as possible to reduce the stage of information message transmission from the enterprise to the final consumer by means of effective use of media space.

Close to the concept of "media space" are the concepts of "information space," "information and communication space," and "hyper-reality." E. Toffler introduced the concept of infonoosphere (authors' comment: "integration of information, digital, technological, and intellectual spheres of human activity")—a transitional stage on the route to transformation of society in future civilizations, from the technogenesis (authors' comment: "the process of transformation of the environment influenced by various types of human technical activities") before anthropogenesis of a human civilization (Toffler, 2004). A. Sokhatskaya marks out three options of information space: a cyberspace, an infosphere, and a noosphere; at the same time the noosphere unites all three spaces (Sokhatskaya, 2012). We suggest considering the concept "media space" as the closest, as it is synonymous with the concept "noosphere."

The wider the market borders are, the more difficult it is to follow the back reaction of consumers and to estimate the efficiency of the enterprise strategy. The need for the formation of a feedback mechanism of use of digital technologies for enterprise strategies has also caused the actuality of this research and problems (Sharko, 2015).

To understand the weak points in the mechanism of interaction of the enterprises and final consumers in the information digital environment, it is necessary to designate the conceptual concepts and basic elements of this mechanism.

The research question is to define tools and methods of interaction of clients of enterprises to ensure the implementation of strategic goals in the era of the digital economy.

The research goals are

1. To define the basic components of the feedback mechanism of use of digital technologies in enterprise strategies
2. To organize research by interviewing potential customers of enterprises in the service sector —users of mobile devices and Internet resources—to identify effective digital technologies for enterprises
3. To summarize the results and make recommendations for businesses that operate in the era of the digital economy

2 BACKGROUND

To ensure the operation of the feedback mechanism of the use of digital technologies in enterprise strategies, it is necessary to use the media space. There are various scientific approaches to determination of content of media space:

- Sociological: set of means of social communication (Berger et al., 1966; Giddens, 2005; Bourdieu, 2007)
- Psychological: set of means of psychological impact on the personality (McLuhan, 2002; Bodriyar, 2006)
- Journalistic: set of mass media and mass media (Matveev, 2012; Plakhty, 2017)
- Philosophical and cultural and anthropological – media space as a multidimensional phenomenon which is considered at various levels and in different manifestations (Baran, 2010; Ilyashenko, 2011; Gritsay, 2012) Ilyashenko

Now there is the sociological approach to definition of the concept "media space," but, except when the aforementioned prevails, philosophical and cognitive and neurophilosophical approaches to the definition of this concept are more and more actively formed.

On the one hand, the media space is an object of attention of media managers, sociologists, culturologists, psychologists, lawyers, etc.; on the other hand the media space itself forms such spheres as social space, culture, the legal field, the right, and so forth. Researchers define the manipulative nature of the processes proceeding in media space, plasticity of media space (defined by these or those social and political structures), and its communication with social space. "Creating new conditions to activity of society, the new reality influences not only its organizational, communication features, but also the character of the public relations which become prompter, media saturated, various and intensive, generating new social and psychological, information and psychological phenomena: information phobias, information loading and tension, aggression and information crimes. Such approach gives us an opportunity to consider definition of the concept "media space" from positions of its multidimensionality, synthetic character, and complexity of modern information and communication space which covers several spheres forming new conditions of public life (Gritsay, 2012):

- A technosphere that is constructed on ICT
- An infosphere that is based on information and network highways
- A socioinfosphere that occupies any flows of information structures that organize and operate them, create them consumption, and influence the condition of social intelligence.

The structural concept of media space is interpreted by different researchers in different ways. But, in our opinion, in all definitions it is a multidimensional superposition or reflection of superposition of three types of spaces: information, virtual, and physical. E. Nem writes about various principles of approaches to a concept of media space (textocentric, structural, territorial, technological, ecological) and suggests three media space dimensions (Nem, 2013):

1. The media space – a space for live transmission of information messages that represents both physical and social aspects; these are media images and media texts as a result of media "mapping" of reality.
2. The mediated space – any type of social space that assumes the use of media and/or comes under their considerable influence, that is, the sphere of distribution of media technologies that change the nature and configuration of spaces (it is possible to speak about a mediatization (authors' comment: "introduction of media technologies in all spheres of human activity") of public and private space, policy, religion, work, rest, shopping, travel, and so forth).
3. Space of media ("old," "new," its convergent forms) – material space of mass media networks and streams (can have both physical and "virtual" geography).

If the media space corresponds to content, mediated – with the sphere of its distribution and consumption – then the media space comprises channels of production and information transmission; these are media and the system of their interrelations. Borders between these three dimensions are very conditional.

Means of exchanging between the enterprises and final consumers are informational digital streams.

Informational digital streams that form media space have the following properties:

- Universality: Information doesn't know frontiers and any other borders.
- Infinity: Information has a cumulative property of continuous accumulation and self-restoration.
- Hierarchy: Information has a hierarchical structure and a tendency to transition to higher levels of hierarchy and to increase in number of new communications.
- The targeting: Information is always connected with some material carrier.
- Orientation: From an object to the subject; continuity.
- Uniqueness.
- Variety.

Understanding the problems of the immersion of the real world to the digital virtual world demonstrates that the formation of the new basic media space is continuing – the re-inovirus. The three-dimensional world of boundless opportunities and properties continues; there is a process of its rapid expansion on the basis of creation of new systems of reproduction of signals and the movements of information streams. At the same time, being the result and the stimulus of the process of globalization, the space eliminates with media all traditional restrictions of physical space and cancels any geographical remoteness; further globalization opens unique information, educational, scientific, and cultural opportunities in front of mankind, forming the person media and a new media civilization (Goncharenko, 2017).

Transformation of media originated in a transition in civilization when an industrial was succeeded by a postindustrial (information) society. It became a new stage of development of society, and the information and computer revolution—process of informatization of all life spheres (societies in general and the person in general)—its quintessence. The latest technologies have extremely powerfully and considerably transformed material and production and social spheres of mankind. At the same time there also were considerable changes of processing, production, and information transfer, the approach to its serving.

The computerization, internetization (implementation of Internet technologies into the company's activities), mediation, and virtualization of information as components of a process of optimization and new submission became driving forces of the designated transformations. For the first time computer text was used to present information, but later it would be able to present any types of information—text, sound, video, or graphic in a digital form. The Internet as virtual space together with the ability of computer memory for instant reproduction, in turn, has promoted a considerable gain of knowledge, enormous expansion of information borders, and the invention of a new information picture of the world. A significantly updated media sphere in general and new media in particular became its main exponents, perhaps the factor in the final transition from a consumer industrial society to postindustrial, with knowledge, information in the center, and then society informational (Wertheim, 2010).

The consumer, as the most important element of the mechanism, is exposed not only to information influence from the enterprises, but also to psychological and emotional influence. The psychoemotional aspect is the main component of the influence mechanism and feedback between clients and the enterprises. The essence of the strategy based on digital communications consists in formation of the consumer desire to buy goods or service (works) of a concrete enterprise, influencing the consumer's psychoemotional state by means of digital technologies in media space. It is important to note that digital methods of impact on the consumer are a little other than classical Internet marketing; as a proof of what has been said we will show its comparative characteristics in Table 1.

According to Table 1, the fundamental difference between Internet and digital marketing consists in borders of the target audience: Internet marketing is aimed at coverage of Internet users, and digital marketing is aimed at involvement of all possible consumers by means of digital media space.

Before adopting any measures, the enterprise has to conduct market research including fields to reveal consumer needs (questionnaires, polls, interviews, a hot interview, analysis of the protocol, etc.). That is, it "addresses" consumers to receive the back information. On the basis of the obtained and processed data the relevant marketing activities are developed. It is not necessary that they concern only policy of advancing a brand, but also any other elements of a marketing mix. However, instruments of marketing communications with the consumer for the purpose of bringing to the consumer information on new offers, sales promotion of already existing goods, formation of system of loyalty, etc. are all the same used further.

Let us consider the structure of media space for formation of a feedback mechanism of use of digital technologies in the enterprise strategy(Figure 1). There are three layers: layer 1 is the enterprise (microenvironment), layer 2 is the mesoenvironment, and layer 3 is the macroenvironment.

Let us consider each element of the structure of the media space from the point of view of participation and a role in the mechanism.

Public authorities have to create a standard and legal base and the normal investment environment for creation and development of the enterprises of a certain sphere. Functions: creative and control.

Structures of the regional level have to provide normal development and formation of an investment climate in the region (conditions in which investors want to invest in companies,

Table 1. Comparative characteristic of Internet marketing and digital marketing.

Features	Internet marketing	Digital marketing
Sphere of impact on target audience	Online sphere	Online + offline
Distribution channels	Internet channels	All types of digital channels (Internet, mobile applications, digital advertising, tablets and video game consoles, digital TV, etc.)
Target audience	Everybody who has Internet access	Everybody who has Internet access + attraction offline—audiences on the online market
Means of communication with audience	E-mail newsletters, landings, websites, advertising (search, banner, tabulated, contextual), etc.	Digital TV, advertising in online games and mobile applications, messengers, interactive screens, POS terminals, local networks of the large cities

Source: Developed by the authors on the basis of Wertheim (2010).

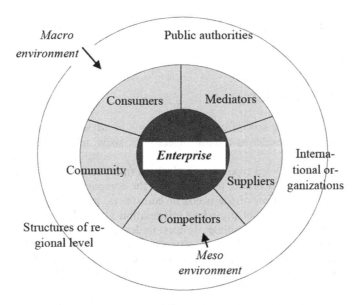

Figure 1.   Structure of media space.
Source: Developed by the authors.

and new companies, both foreign and domestic, want to carry out its activities in this region) and also respect for the principles of perfect competition in the regional markets. Functions: organizational, control, and planning.

The international organizations—if those are present at the concrete market—but they have to use acceptable forms and instruments of competitive activity. Functions: differential and structural (they provide different forms of interaction of participants of the market taking into account the international features). The macroenvironment is formed by external factors and the company cannot influence it; the managers can only develop strategies and policies to reduce the negative or enhance the positive impact of the macroenvironment. Within the

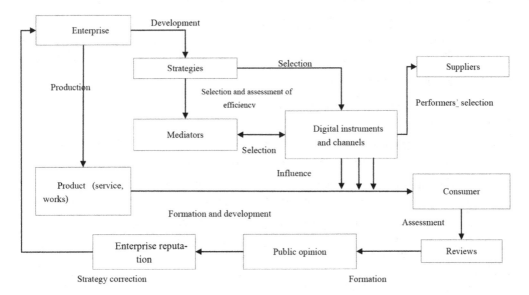

Figure 2.   Mechanism of interaction of media space elements in the mesoenvironment.
Source: Developed by authors.

enterprise media space is being formed to affect the mesoenvironment and connect with it. Next we will illustrate the mechanism of interaction of elements in a mesoenvironment (Figure 2).

3   METHODS AND DATASETS

A large number of marketing specialists fail in attempts to advance a brand on the Internet because they mistakenly consider as a main objective of social media the stimulation of the growth of sales of goods.

Actually, the purpose of such a type of media is the formation of interest among consumers, creation of positive image of a brand, positioning of the company for which a priority is not receiving profit, and satisfaction of the needs of consumers, gaining their trust and attachment. To determine the features of providing information about the company, which is displayed on the Internet and on mobile devices under modern conditions, studies were conducted among 157 consumers of various services (potential customers of enterprises in the service sector—users of mobile devices and Internet resources—respondents were interviewed anonymously). A Google Form was created for this and filled out by 157 various respondents, and the results were tabulated in Excel form and processed.

Results of consumer polls concerning use of the Internet and mobile devices at the choice of the enterprise are given in Table 2.

Table 2.   Results of consumer polls concerning use of the Internet and mobile devices at the choice of the enterprise.

Questions for consumers	Number of affirmative answers	Affirmative answers, %
How the information about the enterprise is captured on the Internet:		
The reviews about the enterprise are read	122	77.7
The website of the enterprise is analyzed	35	22.3
How the information about the enterprise is perceived in blogs:		
Is read and taken into consideration	109	69.4
Influences decision-making	21	13.4
Isn't used	27	17.2
How the information from the sites is read:		
Is considered consistently and in detail	16	10.2
Only the headlines and inserts are read; if necessary the information is considered in detail	141	89.8
What influences information perceiving of the enterprise's site:		
The availability of video content	10	6.4
The availability of detailed information about the enterprise and its services	89	56.7
The availability of comfortable functioning	14	8.9
The possibility of interactive communication with the enterprise	27	17.2
The comfort of information presentation	17	10.8
There is an interest to mobile applications by the cooperation with:		
There is an opportunity and wish to use	104	66.2
There is a wish but no opportunity	41	26.1
Not plan to use	12	7.6

Source: The authors' own elaboration.

4 RESULTS

The research shows that the Internet gives new opportunities for consumers to obtain information on the enterprise from various sources, increases information processing speed, and promotes the emergence of new requirements for information.

About 80% of respondents have answered that before ordering products (services, works), they carefully study information on the enterprise in the network, analyze its website, and read responses blogs that influence decision-making. It is connected with the fact that services are nonmaterial and processes of production and receipt are synchronized. It is possible to estimate services only after they are received or to learn about their assessment by others by means of information on the Internet. The policy of transparency increases chances of the enterprise to gain trust and cooperation with consumers and other stakeholders. Therefore communication of enterprise personnel in a blogosphere has to be the most open (Nikiforova, 2013).

Research has shown the relation between information content on the website and its representation has changed. Most of the respondents have answered that they read information on headings and cuttings, and if necessary consider information in detail. Existence of video content, convenient functionality, and opportunities for interactive communication considerably influence the concentration of attention on the part of consumers. Interest in mobile applications that allow support of communications with the enterprise is considerable, to monitor movements of freight and to correct necessary actions. It is connected with the distribution of smartphones that provide new opportunities for communication.

5 CONCLUSION

An important answer to the research question is the fact that for enterprises today the media space in the era of the digital economy plays an important role in attracting customers and this fact cannot be ignored in the development of strategies for the future. But because there are many tools and methods of information transmission and display in the media space, the authors' research has shown what information this should be and how it should be displayed in the media space in order to strengthen its competitive position in the market.

Also, as a result of the research, an important conclusion was obtained: that it is necessary to regularly interact with users for feedback—to receive feedback and opinions about your image and reputation in order to recognize the risks of losing your competitive positions in time.

Changes in communication processes contribute to the development of enterprises under conditions of rapid changes connected with the fact that digital communications allow enterprises to be present constantly at the virtual sphere, and it provides their availability. Availability to consumers and other stakeholders is crucial for creation of strong communications and creation of trust relationships. As clients get access to content on several devices, an integrated approach that increases coherence in interactions with consumers and other stakeholders becomes an urgent need.

For the constant involvement of consumers, enterprises need to create convenient website functionality and modern design that conveys a feeling of professionalism, convenience, and pleasantness to the consumer. It involves both the use of certain colors and creations of qualitative content and possibility of interactive communication. In speaking in terms of marketing of trust, the question is how, for example with the help of color, to attract consumers to the website, to create feelings of trust, to induce them to take certain actions, and to create an emotional connection with the enterprise brand.

As a direction of further research in this area, we can highlight the improvement of forms and methods of obtaining and processing information obtained through the feedback mechanism with consumers by the main points: reliability, efficiency, and objectivity.

During the digital era, marketing communications have become more and more integrated because of the lifting of restrictions that had existed earlier, as a consequence of fast technologies development. Means of establishing contacts with consumers have moved to qualitatively new levels and give the enterprise the opportunity to introduce communications constantly by means of websites, mobile communication, and social networks. Digital technologies also allow analytical work to be carried out, to trace how consumers use the digital opportunities, and to integrate the obtained data with the existing sets of traditional data from customer relationship management systems.

ACKNOWLEDGMENTS

This article was prepared as part of the project section of the government contract as requested by the Ministry of Education and Science of the Russian Federation on the subject formulated as "Development of Methodological Principles and Organizational Economic Mechanism of Strategic Management of Economic Security in Russia" (Assignment No. 26.3913.2017/ПЧ).

REFERENCES

Baran, G. 2010. Latest tools of Internet marketing. *Economy of Crimea*, 4(33):328–331.

Berger, P., & Luckmann, T. 1966. The Social Construction of Reality: A Treatise on Sociology of Knowledge. London: Penguin Books.

Bodriyar, Zh. 2006. *The Consumer Society: Its Myths and Structures*. Translation by E. Samarskaya. Moscow.

Bourdieu, P. 2007. *Sociology of Social Space*. Translation by N. Shmatko.. Moscow: Institute of Experimental Sociology.

Giddens, E. 2005. *Sociology*. 2nd ed. Moscow.

Goncharenko, L., Sybachin, S., & Ionkin, S. 2017. The state of the research complex of Russia as a fundamental basis for the transition to an innovative type of development of the national economy. *National Safety/Nota Bene*. 2:109–119.

Gritsay, C. 2012. Definition of "media" from the standpoint of interdisciplinary approach. *Bulletin Kharkov*, 36:235–243.

Ilyashenko, S. 2011. Modern trends in the use of Internet technologies in marketing. *Marketing and Innovation Management*, 2(4):64–74.

Matveev, M. 2002. Viral marketing in the international market of banking services. *Journal of European Economy*, 11(3):360–368.

McLuhan, M. 2002. *The Medium Is the Massage: An Inventory of Effects*. New York: Ginko Press.

Nem, E. 2013. Media: Main directions of research. *HSE: Business. Society. Authority*, 14. Available at: https://www.hse.ru/mag/27364712/2013-14/83292427.html

Nikiforova, S., & Sovershaeva, S. 2013. The effectiveness of marketing communications in the digital environment. *Problems of Modern Economy*, 2(46). Available at: http://cyberleninka.ru/article/n/effektivnost-marketingovyh-kommunikatsiy-v-didzhital-srede

Order of the Government of the Russian Federation. 2017. About the approval of the program "Digital economy of the Russian Federation." 1632. Available at: http://government.ru/rugovclassifier/614/events

Plahtiy, I. S. 2017. Development of media in modern society. *Young Scientist*, 17:204–207.

Sharko, E. R. 2014. Approaches and methods for selecting BSC indicators for the enterprise, taking into account the specifics of the business. *Modern Fundamental and Applied Research*, 2(13):147–152.

Sokhatskaya, A. 2012. Monetization of social media in the global information space. *Journal of European economy*. 2012. 11(1):104–d114.

Toffler, A. 2004. *The Third Wave*. Moscow.

Wertheim, K. 2010. Digital marketing: How to increase sales using social networks, blogs, wikis, mobile phones and other modern technologies. Moscow.

Wipperfurth, A. 2007. Brand engagement: How to make the buyer work for the company. St. Petersburg.

Inclusive Development of Society – Lumban Gaol (eds)
© *2020 Taylor & Francis Group, London, ISBN 978-1-138-33476-2*

Influence of the social environment on the development of scientific potential in Russia

L. Goncharenko, D. Gorin, S. Sybachin & I. Sokolnikova
Plekhanov Russian University of Economics, Moscow, Russia

A. Novitskaya
South Federal University, Rostov-on-Don, Russia

ABSTRACT: This article defines the essence and main tendencies of the influence of the social environment on the development of scientific potential in Russia. An increasing number of countries focusing on innovative means of development makes it necessary for Russia to integrate into this worldwide trend, keeping the most intelligent human capital and, above all, the scientific potential of researchers. As a result, Russia needs both to ensure the most effective use of available scientific potential and to stimulate its capitalization and commercialization. At the same time, it is important to invest in innovative development of the scientific potential of researchers who are the main generators of innovative ideas, which allows to dominate in business, based on an innovation sphere. Analysis of the impact of the social environment on the development of scientific potential allows forecasting certain trends, including the number of researchers, research organizations, postgraduate students, and other indicators.

1 INTRODUCTION

The institutionalizing of modern science, functioning of the scientific organizations, integration of scientific communities, and professional and public assessment of results of scientific activity comes under the influence of varied social factors that influence substantially the conditions of development of the scientific potential of modern researchers, a set of its values, beliefs, and research practices.

Scientific potential of the modern researcher includes a set of possibilities, competences, and resources that society provides for the implementation of research activities. Scientific potential develops within the bounds of the social environment, public infrastructure, the system of values that prevail in society, level of economic development, state of labor resources, quality of human capital and educational system, and governmental support of science, business, and civil society. The purpose of the article is to analyze the influence of the main factors of the social environment on trends and dynamics of scientific potential of modern researchers in Russia. Such analysis is necessary for better support of researchers who work in various sectors of science. Identification of social factors for scientific potential development is necessary for the activation of innovation development.

2 METHODOLOGICAL RESEARCH

The problem of character and extent of influence of the social environment on development of scientific potential is debatable and is shown in disputes between internalists (A. Koyre, R. Hall, J. Agassi, etc.) and externalists (R. Merton, A. Krombi, G. Gerlak, E. Tsilzel, J. Nidam, S. Lilly, etc.). However, the generally recognized fact is that influence of social factors is related not only to establishment of restrictions and granting opportunities to the scientific community working in certain cultural, social, economic, and political conditions but also to distribution of a scientific discourse that is beyond the close professional environment in various segments of the media sphere in modern societies.

Moreover, social processes thanks to which there are production, discussion, assessment, and transfer of scientific ideas are considered through the prism of the scientific ethos (Merton, 2007), ideological connotations of societies of a modernist style (Habermas, 2007), concepts of knowledge power and scientific discourses (Fuko, 1996), and the critical analysis of science in the theory of a deconstruction (Derrida, 2000). These and other authors prove that the argument and rhetorical receptions that are involved in scientific discussions have obvious social and cultural connotations. The role of such connotations amplifies in the process of entry of a scientific discourse into the public sphere and mass media. For example, on the basis of the analysis of publications in the most popular American periodicals (including *Time*, *Newsweek*, and *Popular Science*), J. Alexander came to a conclusion that scientific and technical innovations in mass culture are comprehended in the context of traditional religious opposition of rescue and an apocalypse that creates overestimated expectations and unjustified fears of rather new scientific achievements in society (Alexander, 1992).

The influence of social factors on development of scientific potential is considered in this article in three interrelated aspects. The first includes a direct consideration of the researcher, his or her communications, interactions, and relations which are defined by the nature of social system, and institutes that exist in society and the institutional environment. This aspect allows cognitive, social and psychological, professional, administrative, and managerial interactions to be carried out and they influence the functioning of research groups, laboratories, departments, and schools of sciences. The second aspect of analysis focuses on research on the sociocultural conditions and factors that provide interactions of researchers with the close scientific directions that are shown in activities of professional associations and communities, communications around professional magazines, cooperation of the research establishments and institutions of education involved in the process of formation, and individuals with research potential. The third aspect is presented by the conditions and factors all the features of the basic institutes defining structure of the statuses, characteristic of scientific community, and roles, organizational and administrative infrastructure of research activity, criteria of its assessment and recognition.

The research on the environment of formation and realization of scientific potential is based on the idea of a communicative community of research activity that is carried out under direct or indirect influence of the social factors determining the content and the nature of research activity and criteria of public recognition of its results.

For optimization of administrative impact on formation and realization of scientific potential the main point of the social environment is conceptualized in this article and its influences on reproduction of scientific potential are operated. For predicting the main tendencies of the Russian scientific environment mathematical and statistical methods were used, including linear regression and extrapolation, the weighed assessment of criteria, and the multicriteria analysis.

3   RESULTS AND DISCUSSION

The social environment can be considered as a set of the factors promoting or interfering with formation of human, social, and cultural capital—as forms of capital that under certain conditions are capable of being converted to each other and into development of scientific potential. This approach is justified by the fact that the concept of "the scientific potential" reflects the set of means, stocks, the sources available and capable to be actual and used for achievement of development aim of the science sphere and innovations.

The analysis of influence of the social environment on the development of scientific potential assumes a research of the designated perspective in three interconnected and interdependent aspects: institutional, organizational, and individual and personal.

The first aspect is institutional, which represents results of an institutionalization of science in modern societies. Science, interacting with other institutional spheres, is influenced by basic institutes and institutional transformation of society.

In the analysis of institutional aspects of influence of the social environment on development of scientific potential the appeal to the concept of "a package of institutes" that is used in scientific literature that defines a set of the norms and institutions that have historically developed as uniform complexes makes sense. Traditions, morals, legislative designs, and types of relations between regulation from imperious hierarchy and processes of social self-organization can be the components of the package of institutes. Some elements of a package of institutes can promote positive dynamics and provide sustainable development, and others can block this development. In a package of institutes, the so-called "basic institutes" representing the set of the basic rights and freedoms of the person and citizen, the norms and law—enforcement structures guaranteeing respect for these rights and freedoms—are especially allocated. In each concrete society there is "mix" of effective and inefficient institutes. One will encourage investments and innovations, and others will fight for privileges and advantages; one will promote expansion of space of mutually advantageous exchange and cooperation and the others will restrict it. It is possible to assume that development or braking of development of scientific potential is defined by a ratio between the first and second that is shown on each of the described above than the levels.

One of the indicators reflecting quality of the institutes contributing or interfering with innovative development is the index made by the World Intellectual Property Organization (WIPO) of the UN— The Global Innovation Index (The Global Innovation Index, 2017). We will consider the dynamics of positions of Russia in this index during 2013–2017 (see Table 1).

Russia shows stable growth on indicators of development of scientific potential, including internationalizations of science and education. If in 2013 Russia was in the second half of this rating, then in 2017, it got to its first half. According to the rating, in 2014 Russia made a noticeable jump, on 13 points having strengthened the positions. It occurred as a result of improvement of a social and economic situation that positively affected the quality of the educational environment and prestige of the research activity. A positive role was played also by insignificant increases in financing of research activity and expansion of use of the information technologies providing acceleration of the circulation of information. In the next years Russia strengthened its position in the rating and it rose on 1 point in 2015 and on 5 points in 2016. However, in 2017 the position of Russia decreased on 2 points because of higher rates of development of the competing countries (in 2017 in Russia the small growth of the total points reflected the rating of innovation on 0.3 points relative to 2016, with 38.5 to 38.8 observed). Regardless of how stable the accumulation of scientific capacity of Russia looks, 2017 showed that these efforts aren't enough to surpass growth rates of the rival countries. The leading position in rating is kept by Switzerland, followed by Sweden, the Netherlands, the United States, and Great Britain. Singapore in 2017 was located on the seventh line, and China on the 22nd. In 2017 Russia gave way to Turkey, and Ukraine took

Table 1. Position of Russia in the rating of the innovative countries of the world.

Year	Position of Russia in GII
2013	62
2014	49
2015	48
2016	43
2017	45

Source: The Global Innovation Index 2017. Innovation Feeding the World. S. Dutta, B. Lanvin, & S. Wunsch-Vincent (eds.). Available at: https://www. globalinnovationindex.org/

50th place. The Togo Guinea and Yemen close the list. The places next to Russia in the rating are held by Greece (44) and Chile (46). As the originators of the rating note, the weakening of Russia in the rating is connected with deterioration in positions of the universities of the country in the international ratings and a reduction of number of the quoted works and numbers of submitted patent applications. In a segment of the countries with an average level of income Russia took third place, having conceded to China and India (Gatinskiy, 2017).

The second aspect of the analysis of influence of the social environment on development of scientific potential is organizational. It represents the result of the choice of organizational forms in which control of the development of scientific potential is based on the integration of the interests of all participants in this process (Koshkin, 2016). This choice is essentially defined by the character of the institutional environment. The development of scientific potential thus can be considered as a result not only of the actions of internal factors on research activity, but also transformation of the institutional environment.

We will consider a number of tendencies characterizing organizational forms of reproduction of scientific potential in Russia. First of all, is the dynamics of the number of the scientists and the organizations that are carrying out research and development. In 2002–2017 a decrease in the number of staff occupied with scientific research is observed. Reduction of demand for the goods and services demanding provision of research is the reason. Production of hi-tech products decreases; therefore, number of staff occupied with scientific developments is reduced. According to our prognosis until 2020, the decrease in quantity of research associates is stabilized (Figure 1).

At the same time as the number of scientists decreased, a growth in the number of organizations that are engaged in scientific research (including the design, construction, and survey organizations, experimental plants, organizations of higher education, industry organizations having research and design divisions) is observed (Figure 2).

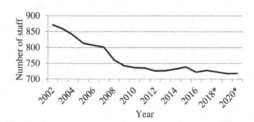

Figure 1. Number of staff occupied with research and development in Russia (thousands of people).
Source: The graph was built by the research team on the basis the federal state statistics service. Available at: http://gks.ru
*Forecast meaning.

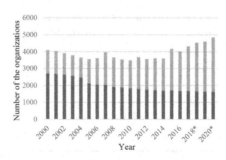

Figure 2. Number of organizations in the sphere of science and education in Russia (units).
Source: The graph was built by the research team on the basis of the federal state statistics service. Available at: http://gks.ru
*Above lines: number of organizations in total; under lines: research organizations.

142

The third aspect of the analysis of the influence of the social environment on the development of scientific potential is individual and personal. It reflects the individual choice of a personal trajectory of training of the researcher that is carried out within forms and is possible under the conditions of a certain social environment. One of the most important indicators is the dynamics of the number of graduates of higher educational institutions. During the period from 2005 to 2012 in Russia a stable growth of this indicator was observed; however, since 2013 the situation has changed cardinally and has begun to decrease in number of graduates of programs of a bachelor's degree, specialist and magistracy. This tendency is caused by the following factors:

- Changes in the need of graduates of schools of higher education
- Reduction of the need for experts with higher education and increase in demand for working professions
- Outflow of the best scientists to foreign higher educational institutions

The decrease in the number of students who graduated in Russia is represented as stable, which is predicted to continue for 2018–2020. By 2020 the number of graduates of higher educational institutions will be about 888,000 people (2005 there were 445,000 people) (for example see Figure 3).

An important quantitative index is also the number of persons continuing to be engaged in scientific activity after obtaining an expert or master diploma, including, coming to postgraduate study.

Data on the number of students who finished postgraduate study in Russia are given in Figure 4. As it is possible to notice, the main tendency is the decrease in the number of graduates of postgraduate study, which is extrapolated also to a prognosis period until 2020.

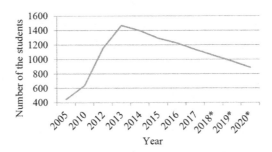

Figure 3.   Number of students who graduated in Russia (thousands of people).

Source: The graph was built by the research team on the basis the federal state statistics service. Available from: http://gks.ru
*Forecast meaning.

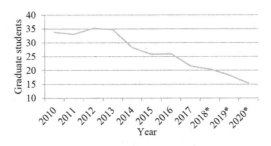

Figure 4.   Number of students who finished postgraduate study in Russia (1,000 people).

Source: The graph was built by the research team on the basis of the federal state statistics service. Available from: http://gks.ru
*Forecast meaning.

It is possible to assume that this decrease depends on both problems with financing of science and decrease in human capital (number of students). The graphics show an increase in the number of graduate students of 6.28% in 2011–2012 and of 0.64% in 2015–2016. Growth in 2011–2012 was connected with additional measures of the government of the Russian Federation. In particular, within three years 12 billion rubles had been allocated to attract scientists to the Russian higher education institutions (including, in 2011, 5 billion rubles; in 2012, 4 billion rubles). In 2015–2016 the small gain of 0.64% can be explained by the measures directed to additional financing of scientific activity (in particular, presidential grants to the young scientists and graduate students who are carrying out prospective research and development in the priority directions of modernization of the Russian economy in 2015–2017 (Grant Council of the President of the Russian Federation, 2017). Expected values of number of graduate students reflect its bigger reduction of approximately 39% until 2020.

In the Russian system of preparation of scientists, continuation of scientific activity of the graduates of a postgraduate study who have defended dissertations for a degree of the candidate of science in doctoral studies where researchers prepare theses for a degree of the doctor of science is supposed.

In an analysis of the data characterizing doctoral training, we notice that their number is unstable and is subject to considerable fluctuations. In 2010–2012 the number of graduates with a doctor of science degree in Russia increased by 9%, followed a small decrease that was replaced by new growth. According to the forecast until 2020 a decrease in the number of of graduates with doctor of science degrees is expected (Figure 5).

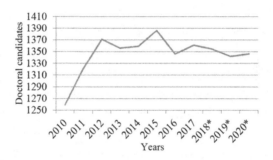

Figure 5. Number of doctoral candidates in Russia (people).

Source: The graph was built by the research team on the basis the federal state statistics service. Available from: http://gks.ru

*Forecast meaning.

4 CONCLUSION

The social environment has a significant impact on the formation of scientific potential and the effectiveness of its use. The analysis of this influence on institutional, organizational, and individual-personal levels allowed a forecast of certain trends in scientific potential development, including the index of social development; the number of researchers, graduate students; and students in higher education; and other indicators.

The analysis of the main tendencies of development of scientific potential in Russia has allowed establishing that dynamics of the number of students has graduate students, doctoral candidates, and the staff of the organizations working in the scientific sphere, dynamics generally descending. It is necessary to pay attention that a reduction in the growth of the number of researchers and number of organizations in the sphere of science and education is observed. It demonstrates that in the conditions of no formation of steady institutional conditions of development of scientific potential, an increase in investment into the sphere of science and education does not provide the expected transition of this sphere from stagnation to growth.

The social environment proves itself as the set of formal and informal structures. The first reflects the formalized relations that the researcher in the course of the research activity enters. Formal structures regulate research activity by precepts of law and other normative documents for its integration with the interests of official institutions of science and education. Informal structures reflect initiatives in informal aspects of activity of researchers and their professional interactions. For development of scientific potential, the basic value has interaction of the formal and informal structures defining quality of the social environment at each level—from the primary microlevel of direct professional interactions to the macrolevel at which incentives and conditions of recognition of research activity from society are provided. Interaction of formal and informal structures has several aspects, including activity and communicative and social and psychological.

The influence of the social environment on formation of scientific potential is a current trend, reflecting fundamental changes of modern societies. Modern production is innovative and knowledge-intensive, demonstrating an increase of the scientific role in contemporary society. However, as a result of globalization of the economy and production there is a transition from hierarchical and territorial and organized structures to network and exterritorial ones. The competition becomes global, which leads to depreciation of research potential if it doesn't join in global network communications. In these conditions processes of degradation of a number of regional schools of sciences that lose real opportunities of positioning in the global world are observed.

ACKNOWLEDGMENTS

This article was prepared with the assistance of the Russian Foundation for Basic Research (RFBR) (Project No. 17-23-03003-OGN).

REFERENCES

Alexander, J. 1992. The promise of a cultural sociology: Technological discourse and the sacred and profane information machine. In R. Münch & N. J. Smelser. *Theory of Culture* (pp. 293–323). Berkeley: University of California Press.

Derrida, J. 2000. *About Grammatology*. Moscow: Ad Marginem.

Fuco, M. 1996. *Archaeology of Knowledge*. Kyiv: Nika-Center.

Gatinskiy, A. Russia has fallen in the ranking of the most innovative countries. RPC. Available at: https://www.rbc.ru/economics/15/06/2017/594271b19a79473ed86548d0/

Grant Council of the President of the Russian Federation. Available at: https://grants.extech.ru

Habermas, Yu. 2007. *Engineering and Science as Ideology*. Moscow: Praxis.

Koshkin, A., Yablochkina, I., Kornilova, I., & Novikov, A. 2016. Integration of interests at university. *Interchange*, 48(3):231–255.

Lisin, V. S., & Yanovskiy, K. E. 2011. *Institutional Constraints of Modern Economic Growth* (pp. 15–17). Moscow: Delo.

Merton, R. 2007. Science and social order. *Questions of Social Theory*, 1:197–207.

The Global Innovation Index. 2017. *Innovation Feeding the World*. S. Dutta, B. Lanvin, & S. Wunsch-Vincent (eds.). Available at: https://www.globalinnovationindex.org/gii-2017-report

Inclusive Development of Society – Lumban Gaol (eds)
© 2020 Taylor & Francis Group, London, ISBN 978-1-138-33476-2

Transparency and effectiveness of a national purchasing system

K.A. Belokrylov & E.I. Firsov
Southern Federal University, Russian Federation

V.P. Kuznetsova
Herzen State Pedagogical University of Russia, Russia

ABSTRACT: Institutional modernization of the system of the Russian government and consistent goal-oriented municipal procurement have increased its transparency in recent years. The measures taken provided creation of the most transparent information system in the world presented by the All-Russian website of government procurements by using a systems approach (www.zakupki.gov.ru). However, it demands quantitative verification through an assessment of the influence of transparency on purchase efficiency. Processing of the survey results of people who are studying at the pilot center of government procurements of the Southern Federal University, with use of the SPSS program, revealed a strong dependence between efficiency of electronic auctions and receiving information on results of the held auction (i.e., transparency) by all interested persons, and also other dependences.

1 INTRODUCTION

M. A. Eisner gives the following grounds for realization of new forms of state interference into market mechanisms under the current conditions of high global instability and imperfection of the market: "The state and economy do not evolve in isolation. Rather, the two are best viewed as evolving together" (Eisner, 2011). One of the "functional economic systems" that is able to harmoniously manage and regulate socioeconomic processes and to provide sustainability of the economy is the government procurement system which guarantees the fulfillment of most state functions. However, these effects, as well as an increase in efficiency from the use in 2014 of sequestered state costs can be achieved only thanks to informational transparency of all stages of government procurement: planning, implementation, and enforcement of government contracts and their realization.

Increase in efficiency from use of budgetary funds in the conditions of transparency of procurement procedures can be achieved through its direct influence on the procurement prices. How may the transparency of information influence the prices? On the one hand, making the information more transparent may lead to lower prices. First, the availability of information makes it easier for firms to participate in the procedure, increasing the competition and hence lowering the prices. Second, information transparency decreases monitoring costs [3, p. 438; 4, p. 522] and "makes it easier for controlling parties to reveal the facts of opportunistic behavior of bidders and procurers. Consequently, information transparency will increase the participation of "honest" firms by signaling trust in the process" and result in lower price [3, p. 438]. Proper level of transparency [5] of procurement is one of the key indicators of methodology of evaluation of EBRD procurement systems, included in its basic principles [6].

The updated Model Law on Procurement, which was renewed by the UN Commission on International trade rights (UNCITRAL), side by side with stage-by-stage realization of basic principles of procurement organization (Open and effective competition, Accountability and due process, Fairness, stated in a number of international documents—the EU Treaty, the EU Procurement Directives, Multilateral Treaty on government procurement in the framework of

WTO, documents of Asian-Pacific Economic Cooperation etc.) specified the key principle of government procurement. These are transparency, pointing out to its contents-related elements: revealing information and clarification of the rules of procurement realization, institutionalized in national law, publishing in proper time and beforehand of information in regard to procurement plans, preciseness in naming of procurement object and true-to-life requirements, transparent and fully corresponding to the law on realization of procurement when officers must be enforced to implement it, availability of control system, and opportunity to file a protest against the course and results of procurement [7]. All these structural components of transparency principles were embodied in the federal law dated 5.04.2013 № 44-FZ "About contract system in the sphere of procurement of products, works and services to satisfy state and municipal needs" with little extension of them by engagement of the institutes of civil society—in other words, an openness of the government procurement system for the public. It must be emphasized that public control becomes a common rule in different spheres of state management and regulation is implemented through different forms (institutes), which perform some functions: public or expert commissions cooperating with state bodies, public monitoring and debates on significant issues, public investigation, checks and surveys, and activity of NCOs and others that correlate with open government (http://большоеправительство.рф/).

Apart from the aforementioned effects of government procurement as a market mechanism to satisfy social needs, in our opinion, openness in the sphere of government procurement and its transparency directly influence the efficiency of government costs.

That is why the aim of this study is empirical analysis of the dependence of efficiency of government costs on transparency of the government procurement market. The government procurement market as an independent institution includes entities (society, power bodies, producers, and NCOs), objects (products, works, services) and processes taking place in this system [9].

The importance of this study is also determined by mass spread of a specific model of state officers' behavior (Russia) that has become total institutional practice and is called "concealment or distortion of information about procurement" ("blind procurement") In estimates of 2012 their number was not less than 2,500 and partially blind—9,000 which caused harm to the state in the amount of 874 million rubles [10].

2 MATERIALS AND METHODS

The key method of the study is analysis of the results of regular sociological surveys based on interviews (performed through questionnaires) of state officers, which study in pilot center of the government procurement of Southern Federal University with the aid of the SPSS program.

What is the degree of openness of information about bidding?

y_1 = opened fully
y_2 = rather opened but there are some reserves to increase this openness
y_3 = low openness
y_4 = information inaccessible

The results of observations over n state officers are provided in Table 1.

Table 1. Correlation of attributes.

	y_1	y_2	y_3	y_4
x_1	n_{11}	n_{12}	n_{12}	n_{14}
x_2	n_{21}	n_{22}	n_{23}	n_{24}
x_3	n_{31}	n_{32}	n_{33}	n_{34}
x_4	n_{41}	n_{42}	n_{43}	n_{44}

In Table 1 the variable n_{ij} denotes the frequency of simultaneous occurrence of events x_i and y_j; in other words, event $x_i y_j$, $i = \overline{1,4}$, $j = \overline{1,4}$. Thus, n_{11} is a number of state officers participating in the interview who together chose the answer x_1, "Question 1," and answer y_1, which was in "Question 2."

Let us put forward zero hypothesis H0 about the absence of dependency between attributes X and Y in regard to alternative hypothesis H1 about availability of such dependency. It is obvious that attributes X and Y are dependable if the condition $P(x_i \cdot y_j) = P(x_i) \cdot P(y_j)$ $P(x_i/y_j) = P(x_i)$ is true for all pairs of x_i, y_j The frequency of attribute x_i is equal to $N_i = \sum_{j=1}^{4} n_{ij}$, and attribute y_j is equal to $N_i = \sum_{i=1}^{4} n_{ij}$. The number of state officers participating in interviews is equal to $N_i = \sum_{i=1}^{4}\sum_{j=1}^{4} n_{ij}$ Probability of choosing attributes x_i and y_j:

$$P(x_i) = P(x_{i1}) + P(x_{i2}) + \ldots + P(x_{i4}) = \frac{\sum_{j=1}^{4} n_{ij}}{N},$$

$$P(y_i) = P(y_{1j}) + P(y_{2j}) + \ldots + P(y) = \frac{\sum_{i=1}^{4} n_{ij}}{N},$$

The last value of the variable $x_{i\alpha}$, $\alpha = \overline{1,4}$ denotes the following event: respondent will choose attributes x_i and y_α simultaneously. As mentioned earlier, we denoted the number of such choices through $n_{x\alpha}$ and provide them in Table 1. The variable $y_{j\beta}$, $\beta = \overline{1,4}$ denotes the event of simultaneous choice of attributes y_j and x_β. Their number is denoted in Table 1 through $n_{j\beta}$. The product of probabilities $P(x_i) \cdot P(y_j)$ is the expected relative frequency. The observed relative frequency of simultaneous occurrence of events x_i and y_j is the probability of the product $P(x_i \cdot y_j)$ and is found in the table by the formula $P(x_i \cdot y_j) = \frac{n_{ij}}{N_i N_j}$. As we mentioned earlier in the case of refusal of zero hypothesis H0 about the absence of dependency between attributes x_i, y_j and choosing alternative hypothesis H1 about the existence of a stochastic relationship, the condition $P(x_i \cdot y_j) = P(x_i) \cdot P(y_j)$ must be true, in accordance with which the observed n_{ij} and expected $\sum_{i=1}^{3} n_{ij} \cdot \sum_{j=1}^{6} n_{ij}$ must be equal to each other: $n_{ij} = \sum_{i=1}^{3} n_{ij} \cdot \sum_{j=1}^{6} n_{ij}$.

3 MAIN PART

Results of interviews were aggregated by us into cross-tables that were built with the aid of the program SPSS Statistics ("Statistical Package for the Social Sciences") and this allowed us to show the degree (strength) of relationship between the analyzed parameters [11]. The answers to the first question to 200 state officers in the first half of the 2013 questionnaire about efficiency of government procurement, "What is efficiency of electronic auctions?" were distributed as depicted in Figure 1.

The correlation of the answers with the answers to the question "Does the current system of budgetary government procurement allow all stakeholders to obtain information on the auction's results?" (this question characterizes transparency of the government and municipal procurement system) was analyzed; see Figure 2.

The analysis of dependency between efficiency of electronic bidding and obtaining by all stakeholders information about the auction's results proved the thesis about high efficiency determined by transparency of this procedure and reduction of costs (Figure 3). In accordance

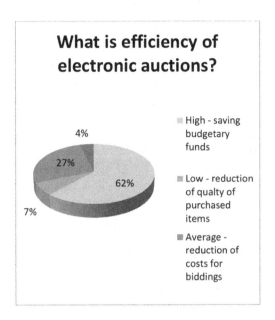

Figure 1. Distribution of answers of state officers to question 1, "What is efficiency of electronic auctions?".

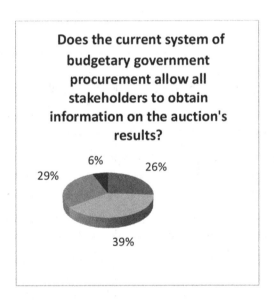

Figure 2. Distribution of answers of state officers to the question, "Does the current system of budgetary government procurement allow all stakeholders to obtain information on the auction's results?".

with our hypothesis it was necessary to identify the dependency between the variables "Question 1" and variables "Question 2." Because these are nominal variables the pure logit-analysis cannot be used but we can use cross-tables in which observed frequencies (Count) are compared with expected frequencies (Expected count) and the "unstandarted residual" is evaluated. The value "unstanarted residence," which is calculated as the difference between expected frequency and observed frequency, is of special significance for this analysis.

The higher the unstandarted residence is, the stronger is the dependency between values. The sign before unstandarted residence does not matter (only "module value" is important), and the

149

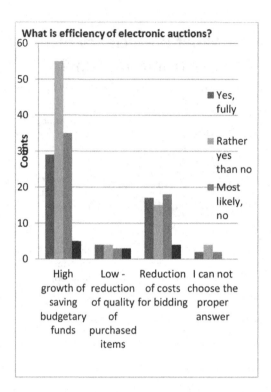

Figure 3.   Relationship between openness and efficiency of electronic auctions.

strength of dependency is estimated by the expert. Residence value = 5, as the threshold from which the relationship between the chosen indicators is beginning to be observed, is chosen by us to optimize the results The higher the unstandarted residence is, the stronger the relationship between questions, and it is considered by module, which means that every question is considered and the value of unstandarted residence that is more than 5 by its module is chosen.

Availability of dependency demands that observed frequencies must significantly differ from expected ones, which is shown by unstandarted residuals—the difference between observed frequencies and expected frequencies. These residuals are evaluated by experts on the basis of their experience.

Using the method of cross-tables the questions from the blocks "Efficiency" and "Transparency" are compared with each other, which allowed us to obtain corresponding dependencies. We used an equal number of answers (100%) to determine dependency. Analysis of the results shows that the efficiency of electronic auctions greatly depends on the transparency of the results of the bidding.

The next step of our study is testing of the dependency between answers to the questions about efficiency of electronic auctions and evaluation of the opportunity for all stakeholders to participate in bidding and get the orders. In our opinion this reflects the relationship between saving of funds spent on carrying out of auctions (their efficiency) with transparency in the context of access to biddings. Thanks to it, growth in saving budgetary funds will occur, which characterizes the efficiency of this way of placing orders. By this a strong dependency between efficiency of electronic auctions, and their role in the increase of transparency of the system of government procurement was found because most respondents pointed to big savings and a substantial increase in transparency of procurement procedures.

This thesis is also proved by a strong relationship between answers to the question about the economy of budgetary funds and agreement of respondents with a high degree of openness of information about budgetary procurement. Thus, interviews demonstrated that there is a dependency between saving and obtaining information about the results of realization of contracts. But in this case the relationship is not strong and multidirectional, which shows a difference in opinions of state officers in regard to this question.

Thus most respondents considered the open auction the most efficient and transparency way to place the order (dependency is shown in cross-tables). But some respondents believe that transparency in procurement is completely absent and open auction in electronic form as the most common and efficient way of procurement determines this nontransparency. The others believe that open auction in electronic form is not only efficient but also increases transparency and saves budgetary funds. This dependency is confirmed by the answers to the question "Did most open auctions in electronic form save money for your organization?" Answers such as this dominated: "Use of open auction and electronic bidding increase transparency in the sphere of government procurement and increase efficiency of spending of budgetary funds in the framework of government procurement."

4 INFERENCE

A formed system of state and municipal procurement provides an increase in efficiency of spending of budgetary funds but only if all stakeholders obtain information about bidding even partially. This determines broadening of the information component in the framework of the formation of the contract system. Thus, if in correspondence with 94-FZ (Law on Public Procurement, acted until 2013) only information about the contents of the order was placed on the Russian website, in accordance with 44-FZ, full access to information about government procurement will be granted—about implementation and the result. In our opinion, the result of it is "through" (complete) informational follow-up of the procurement in a unified information system. But conceptual principles of such a system have not been found yet, first of all, in terms of how this system will correlate with an already functioning Russian website that includes more than 200 customers who participate with the use of electronic signatures and key widgets, which are oriented to consumer groups: general statistics, statistics of procurement group, questions to suppliers, prices. In 2011 procurement in electronic form amounted to 223 billion rubles, in 2012, 224 billion rubles, and for 8 months of 2013, to 202 billion rubles [12].

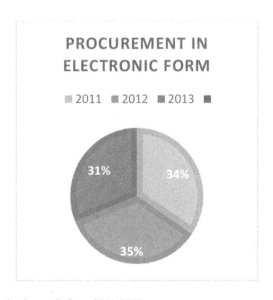

Figure 4.   Procurement in electronic form 2011–2013.

5 CONCLUSIONS

The efficiency and transparency of functioning of the Russian government and the municipal procurement website allows the conclusion that the unified information system of government procurement must not be started from scratch but must use current websites as a base to which other systems will be connected.

REFERENCES

Boehm, F., &Olaya, J. 2006. Corruption in public contracting auction: The role of transparency in bidding processes. *Annals of Public and Cooperative Economics*, 4(77):431–452.

Eisner, M. A. 2011. *The American Political Economy*. New York: Routledge.

Evaluation of restricted access to participation in government procurement determined by distortion of information about procurement. Date Views 18.11.2013. Available at: http://naiz.org/arhive/documents/

Kolstad, I., & Wiig, A. 2009. Is transparency the key to reducing corruption in resources-rich countries? *World Development*, 3(37): 521–535.

Model Law UNCITRAL on Procurement, 2012. Moscow: NAIZ. (in Russian).

Nasledov, A. 2013. BM SPSS Statistics 20 and Amos: Professional statistical analysis of data. St. Petersburg: Piter.

Potential and risks for new government procurement system, Date Views 2.12.2013. Available at: http://большоеправительство.рф/events/5508931/

Procurement Policies and Rules, 2010. London: The European Bank for Reconstruction and Development. Date Views 16.08.2013. Available at: http://www.ebrd.com/downloads/procurement/ppr10.pdf

Public procurement assessment. Review of laws and practice in the EBRD region, EBRD, 2011. Date Views 12.05.2013. Available at: www.ebrd.com/downloads/legal/procurement/ppreport.pdf

Smotritskaya, I., & Chernykh, S. 2010. Government procurement in the system of relationship between the state and market. *Society and Economy*, 1:78–87.

UNCITRAL Model Law on Procurement of Goods, Construction and Services with Guide to Enactmend, 2012. Date Views 18.11.2013. Available at: http://www.uncitral.org/pdf/english/texts/procurem/ml-procurement-2011/ML_Public_Procurement_A_66_17_E.pdf

Inclusive Development of Society – Lumban Gaol (eds)
© 2020 Taylor & Francis Group, London, ISBN 978-1-138-33476-2

On the application of a cost-based approach to the evaluation of intangible assets

L.I. Khoruzhy
Russian Timiryazev State Agrarian University, Moscow, Russian Federation

I.N. Bogataya
Rostov State University of Economics RSUE (RINH), Rostov-on-Don, Russian Federation

R.U. Simionov
Rostov Branch of the Russian Customs Academy, Rostov-on-Don, Russian Federation

A.U. Panakhov & N.Yu. Lebedeva
Southern Federal University, Rostov-on-Don, Russian Federation

ABSTRACT: Modern financial accounting rules refuse to identify a number of intangible assets including them into the composition of goodwill. In the article it is considered that in management accounting some of such assets can be identified and subjected to cost-based evaluation. For the purpose, the methodological provisions of cost center accounting, Time-Driven Activity-Based Costing and Root Cause Analysis can be applied. Alternative options and perspective problems for cost accounting and capitalization of intangible assets are proposed for discussion.

1 INTRODUCTION

At the present stage the concept of intangible assets in accounting theory is closely related to the goodwill concept.

In economic literature, it is widely believed that goodwill is an indivisible set of assets each of which cannot be identified and subjected to historical (i.e., costly) valuation separately: such a viewpoint is dominant from the middle of the last century and amply considered in the works of such authors as T. Lee (1971), A. Seetharaman, M. Balachandran and A. Saravanan (2004), K. Ramanna (2008), Y. Ding, J. Richard and H. Stolowy (2008), R. Ratiu & A.T. Rudor (2012). However, the spread of modern management accounting approach leads us to the assumption that a number of assets included in goodwill can be identified, and a cost approach to evaluation such assets can be applied. In particular, the possible identifiability of goodwill is considered by S. Gupta, Y. Levant and Ch. Ducrocq (2003), R. Reilly (2015).

To analyze goodwill, it is important to consider the different economic and institutional nature of its constituents, which were considered, in particular, in works of J. Courtis (1983), V. Kam (1990).

Indeed, one of the goodwill constituents is the synergistic effect of a combination of other assets that were already identified in financial accounting (including tangible assets) and reflected in other items of the balance sheet. The part of the goodwill, which is the difference between the value of the combination of such assets and the sum of their individual values, definitely cannot be a subject of cost-based evaluation.

Also, goodwill can arise from changes in the external environment that are not created by the enterprise. For example, this can be due to climate change at the location of the enterprise which allows to produce goods with lower costs than would be required if the similar resources, competencies and technologies were combined and used by exactly the same way in any other location, respectively giving a competitive advantage to the enterprise. The impact of such external factors also cannot be evaluated on the basis of costs since the enterprise resources have not been purposefully spent to obtain these effects.

However, goodwill also includes results of the firm's purposeful actions such as business process reengineering, applying or creating new software, training, development of new management subsystems (organizational, budgeting, motivation and personnel management and other systems), finding new employees, suppliers and sales markets, designing firm style etc. To carry out these actions the firm bears costs that can be identified. Such costs are usually related to:

- use of labor of employees;
- use of other resources controlled by the firm;
- acquisition of new resources;
- alternative costs associated with the diversion of controlled resources (including labor) from other, former activities.

Today we know that one accounting object can have different types of value simultaneously. But not all types of value are associated with the possibility of selling the object. Contrariwise, concept of assets in economic is primarily associated with an opportunity to bring benefits. Identification of assets by the criterion of their opportunity to be sold, applied in financial accounting theory (and also assumed in IAS 38 as a separation criterion), does not satisfy the needs of management accounting. In the post-industrial economy, there is a growing interest in the historical valuation of intangible assets; furthermore, we should consider as assets even those resources that do not have an independent sales value and cannot be separated from enterprise: this viewpoint is expressed, in particular, by V. Paliy (2009).

In this regard, we raise the following research question: is there an actual potential of the methodology of the cost-based evaluation of intangible assets?

To answer the question, we intend to achieve the following research goals:

- to determine whether there is an economic justification for the cost-based evaluation of intangible assets that cannot be identified separately in accordance with modern financial accounting rules;
- to identify specific accounting techniques that could allow to measure the costs of creating an intangible asset;
- to consider promising challenges related to further development of the methodology of cost accounting for intangible assets.

2 METHODS

The traditional approach to evaluation of assets in financial accounting is based on taking into account the fact of the asset has and independent sales value.

In the article we provide a possibility of especially cost-based evaluation with identification of an intangible asset which is the result of targeted actions performed within the firm even if the asset does not have an individual sales value.

The methodological basis of the discussed approaches involves cost center accounting, project costing, Time-Driven Activity-Based Costing with application of standard cost accounting tools and Root Case Analysis.

3 RESULTS

3.1 *Outline of the theoretical example*

Suppose, we analyze the financial indicators of two firms, Firm A and Firm B. For a number of previous years, their activities were completely identical. They had the same resources in the same volume, the same external environment, produced the same quantity of the same products and sold them to the same number of buyers at the same prices. Their costs, revenues and profits were identical. Moreover, they operated in an imaginary un-changing environment in which all their indicators still the same every year and did not change over time.

However, in one of the years (we will conditionally mark it as Year $_1$), changes began to occur. Firm A received relatively less revenue, and its costs were higher. The following year (Year $_2$), the costs of firm A also exceeded the costs of firm B, while the revenue grew although remained less than the revenue of firm B. In Year $_3$, the costs of firm A decreased and were lower than that of firm B, while the revenues of the companies have become identical again. This state of affairs persisted in the following years.

In this case, the only change that has occurred during the three years is the following: in the Year1, the Firm A began to design and implement a new system of corporate budgeting that began to function fully correctly after the completion of the project at the end of the Year2. That is, additional costs and lower revenues for the first two years are associated with the obtaining of this intangible system, and cost reduction since year N3 is the benefits (positive effect) of its use.

There were no other changes related to other circumstances in the activities of the firms.

3.2 Modern financial accounting approach

The general financial accounting approach (which is applied in particular by companies that follow IFRS) suggests following sight.

In the Year $_1$ and the Year $_2$ accountant recognize the lower financial results of Firm A; in later years, when the higher profitability of Firm A activity became noticeable, goodwill can be taken into account.

Obviously this contradicts the methodology for considering long-term tangible assets - for example, if a firm acquires a fixed asset, the costs of its acquiring do not reduce financial results directly (the decline will occur only through depreciation in future). However, from the point of view of going concern principle (i.e. while we can consider 'asset' conception not from a liquidation position but from the point of view of usefulness), the role of the budgeting system is analogous to the role of the fixed assets: it is an asset that was acquired through targeted costs and which can bring benefits or not bring them depending on how effectively it will be used in future business activities.

However, without timely identifying the implemented budgeting system as an asset, financial accountants will link the increase in the efficiency of business activities of Firm A starting from the Year $_3$ with certain events that were supposedly occurred in Year $_3$, but cannot be determined, and therefore goodwill should be recognized subsequent years.

3.3 Management accounting approach

Firstly, we list all resources controlled by firm. Then we list the synthetic produced resources that can be produced in firm for internal consumption in link with cost centers that produce such resources.

A synthetic resource is usually a service that is consumed in business processes. For example, a workshop worker produces a service of production labor that is not equivalent to his personal labor for which he receives wages (his personal labor is only a part of such a service). Synthetic resource 'production labor' is formed in combination of:

- worker's labor;
- use of production equipment;
- use of production premises;
- electricity consumption.

For each synthetic resource (output) at each cost place a 'recipe' for its derivation (input) must be formed. This information can be presented in a matrix form (an example for IT department is given in Table 1, where the value of 1 means that a resource is required, and the value of 0 means that it is not required). Different types of services there can be, for example, the following: programming, participation in meetings with counterparties (abbreviated as "Meetings"), preparation of paper documentation (abbreviated as "Documenting"), repair of IT infrastructure (abbreviated as "IT-repair") etc.

In practice such a table can be supplemented with a wide range of resources, depending on the specifics of business processes within each department.

Further we can apply the methodology of time-driven activity-based costing (TD ABC) considered by such scientists as Kaplan & Anderson (2004), Gervais, Levant and Ch. Ducrocq (2010), Tanis & Ozyapici (2012), L. Siguenza-Guzman with co-authors (2013).

We should determine the rules for allocating different resources to other (synthetic) resources on the basis of normal capacity. For example, for a type of synthetic resource, followed rules can be defined:

- if the resource includes employee's labor, for each unit of the resource (service hour) the distributed established share of the monthly cost of labor (based on the average salary of the employee of the relevant qualification, the average market value of similar labor);
- if the resource requires the use of fixed assets, for each unit of the resource (service hour), the allocated established share of the monthly cost of the equipment (based on the amount of depreciation, leasing agreement or prices for the lease of similar equipment);
- if the resource requires the use of electricity, for each unit of resource (service hour), the assigned amount of electricity is distributed, which is calculated on the basis of the capacity of the equipment used.

The unused amount of resources (if their monthly use is lower than normal capacity) should be considered as a period expense and not be allocated.

Further, in management accounting we introduce a dimension of processes. This dimension should be used to account for each business transaction and also can include projects (as particular types of intra-firm processes). Relatively, in the dimension we should introduce the project "Design and Implementation of the Corporate Budgeting System" (abbreviated as "Budgeting system implementation").

Further, in the course of economic activity, hourly records of the use of labor resources should be maintained. All hours of labor used in the framework of budgeting system implementation should be related to this project by using the process dimension.

Thus, we have at our disposal all information about the non-financial resources that were in firm possession before the project and were spent in the project. To calculate the related costs of project, it is enough to multiply the number of each synthetic resource (services) units by cost of each unit of each resource that was required to generate the service. Since the cost of resource can change with time, we can carry out the calculation with a certain step (the monthly periodicity seems optimal).

3.4 Cause analysis

To account for the use of resources that have been purchased under the project, but the use of which in the company may go beyond the scope of the project (for example, if their useful

Table 1. Matrix form of list of produced resources with reference to required resources.

Cost place	Output resource	Input			
		labor	equip.	PC	electricity
IT	Programming	1	0	1	1
IT	Meetings	1	0	0	0
IT	Documenting	1	0	0	0
IT	IT-repair	1	1	0	1

terms in the company exceeds the project implementation period), two different approaches can be used.

For example, the enterprise purchased a certain server (Server Y) which is required for automation tools of the budgeting system that is being implemented.

The first approach is to take into account, as for already controlled resources, only a part of the expense (for example, in the form of depreciation) in proportion to the cost driver when using them within the project. In this case, the Server Y will form the project costs only when it is used in this project. In the rest of the time its use (in the form of depreciation) will be attributed to other projects or will be written off for the financial results of the period. Other transactions (such as revaluation) will also be considered as general business actions without reference to this project.

The second approach is to try to analyze the entire chain of processes associated with the turnover of the resource in the organization in conjunction with the project. This can be interesting, since the entire resource turnover (started with its purchasing) was initiated by the project.

Here we can use a 'final cause' concept, which was introduced by Aristotle, according to philosophical research, presented, for example, by M. Alvarez (2009). Indeed, the project was both the goal and the reason for buying the Server Y. In turn, the purchase of the server will cause in the future costs (depreciation), and, possibly, revenues (for example, if the server is sold more expensive than purchased). If at the end of the project the server remains at the disposal of the firm, the project will also prove to be the reason for the availability of this asset in the future.

To analyze this situation, we can consider "input" to the project and the "out" from the project. As an input, the "negative" aspect of the purchase transaction (occurrence of accounts payable or the immediate payment) can be considered as project costs. Further we can consider two alternative cases.

The first case. Suppose, six months after the Server Y was purchased, it was decided to use it for other needs, and not use it in the budgeting project. At this point, the resource is "extracted" from the project, and at the moment the resource is estimated. This estimate is considered the income earned by the firm from the project.

The second case. The budgeting system implementation project is successfully completed, but after that it is decided to use the Server Y for other needs. Nevertheless, managers continue to analyze the cause for this resource. In this case, all further operations, up to the server's retirement from the firm, will be taken into account in the process dimension related to 'Budgeting system implementation' project.

In this connection, an important methodological conclusion follows: the measurement of "processes" can be applied not only to economic operations, but also to a resource as an accounting object.

Once the Server Y is registered as a fixed asset, the binding it to the project must be assigned, and then this relation must be automatically used in all business transactions associated with this server (until managers will break the binding).

However, two cases do not contradict each other and imply a common requirement for the methodology of management accounting. In managerial accounting, it should be possible to analyze the balance of resources (in-kind) in terms of projects, and the ability to write off the resource from the project with the determination of the corresponding deviation in its cost. An example of requirements for management accounting information is given in Table 2 and Table 3.

Table 2 reflects information before resource extracting.

Table 3 reflects information got after resource extracting.

Thus indicators of "Costs" is used identically accounts "Financial results" in the modern financial accounting theory, however assumes a possibility of further capitalization of the saved-up sums as a part of an asset of balance sheet (this corresponds to cost approach to assessment of the project results, i.e., the implemented budgeting system, in management accounting).

The provision of such requirements within the double-entry system remains open.

Table 2.   Matrix form of list of produced resources with reference to required resources.

Resource	Project	Value
Server Y	Budgeting system implementation	n

Table 3.   The balance of the project "Budgeting" and the resource "Server Y" after extracting the resource from the project (an example of requirements for management accounting information).

Indicator	Project	Value
Server Y	(none)	n
Costs (account)	Budgeting system implementation	(n-m)

3.5   *Evaluation*

Now back to determining the value of the resources that were already available to the firm used in the project.

In reality, there several approaches can be applied based on the solving of different tasks.

The first task is to determine economically justified value of creation of the intangible asset based on the resources technologically required to create it (by analogy of the analysis of the resources technologically required to produce material goods in the industrial economy). The value is, in particular, important for forecasting and can be the basis for correct planning the cost of similar projects in future. In our opinion, to solve the task it is necessary to identify the physical volume of use of each of the enterprise's resources within the project. Further, to convert these quantities into value terms it is necessary to multiply them by a standard cost of each unit of resource use. Accounting for the casual effects (such as the speculative gains from the resale of the Server Y) with their binding to the project is not required in this case.

The second task is to estimate the overall actual result of the project by comparison of the costs incurred by the company due to the project implementation with the benefits that it received from the obtained intangible asset. Since such a task is associated with the What-If Analysis (which is in comparing the actual state of affairs with that which would have arisen if the project had not been implemented), in this case it is important to analyze only the additional costs that would not have occurred if the project had not been carried out.

However, it is difficult to measure the costs and benefits that were related to the project implementation indirectly as well as to exclude from the project evaluation the cost of resources consumed within the framework of its implementation, however, already paid for (so that additional monetary costs due to the use of these resources within this project did not arise). With such an analysis, we should identify within the enterprise economy a "free-rider effect" studied in economic theory (for example, this effect is described by G. Jones (1984)): thus, the project, as it were, uses re-sources that are already "financed" by other processes (that is, attributed to financial results from other activities). This means that we need to analyze each unit of costs for communication with the project – that is, in effect, conduct a continuous Root Cause Analysis. This approach should not be recommended for mass application since economic at all consists of a conventionally innumerable set of actions each of which is indirectly caused by a conditionally uncountable set of causes that "layer" one another ("put in each other") that was outlined, in particular, by Sharovatova, Omelchenko (2015).

Moreover, What-If Analysis can be conducted by real costs only or can also take into account implicit costs. So, if the worker did not perform any of his direct duties for several hours due to the fact that in his workplace new equipment for production operation control (as part of the budgeting system) was being installed, these hours of his downtime may also be included in the project costs. In our opinion, in this case they can be evaluated based on the standard cost of each hour.

Relatively, it should be noted the problem of capitalization of opportunity costs in management ac-counting.

On the one hand, these costs do not increase the value of an intangible asset in terms of its useful-ness. They apparently will not increase the benefits of the enterprise in the future, but, on the contrary, are definitely related to the period in which they arose and should reduce the financial results of that period. From this point of view, they can not be reflected in the balance sheet.

On the other hand, these costs are part of the cost of the project that should be recompensed by its results and can be compared with them only in future, when the intangible asset will be used.

Perhaps, off-balance sheet accounts could be used to store information about opportunity costs for further analytical estimation of the project's efficiency, and the study of this issue is an important direction for further research.

4 CONCLUSIONS AND DISCUSSION

Certain intangible assets such as an implemented budgeting system which do not have an independent sales value but are the result of targeted intra-firm activities during which the firm resources are consumed can be identified in management accounting and subjected to cost-based evaluation.

The evaluation of the asset can be made on the basis of the calculated costs of the project of its creation. Generally, the calculation can be carried out on the basis of distribution of the cost of resources in proportion to the volume of their use during the project and/or by tracking the cause–effect relationship between the project implementation and operations of purchase and sale of resources.

The analysis of the use of resources can be provided by the application of Time-Driven ABC methodology A and determination of standard cost of each unit of the volume of resource use in the enterprise, while the analysis of the cause-effect relationships requires the attribution of both positive and negative effects that have become a direct or indirect consequence of the progress of the project.

Accordingly, further development of the cost-based approach to intangible asset evaluation seems to be related to solution of a number of perspective problems that are largely associated with the choice of one of these approaches.

The first problem is the need to find an answer to the following question: if the company uses during the project the resources that are already in its possession and that would stand idle otherwise, should the cost of using them be attributed to the cost of the project?

The second problem is related to the need to deter-mine the boundaries of the cause-effect analysis since it is obvious that an attempt to identify absolutely all the consequences of project implementation is impractical.

The third problem is related to the issue of capitalization of implicit costs in management accounting.

We offer these issues to the further discussion. Meanwhile, at the present stage of development of the management accounting methodology, in our opinion, the evaluation of intangible assets based on the actual amount of resources consumed within the project of creation (or otherwise obtaining) of these assets using the standard-costing tools is the most suitable for mass practical application cost-based approach.

REFERENCES

Alvarez, M.P. 2009. The four causes of behavior: Aristotle and Skinner. *International Journal of Psychology and Psychological Therapy* 9: 45–57.

Courtis, J.K. 1983. Business Goodwill: Conceptual Clarification via Accounting, Legal and Etymological Perspectives. *The Accounting Historians Journal* 10: 1–38.

Ding, Y., Richard, J. & Stolowy, H. 2008. Towards an Understanding of the Phases of Goodwill Accounting in Four Western Capitalist Countries: From Stakeholder Model to Shareholder Model. *Accounting, Organizations & Society* 33: 718–755.

Gervais, M., Levant, Y. & Ducrocq Ch. 2010. Time-Driven Activity-Based Costing (TDABC): An Initial Appraisal through a Longitudinal Case Study. *Journal of Applied Management Accounting Research* 8: 1–20.

Gupta, S. & Lehmann, D.R. 2003. Customers as assets. *Journal of Interactive Marketing* 17: 9–24.

Jones, G. 1984. Task visibility, free riding, and shirking: Explaining the effect of structure and technology on employee behavior. *Academy of Management Review* 9: 684–695.

Kam, V. 1990. *Accounting Theory*. 2-nd ed. John Wiley & Sons, Inc. 581 p.

Kaplan, R.S. & Anderson, S.R. 2004. Time-Driven Activity-Based Costing. *Harvard Business Review*: 131–138.

Lee, T.A. 1971. Goodwill – further attempts to capture the will-o'-the-wisp. *Accounting and Business Research* 24: 79–91.

Paliy, V.F. 2009. Cost of intangible asset in post-industrial economy. *Accounting* 22: 52–56.

Ramanna, K. & Watts, R.L. 2008. *Evidence from Goodwill Non-impairments on the Effects of Unverifiable Fair-Value Accounting*. Boston: Harvard Business School.

Ratiu, R.V. & Tudor, A.T. 2012. The Definition of Goodwill - a Chronological Overview. *Revista Romana de Statistica – Supliment Trim* IV: 54–59.

Reilly, R.F. 2015. Goodwill Valuation Approaches, Metods and Procedures. *Insights* 3: 10–24.

Seetharaman, A., Balachandran, M. & Saravanan A.S. 2004. Accounting treatment of goodwill: yesterday, today and tomorrow: Problems and prospects in the international perspective. *Journal of Intellectual Capital* 5: 131–152.

Sharovatova, E. & Omelchenko, I. 2015. Methodology of management accounting of costs in the framework of application of the ordering method at the industrial enterprise. *Auditorskie Vedomosti* 6: 62–76.

Siguenza-Guzman, L., Abbeele, A., Vandewalle, J., Verhaaren, H. & Cattrysse D. 2013. Recent Evolutions in Costing Systems: A Literature Review of Time-Driven Activity-Based Costing. *Review of Business and Economic Literature* 58: 34–64.

Tanis, V.N. & Ozyapici, H. 2012. The Measurement and Management of Unused Capacity in a Time Driven Activity Based Costing System. *Journal of Applied Management Accounting Research* 10: 43–55.

Inclusive Development of Society – Lumban Gaol (eds)
© *2020 Taylor & Francis Group, London, ISBN 978-1-138-33476-2*

Russian universities' organizational development: Models, factors and conditions

N.V. Pogosyan & O.S. Belokrylova
Southern Federal University, Rostov-on-Don, Russian Federation

E.A. Tropinova
Saint-Petersburg State University, Saint Petersburg, Russian Federation

L.P. Bespamyatnova
K.G. Razumovsky Moscow State University of Technologies and Management (the First Cossack University), the Don Regional Branch - State Institute of Food Technology and Business, Rostov-on-Don, Russian Federation

ABSTRACT: Permanent reforms of the higher education system in Russia, the on-going transition from the entrepreneurial university model to efficient inclusion of Russian universities in the worldwide market of educational services through ascension into the group of top universities in various rating systems act as the real driving force of non-stop restructuring in the corporate labour market and formation of university innovation infrastructure. Therefore, our research was intended to analyse the chronological transformation of various university models based on the global experience, and to estimate the potential of their innovation infrastructure during every phase of development. Historical and genetic approach, and benchmarking methods were employed in the research as efficient tools of education marketing intelligence.The results obtained allow a justified conclusion that transition of the models and mission of modern universities is mediated by social trends related to the development of knowledge economy, digital economics, and expansion of Industry 4.0.

1 INTRODUCTION

The pace of social and educational changes during digital era increases exponentially. Rapid developments of the 4th Industrial Revolution impart unparalleled significance to the infrastructure component and its role in the acceleration of the transition to innovative development path in the economy as a whole and in the national system of education. The situation predetermines domination of services in the national GDP structure, growing share of intangibles and intellectual capital in comparison with material assets of companies, increasing importance of institutional support for sustainable development and, in the end, general priority of the role of human capital with a set of specific competences both in material production and in the educational system. Still being tested in all countries, these transformational changes determine the need for deeper theoretical research of their influence upon the formation of conditions and recognition of the factors affecting the transformation of modern university models as well as assessment of the potential and role of innovation infrastructure in their development. Indeed, while the role of infrastructural component remains extremely significant as regards ensuring innovation-driven development of educational system, its influence on the formation of soft infrastructure, institutional support and specific human capital in the national system of education in general also become increasingly important. Consequently, both the national innovation policy and its shaping and implementation on the level of universities as autonomous subjects of the higher education system define the principles and the appropriate tools for development of innovation infrastructure, determine the arrangement of key objects in the national innovation system of modern Russia, where universities preserve their paramount importance as centres of the innovation generation space.

2 RESEARCH METHODOLOGY

Changes of the entire paradigm for strategic management of innovation campus in a modern university seem to be determined both by external and internal corporate reasons, by objective civilization-scale changes of public policy, culture, values, and by formal stimuli and signals projected by key regulators of innovation processes and the participants: administrations of higher educational establishments, professors, researchers, and students.

Positioning of higher educational establishments as the innovation generation sites implies their deep integration in the system of local production industries and technology channels, mastering of a conceptually new innovative entrepreneurship mission on their part. Researchers point out that the influence of science and education on economy is not linear, albeit significant.

A. Salter and B. Martin define six basic vectors of scientific influence on modern socio-political and economic systems, which are as follows: "increasing the stock of useful knowledge, training skilled graduates, creating new scientific instrumentation and methodologies, forming networks and stimulating social interaction, increasing the capacity for scientific and technological problem-solving, creating new firms" (Salter A.J., Martin B.R.). The research mission adopted and accomplished by the European and American universities in the late XIX – early XX centuries facilitated shaping of stable relations with the state, which according to B. Readings perceived them as testing grounds for implementation of their plans (Readings B.). Developing connections with the government enabled universities to reach out to external markets, diversify the sources of financing for their projects, and involve general public in the assessment and diffusion of innovation. As a result, *innovation infrastructure* became a necessity as a special system of entities carrying out innovation work or implementing the connections related to the circulation of its products. However, despite the broadening ties with governments and major corporations in order to satisfy the state demand, universities 1.0 were mainly focused on dissemination of knowledge, its propagation, increasing of cultural and educational level of various social layers and strata, saturation of labour market with qualified staff capable of active mastering of new mechanised tools.

Later on, modification of the first research universities and emergence of the "technocratic" university 2.0 was ignited by the restructuring of social and economic architecture of the "third wave" society (post-industrial), where ideas, technology and capital become basic resources according to D. Bell and A. Toffler (Bell D.). Burton Clark, the author of the "entrepreneurial university" concept, has explained that campuses become the sites for development of innovation culture and science-intensive technologies leading students up to creating their own businesses (Clark B.).

Of course, such numbering of university models is largely conventional; however, it reflects specific changes in the environment and organizational culture manifested in the historical transition of university models. Transdisciplinarity becomes a characteristic aspect of the research mission in an entrepreneurial university, and network interaction is actively employed in the educational process. Such changes provide opportunities for breakthrough innovations forming entire production and technology spaces (clusters), such as the Silicon Valley (USA), technopolitan cities emerge. Business becomes actively involved in the research and innovative projects of universities, innovation infrastructure of higher educational establishments expands through addition of fundamentally new, mostly communication-centered sites, such as industrial parks and business incubators. The "multiversity" phenomenon suggested by C. Kerr appears (Neborsky E.V.).

Accelerating pace of social progress by the end of the XX century, radical technological advancements, strengthening of the connection between an idea and innovation, and availability of an open innovation cycle for the economic system determined the next upgrade of the university model 3.0 preserving the research component but reinforcing the entrepreneurial focus. Competition becomes particularly important for development of universities in the social and economic system. "Academic capitalism" demands from higher educational establishments designing of commercial strategies in order to raise development funds (Talagaeva D.A.).

Clear borders between the private and public sectors are gradually blurred in education while competition is becoming more intensive on the institutional and micro levels. Under the influence of fundamentally new goals, innovation infrastructure is modified towards greater availability and creation of communication sites both within its base (industrial parks, business incubators) and in urban environment (open lecture halls) or local production clusters (specialized departments collaborating with corporations, laboratory and testing departments of universities at companies). The architecture of information streams and communications plays a significant role, most services ensuring interactions within a national innovation system become open, remotely accessible, and digital.

Support of the existing communication intensity, combining requests and initiatives, creating the conditions necessary for functioning of creative alliances between the state, business, and educational establishments become a fairly important task for the latter (the "Triple Helix" concept (Etzkowitz H.)).

Objectively accelerating dynamics of communications within the Triple Helix is a consequence of the new (fourth) industrial transformation unfolding around the fundamental "standardization-individualization" backbone (Idrisov G.I., Knyaginin V.N., Kudrin A.L., Rozhkova E.S.). It is focused on the application of advanced technology, platform-based management solutions, and digital transformation. In addition, the new phase of innovation-driven development within the fourth industrial transition implies interaction that occurs not so much between the innovation process stages as between the entities participating in it (universities and business), their mutual influence, and interdependence primarily based on commercial motives. As V. Tambovtsev has rightfully emphasized, lack or insufficient development of any component in the Triple Helix results in zero outcome (Tambovtsev V.L.).

3 METHODS OF TESTING AND ANALYSIS OF RESULTS

Research methodology is based on the of analytical of legal frame and political decision making process and reforms in high education institution.

Thus, the potential for participation of universities in the economic development of Russia is determined by the quality of collaboration between the innovation process participants, as pointed out by the Open Innovations Forum 2016 (Skolkovo, Russia) [10].

The concept of university model 4.0 currently discussed in the professional community is essentially a format for prognostic anticipatory collaboration between innovators capable of solving tasks, which modern industry perceives as impossible (Nesterov A.V. 2018).

Science-oriented policy in the Long-term Program of Fundamental Scientific Research in the Russian Federation (2013-2020) provides for "shaping of balanced and sustainable sector of fundamental research, ensuring multiplication of knowledge …. intensification of integration processes in Russian science, production and education, increasing efficiency of research and its application for development of promising technologies necessary to … ensure social and economic development of the country, considering institutional changes of national economy" (Long-term Program of Fundamental Scientific Research in the Russian Federation (2013-2020) approved by the RF Government Decree No. 2237-p of December 3, 2012). The document emphasizes interdependence between the competitiveness of national economy and the efficiency and competitiveness of science, which should be viewed in the context of national security. Achievement of the goal is assumed to be based on several mechanisms.

Key criteria of university success are: the "quality" of human capital represented by the students, scientific research activities of the scholars and public recognition of achievements (the number of scientific research works, publications in international databases, citations), the level of internationalization of educational and innovation activities, the demand for graduates in the labour market, and the dynamics of human capital development manifested in the university fellows (Karpov A.). It must be stressed that the public recognition estimate for the results of university activities is currently dominated by quantitative indicators, which demonstrate only slight changes in quality when examined closely. Thus, the actual standing of Russia in global science is improving during the recent years, publication activity is increasing and in 2016 it has promoted the Russian Federation to the 14th place in the world rating. During that period Russian scientists published 46.23 thousand works in journals indexed by the Web of Science thus bringing the Russian share in the global corpus of publications to 2.67% (Subbotin A.).

One third of the papers published by Russian researchers were created as a result of international collaboration. Citation frequency increases though it still remains below the average global level. While the number of publications grows, impact factor of the journals where most papers are published is not very high and thus determines low influence of Russian science on the research agenda (29th place in the rating of global research fronts) (Subbotin A.).

Apart from public appreciation of university productivity, its position in the national and international ratings is also significant for domestic political institutions. Within the Russian Academic Excellence Project 5-100 [15], 27 Russian universities entered the Times Higher Education (THE) rating in 2018, 24 – in 2017 [16]. On one part, growing core of successful higher educational establishments adapted to international competitive environment confirms success of the roadmaps devised by the state and productive utilization of budget subsidies allocated for their implementation; moreover, it indicates that the domestic educational system has internal reserves to improve its competitiveness. On the other part, arrangement of universities into a pyramid, priority support to the select few within the framework of the 5-100 concept already results in considerable disproportion both in education and in the development of local labour markets and industries depending upon them. Preservation and development of human capital evenly distributed throughout the entire Russian territory, support of quality-based demand for education and innovation in local public and production system must become a strategic goal.

Today, when university innovation infrastructure is shaped in most sectors, the state sets the goal to achieve integrity, emergence and balance of the national innovation system capable of producing and implementing innovations providing for real growth of economy in the country. Benchmark assessment of our country based on the Global Innovation Index (in 2016 Russia actually used 65% of its innovation potential (Vlasova V., Kuznetsova T., Rud' V.)) in comparison with BRICS countries demonstrates lack of direct influence of science and technology on social and economic progress: positive dynamics is observed in the figures reflecting the quality of human capital, growing expenses to develop science and innovation infrastructure and accomplish international recognition of modern universities, patent-related and publication activities, while indicators of research application in science, technology and production demonstrate negative dynamics evidencing insufficient development of the institutions ensuring interaction between actors in innovative environment.

4 CONCLUSIONS

Thus, as the benchmarking results in comparison with foreign universities demonstrate, modernization of the Russian university model should proceed in two directions: searching for new regulation strategies for the national innovation system in order to achieve better efficiency on one part while providing for inclusion of Russia in the system of international division of labour, ensuring its place in new industries and creation of new markets on the other part.

Future directions of research for us are: adaptation of new business models of University management, interaction and consent of all participants of the research and the educational process in achieving strategic goals.

Today, universities exist in the global competitive space of several markets: educational services and products, market of labour and competences, market of scientific research results and innovation. Success of a higher educational establishment in these markets is determined by the degree of its integration with regional or global educational/scientific and innovative business networks, its goodwill, the degree of openness and adaptability of its innovation eco-system (infrastructure) enabling it to become a part of higher order innovation infrastructures, such as megascience projects or on the contrary take on the properties of an innovation node coordinating innovation and production-related activities of members within a local cluster. As analysis has demonstrated, success of such university ecosystems is based on the state support also depending on internal resources of the university, on the inclusion of all participants of scientific and educational activities in the innovation process.

In situation of selective state support, prioritization ranking of universities and targeted innovation projects, many Russian universities face the need for competitive positioning in the system containing barriers of opportunistic member behaviour and undeveloped institutional environment. Apparently, internal reserves have to act in such conditions as the source of development.

REFERENCES

Bell D. New York: Basic Books, 1973. *The Coming of Post Industrial Society: a venture in social forecasting.*

Clark B. *Creating Entrepreneurial Universities: Organizational Pathways of Transformation.* Bingley: Emerald Group Publishing Ltd. 2001.

Decree of the RF Government No.2237-p of December 3, 2012. Online resource. URL: http://www.ras.ru/scientificactivity/2013-2020plan.aspx (access date: Aug. 27, 2018).

Decree of the RF Government No.2237-p of December 3, 2012. *Online resource. URL*: http://www.ras.ru/scientificactivity/2013-2020plan.aspx *(access date*: Aug. 27, 2018*)*.

Etzkowitz H. 2008. *The Triple Helix.* New York: Routledge.

Idrisov G.I., Knyaginin V.N., Kudrin A.L., Rozhkova E.S. *New technological revolution: Challenges and opportunities for Russia.* Voprosy Ekonomiki, 2018, No.4, p.9 (5-25).

Karpov A. 2018. *Modern university as an economic growth driver: Models & missions.* Voprosy Ekonomiki, 2017, No. 37, pp. 58–76.

National rating of universities, (http://unirating.ru/news.asp?lnt=6&id=1165)

Neborsky E.V. *Reconstruction of the university model: transition to the 4.0 format.* Mir Nauki online journal. 2017. Vol.5, No.4. Online resource at: https://mir-nauki.com/PDF/26PDMN417.pdf (access date: Aug. 27, 2018).

Nesterov A.V. 2018. *What Sets University 4.0 apart from University 3.0: Critical Contemplation.* Online resource at: https://nesterov.su (access date: Aug. 27, 2018).

Readings B. *The University in Ruins.* Moscow: Higher School of Economics Publishing House, 2010.

Russian Academic Excellence Project of the Ministry of Science and Higher Education of the Russian Federation. Online resource. URL: https://5top100.ru/universities/ (access date: Aug. 27, 2018).

Salter A.J., Martin B.R. 2001. *The economic benefits of publicly funded basic research: A critical review.* Research Policy, Vol.30, No.3, pp. 509–532.

Subbotin A. *Russians know little about science but believe it.* Poisk. Weekly newspaper of the scientific community. 2018. No. 7. 16.02.2018. Online resource. URL: http://www.poisknews.ru/theme/info sphere/33140/ (access date: Aug. 27, 2018).

Talagaeva D.A. *European Research Area in Action: Horizon 2020.* Polis. Political Studies. 2018. No.1, p.179 (175-183).

Tambovtsev V.L. *On scientific validity of Russian science policy.* Voprosy Ekonomiki. 2018. No.2. p. 5 (5-32).

Vlasova V., Kuznetsova T., Rud' V. *Drivers and limitations of Russia's development based on the evidence provided by the Global Innovation Index.* Voprosy Ekonomiki. 2017. No. 8. p. 34 (24-41).

Inclusive Development of Society – Lumban Gaol (eds)
© 2020 Taylor & Francis Group, London, ISBN 978-1-138-33476-2

Establishment of a common digital space under conditions of Eurasian economic integration

N.N. Yevchenko, T.V. Voronina & A.B. Yatsenko
World Economy and International Relations Department, Faculty of Economics, Southern Federal University, Rostov-on-Don, Russia,

D.M. Madiyarova
Eurasian National University after L.N. Gumilev, Astana, Kazakhstan

ABSTRACT: This research is focused on the establishment of the digital economy in the Eurasian Economic Union (EAEU) states as an objectively conditioned process of forming a common digital space of the integration association. The research shows that legislative solutions adopted in the EAEU are aimed at enhancing the use of the integration association's digital technologies in general. This being the case, it was made clear that national strategies in the field of member states' economy digital transformation demonstrate diverse vectors involving the existing dominance of national interests over the supranational, but institutions and tools for creation of a common digital space (CDS) of the EAEU are being formed. A comparative analysis of world rankings of development of information and communication technologies in the EAEU allows coming to a conclusion that establishment of the integration association's digital space goes faster than the establishment of a common economic space as the dynamics of development of the digital economy sphere of the member states exceeds the world average level.

Keywords: *common digital space, IT technologies, information and communication technologies, digitalization strategies, EAEU, economic integration, national interests*

1   INTRODUCTION

The world states are actively developing modern information and communication technologies (ICT), which form a significant synergetic innovation effect for the economic system as a whole on the basis of information platforms, speed up the exchange of information and contribute to new knowledge creation. The impact of ICT on economic well-being of the population as a result of increased productivity and competitiveness, economic diversification and business activity stimulation is obvious.

Digital economy contributes to the dynamics of economic development as for the emergence of new technology companies and enterprises, so increasing productivity in traditional industries and value added (more than 75%) through the use of the Internet (Digital transformation, 2018). The impact of digitalization on global GDP growth was higher than the impact of world merchandise trade (Key Barriers to Digital Trade, 2016).

Globally, cross-border data flows between 2005 and 2014 increased 45 times, reaching $ 2.8 trillion. The use of digital platforms has a key impact on international trade: creating new markets and customer communities and allowing companies to work globally despite their scale is. (Manyika, 2016).

Research literature analysis shows that ICT development is explored to the maximum extent on technology level (platforms, ways and methods of data transmission, processing, their storage and digital technology compatibility), this being the case, considerably less attention is being paid to organizational and managerial decisions pertaining to the strategy of development of this economy sphere.

Relevance of conducted research is defined by its subject – state and integration level of ICT management in the EAEU – and the need for elaboration of conceptual organizational decisions on creation of a Common Digital Space (CDS) of the EAEU in order to achieve synergetic effect for the national economies of the member states.

2   RESEARCH GOALS & HYPOTHESIS

The goal of the article is to study the phenomenon of «digital economy» as an objective process of forming a single digital space using the example of the Eurasian Economic Union, identifying and analyzing indicators that can measure the efficiency of creating a common digital space of the EAEU.

The hypothesis of the study is the assumption of the possibility of an institutional impact on overcoming the various-speed dynamics of the development of the digital sphere of the EAEU states, which exists due to the unevenness of the national starting conditions.

3   THEORETICAL FRAMEWORK

Currently, a lot of attention is paid to the processes of creation and development of the digital economy at the level of national economies.

Meanwhile, integration associations, development strategies of which provide for the removal of barriers to movement of production factors and formation of a common market of goods and services, are the leading world economy subjects.

Obviously, development of the Eurasian Economic Union (EAEU) as a successful integration association along with the digital economy global leaders (EU and ASEAN) requires digital transformation of economy. For this reason, the research of a digital single market creation within the integration associations (EU and ASEAN), of its performance indicators and establishment problems is now becoming ever more relevant for the EAEU.

A review of the research literature allows coming to a conclusion that there are a large number of publications addressing the essence of the digital economy, its impact on economic growth and creation of a digital single market in the EU and Eurasian Economic Union.

Digital economy research foundations were laid in the 60-70s of the twentieth century by Daniel Bell (1999), who associated the development of post-industrial society with the expansion of the service sector in economy and the increasing dependence of economy on science as a tool for innovation and organization of technological changes in society.

Building on D. Bell's ideas, Manuel Castells (2000) introduces the theory of "network society", which is an integral part of the information society concept, into scientific discourse.

M. Castells (1999) treats information society as a specific form of social organization, in which, by virtue of new technological conditions arising during this historical period, creation, processing and transmission of information became fundamental sources of productivity and power.

"Digital economy" definition is being introduced into scientific discourse with the revolution in computer science and active expansion of digital technologies in business and society. However, there is no single interpretation of this concept.

The majority of authors note that digital economy promotes distribution of goods and development of service sphere through the exchange of digital information and online trade. T. L. Mesenbourg (2001) focuses on the analysis of the three major components of the digital economy including: supporting infrastructure (software, telecommunications, networks, etc.); e-business (business activities and any other business processes through computer networks) and e-commerce (distribution of goods via the Internet).

S. Guo, W. Ding and T. Lanshina (2017) interpret the digital economy as a series of economic activities or social behavior based on the information and communication technologies (ICT) and via the Internet.

In a number of publications the emphasis is made on the role of information and communication technologies (ICT) sector for the world economy development.

According to R. Amit and C. Zott (2001), e-commerce has a direct impact on productivity gain and business activity effectiveness.

Garicano and Kaplan (2001) note that thanks to electronic commerce, business entities are increasing their products' market share and develop new markets.

Thus, Knickrehm, Berthon and Daugherty (2016) forecast world production growth by 2 trillion dollars by 2020 due to use of digital skills and technologies.

A number of publications illustrate the potential of the digital economy for deepening integration in regional integration associations (EU, ASEAN).

Analysts of the World Economic Forum (2016) note that access to the Internet is already insufficient to deepen integration in ASEAN countries. High-priority tasks are harmonization of rules and liberalization of the IT sector, as well as joint efforts to modernize the education systems of all ASEAN states.

In this regard, according to the authors of this article, the experience of the EU is a good example on forming a digital single market for current regional integration associations.

In the context of our research, we believe it is important to highlight the category of a digital single market created within the integration association.

The European Commission (2017) understands a digital single market of the EU as a barrier-free space within 28 countries providing economic entities with access to digital and online technologies and services safely, legitimately and at an affordable price to improve the lives of the population and simplify business. Its structural components include digital marketing, e-commerce and telecommunications.

The European Commission publications (2017) indicate the importance of the information sector for the EU, which generates almost 5% of the EU GDP, it accounts for ¼ of the overall expenses for business, and investment in ICT ensure 50% of the total productivity growth in Europe.

Benefits of a fully functioning in the European Union Digital Single Market are reported. They lie in the fact that Digital Single Market can provide the EU budget with 415 billion euro of additional income per year; create hundreds of thousands of new jobs; save 11 billion euro for consumers shopping online; and for small business savings can be up to 9,000 euro for each market thanks to the common legal regulations and consistent legislation; and improve access of the population of 28 countries to information.

Digital Economy and Society Index (DESI), annual European Commission expert research, is devoted to evaluation of the status and evolution of digitalization of 28 EU countries' economy. The indicators for calculating the composite index include access to broadband Internet, Human Capital/Digital skills, Use of Internet Services by citizens, Integration of Digital Technology by businesses, Digital Public Services and Research and Development ICT.

4 METHODOLOGY

The methodological basis of research is a comparative analysis of the digital transformation in the EAEU, which was made for two management levels: member states (Armenia, Belarus, Kazakhstan, Kyrgyzstan and Russia) and integral supranational level. Complexity of processing of statistical indicators characterizing the national level of ICT development is the peculiarity of the digital economy sphere research.

The authors used a number of indices and rankings of the ICT sphere used in global analytics and in practice, each of which to some extent comprehensively characterizes the research area.

A comparative analysis of dynamics of the digital economy establishment process and use of ICT by the EAEU states was made according to the data of rankings of international organizations and professional associations.

In the course of the research a wide range of empirical data of the World Economic Forum, the WTO, Euroasian economic commission is analyzed.

In the first part of the work, the relevance and novelty of the study for the EAEU countries are formulated, which involve a two-level approach to the analysis of the digital sphere state - national and supranational.

In the second part of the work on the basis of a monographic survey, a review of the scientific literature carried out on following aspects: an assense of the digital economy, its impact on economic growth, the formation of a single digital market in the EU and the Eurasian Economic Union.

The following part substantiates the research methodology and its information and empirical base.

In the fourth part, the authors analyze the activities of the supranational authority of the Eurasian Economic Union (EEC) on the formation of the institutional framework for the functioning of the ECS in the EAEU. Further, following the logic of the research, an analysis of national strategies for the development of the digital economy of the EAEU states is carried out. Based on the analysis of world digital economy indexes, the positions of the EAEU countries in the global digital space are identified. Focusing the study on the EEU ICT integral index, the authors concluded that the EAEU countries demonstrate positive dynamics in the field of digitalization of the economy for 10 years (2007-2017), the excess of the average ICT index, and the increase 5.6 times in 2008-2017.

The fifth part summarizes the research and formulates its conclusions.

5 FINDINGS

5.1 *Institutional framework for the functioning of the CDS in the EAEU.*

EAEU business community (2015) initiated granting the powers to the Eurasian Economic Commission to create a common digital space of the Union, which may become a key to the Eurasian integration processes' deepening and consolidation of efforts on implementation of joint programs and projects on digital transformation and CDS forming.

The business community identified priority areas of the EAEU CDS: development of electronic commerce, common digital infrastructure, digital space ecosystem and use of digital technologies in state regulation and control to improve business environment of the EAEU member states.

In 2016 the Board of the Eurasian Economic Commission (EEC) endorsed the decision of the Board of the EEC on preparation of the EAEU CDS Creation Concept that would include priority areas of cooperation:

- digital modernization of integration processes;
- digital markets;
- digital infrastructure;
- digital economy development institutions.

Integration processes form the system of supranational institutional mechanisms, successful operation of which is impossible without an effectively functioning CDS. Russian Union of Industrialists and Entrepreneurs (RSPP) (2018) proposed the idea of creating a single payment area (SPA) of the EAEU on the basis of the Concept of Creation of a Common Financial Market of the EAEU and Agreement on Harmonization of Legislation in the Sphere of Financial Market being developed at present. According to experts, main principles of transition to a SPA are transparency, coordination of supervisory policies and proactivity of supranational institutions. These basic principles may be used for the preparation of the EAEU CDS Concept.

Thus, EAEU industry research institutions report about the beginning of development of single business environments (financial, communications, etc.), which must rely on effectively operating CDS. Integration effect of the EAEU may be significantly reduced without the operating data transmission and processing technological and management platforms fit for contemporary challenges.

The structure, procedure and time limits of preparation of declared by the Concept single industry and service spaces of the EAEU should probably be created on the basis of the developed and approved CDS Concept as a management and technological platform of their unification. Therefore, it is necessary to accelerate the development of the EAEU CDS.

5.2 *National strategies of the EAEU states' digital economy development*

EAEU states have developed government programs and plans of the digital economy development, although they currently have different ICT sphere development level. Its comparative analysis of the provisions and planned results is given in Tables 1 and 2.

Table 2 shows that the implementation of national digitalization strategies can have a positive impact on the economies of the EAEU member countries: the economic growth rates will accelerate, the quality of human life will improve through the use of digital technologies. However, the formation of a single digital space throughout the Union ensures the achievement of **more significant** results for the countries participating in the EAEU due to the synergy of integration.

So far, each EAEU country is developing national strategies for a digital economy based on its own strength but poorly takes into account integration possibilities (Tables 1 and 2).

Thus, the need for linking national concepts, programs and plans for the establishment of ICT sphere common standards within the integration association becomes topical, since common digital space provides a synergetic effect for development of the EAEU as integration association.

Table 1. Comparative analysis of the institutional framework for digital development programs in the EAEU states *.

	ICT regulator	Basic documents (date of acceptance)
Armenia	Ministry of Communications and Transport; Public Services Regulatory Commission	Strategy "Digital Agenda of Armenia 2030" (2017)
Belarus	The president; Ministry of Communications and Information; Ministry of Information of the Republic of Belarus; Digital Economy Development Council (2018)	Informatization Development Strategy for 2016–2022 (2015); Decree "On the Development of the Digital Economy" (2017)
Kazakhstan	Government; Communications, Informatization and Information Committee of the Ministry of Investment and Development	The program of the Third Modernization of Kazakhstan; The state program "Digital Kazakhstan" for 2018-2022 years (2017)
Kyrgyzstan	Ministry of Transport and Communications; State Communications Agency; State Radio Frequency Commission; State Committee of Information Technologies and Communications	National Sustainable Development Strategy for 2018-2040 (at the stage of adoption); The project of state informatization "Taza Kom". (2017)
Russia	Ministry of Communications and Mass Media; Federal Service for Supervision of Communications, Information Technology and Mass Communications (Roskomnadzor); Federal Communications Agency (Rossvyaz)	National Program "Digital Economy" until 2024 (2017)
EAEU	EEC	Main Directions for the Digital Agenda until 2025 (2017)

* Based on EEC materials

170

Table 2. Comparative analysis of the main directions and planned results of digital development in the EAEU states *.

	The main directions of digital development	Planned result
Armenia	- digital government, - cyber security, - private sector, - institutional framework - digital skills and infrastructure	100% digitalization in the government-business relationship and 80% in the line of services to citizens; by 203 - reduction of public administrative expenses by 50%; - enter the Top 30 in the Global Competitiveness Index; - enter the Top 30 in the Talent Competitiveness Index; - enter the Top 20 in the E - Government Development Index.
Belarus	- conditions for attracting global IT-companies, production of competitive IT-products; - ensuring investment in the future (IT personnel and education); - The introduction of new financial instruments and technologies; - removing barriers to the introduction of new technologies	- to provide 82% of households with access to the Internet; - the number of subscribers of wireless broadband Internet access - 90 units per 100 people by 2020; - growth in export revenue of the IT sector from $ 1 billion in 2017 to $ 4.7 billion by 2030; - growth in the number of people employed in the IT sector from 30 thousand to 100 thousand by 2030
Kazakhstan	- digitalization of industries; - transition to a digital state; - implementation of the digital Silk Road; - development of human capital; - creation of an innovation ecosystem	- the share of Internet users - 82% of the population; - entry into the top 30 most developed countries of the world by 2050; - By 2022, the creation of 300 thousand jobs; automation of 80% of public services; - labor productivity growth in the manufacturing industry by 49.8%; - GDP growth from 2025 to 30%.
Kyrgyzstan	- building a world-class digital infrastructure; - a favorable environment for sustainable innovation development; - digital opportunities for all segments of the population; - involvement of citizens in the management of the country through digital technologies; - the formation of an open digital society; - a regional hub of digital Silk Road for IT-business and IT-innovations	- the proportion of Internet users - 75% of the population; - 00% security of all households Internet; - 100% provision of rural population with Internet access; - enter the Top 50 countries in the UN e-Government Development Index; - enter the Top 50 countries in the Information Society Index (according to the ITL methodology)
Russia	- regulatory regulation; - human resources and education; - the formation of research competencies and technical reserves; - information infrastructure; - Information Security.	- growth of information technologies in management by 2020 by 25%, and by 2025 - by 50%; - the share of households with broadband access to the Internet - by 2020 - 50%, and by 2025 - 97%;

(Continued)

Table 2.   (*Continued*)

	The main directions of digital development	Planned result
		- the share of electronic public services, of the total number of services provided by 2020 - 50%, and by 2025 - 80%;
		- reduce the amount of commercial losses of electric energy in comparison with 2017 by 2025 to 5%
EAEU	- ensuring quality and sustainable economic growth, transition to a new technological structure; - synchronization of digital transformations of the participating countries; - the use of new business processes, digital models, the creation of digital assets; - stimulating and supporting new digital initiatives	- additional GDP growth of the EAEU to 1% per year; - employment growth in the ICT industry by 66.4% and additional growth in total employment by 2.46%; - an additional increase in the volume of exports of ICT services up to 74%.

* Based on EEC materials

5.3   *EAEU in the digital economy world rankings*

Integration cooperation process may be analyzed using the dynamics of ICT sphere development in certain states and average values of the integration association (EAEU) as a whole in comparison with global trends, and world average integral indicators.

A comparative analysis of the resulting indicators of the digital economy establishment process for EAEU member states was carried out based on the data of rankings of international organizations and professional associations for 2007-2017. (Tables 3 and 4, Figure 1).

IDI composite index calculated in 2017 using 11 indicators (access, use, skills) characterizes positive dynamics of the EAEU economy digital sphere in 2007-2017, with signs of a certain unevenness of development. This index is a globally recognized indicator of the state of the digital economy of countries, which allows to study cross-country comparative researches both within the EAEU and association position in the world. States, ordered by index, occupy the corresponding places (Table 3) in the world ICT rankings.

Belarus became the EAEU leader in the ICT sphere from 2012, outperforming Russia (Figure 1).

EAEU states' ranking placement in accordance with the IDI values in 2007-2017 allows coming to a conclusion that digital economy sphere is developed fastest in Belarus (+21 places), Kazakhstan (+18 places) and Armenia (+14 places), which maximized IDI value as shown in Table 3. Note that the total number of states in the list affects individual country IDI (Kyrgyzstan).

Calculation of relative indicators of IDI change characterizes the positivity of dynamics of increase of the EAEU index versus the world average level indicator, as well as increase in the speed of this increase which reached the level of 5.6 times in 2008-2017 (Table 4).

Table 3.   Place of the EAEU countries in the IDI world ranking.

	2007	2012	2017	Place change	Index change
Armenia	89	73	75	+14	3.10
Belarus	53	43	32	+21	3.78
Kazakhstan	70	53	52	+18	3.62
Kyrgyzstan	96	107	109	−13	1.85
Russia	46	41	45	+1	2.94
Total number of states	159	157	176		

Table 4. Comparison of IDI of the EAEU and world average level.

	2008	2010	2011	2012	2013	2015	2016	2017
Average IDI in the EAEU	3.53	4.60	4.94	5.46	5.71	5.88	6.11	6.31
Average world IDI	3.32	3.94	4.15	4.60	4.77	5.03	4.94	5.11
IDI deviation (EAEU) from the world average level of IDI	0.21	0.66	0.79	0.86	0.94	0.85	1.17	1.20
IDI deviation (EAEU) from the world average level of IDI (%)	6.45	16.75	19.13	18.74	19.62	16.98	23.64	23.44

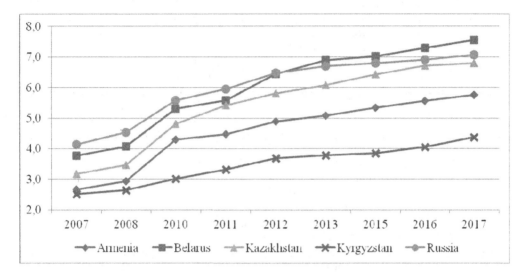

Figure 1. ICT development indices (IDI) of EAEU member states in 2007-2017.

Dynamics of growth of the EAEU ICT sphere is confirmed by the fact of increase in IDI in 2012-2017 occurring against a backdrop of slowing economies. In 2010-2017 the share of the EAEU members in Global GDP fell from 4.01% (2010) to 3.66% (2017/12/31), which indicates a low rate of growth of national economies (Member countries, 2018).

Monitoring of the status of national digital economies is based on the system of indices and rankings. Integration association CDS establishment may be analyzed using the indicators

Table 5. Indicators of the CDS status for the state and integration association.

State Indicators	Integration association Indicators
ICT Development Index	Availability of ICT development regulatory and legal framework – approved legal acts, concept, programs, plans
Networked Readiness Index	CDS creation investment program
Global Innovation Index	Degree of uniformity of national levels of ICT development of the integration association states
Global Cybersecurity Index	Degree of association of supranational (integration) and national interests of ICT sphere development
Ease of DoingBusiness Ranking	Priority of stages of CDS establishment by industries and regions
World Digital Competitiveness Ranking	

characterizing key areas of conflict of national and supranational interests proposed by the authors (Table 5).

The use of the indicators proposed by the authors will allow speeding up the process of creation of a CDS in the EAEU.

6 CONCLUSION

The conducted research identified the priority and importance of the modern information technology strategy development on the meso- and mega-levels of management implemented in national states and supranational integration associations (EAEU and EU).

In comparison with the world values, average ICT Development Index (IDI) of the EAEU demonstrates growth rate exceeding the world average that characterizes the EAEU states as dynamic and successfully creating digital economies.

Priority attention to ICT sphere and recognition of its importance in the integration processes is characteristic of the EAEU – a supranational integration association.

A comparative analysis of world rankings of development of information and communication technologies in the EAEU allows coming to a conclusion that establishment of the integration association's digital space goes faster than the establishment of a common economic space as the dynamics of development of the digital economy sphere of the member states exceeds the world average level.

In the first stage of establishing the foundations of conceptual and regulatory and legal framework of the EAEU digital space a number of inconsistencies are found, comprising: 1) essential individuality of the paths of member states' national digital spaces; 2) absence of a single concept and mechanism of creation of the integration association's CDS; 3) need to recognize the priority of creating a CDS over sectoral integration processes (creation of a common financial space, communications space, etc.).

Overcoming these contradictions stipulates: creation of the EAEU CDS Concept based on the common integration policy of the Eurasian Economic Commission; development of harmonized ICT infrastructure development plans and their investment support programs; training of specialists for implementation of the decisions made.

REFERENCES

Amit, R. and Zott, C. 2001. *Value Creation in E-Business.* Strategic Management Journal. 22, pp. 493–520.

Bell, D. 1999. *The Coming of Post-Industrial Society: A Venture in Social Forecasting.* Basic Books.

Biznesmeny predlozhili sozdat' edinoe cifrovoe prostranstvo EAES. 2015. [Businessmen suggested to create a common digital space of the EAEU]. https://www.kt.kz/rus/economy/biznes meni_predlozhili_sozdatj_edinoe_cifrovoe_prostranstvo_eaes_1153613222.html. (rus)

Castells, M. 1999. *The Rise of the Network Society: New post-industrial wave in the West. Anthology.* V. Inozemtsev (Ed.), Moscow: Academia.

Castells, M. 2000. *The Information Age: Economy, Society and Culture.* O. Shkaratan (Transl.), Moscow: GU VShE.

Chistyakova, E.A. 2018. *Cifrovizaciya v ramkah EAEHS, Sovremennye aspekty mezhdunarodnogo biznesa,* Saratov, pp. 29–32 (rus).

Digital Economy and Society Index. 2017. European Commission.https://ec.europa.eu/digital-single-market/desi.

Digital Economy has the Potential to Drive ASEAN Integration. 2016. World Economic Forum. https://www.weforum.org/press/2016/06/digital-economy-has-the-potential-to-drive-asean-integration.

Digital Kazakhstan state program. 2017.//https://primeminister.kz/rupage/view/gosudar stvennaya_ pro gramma_ digital_ kazahstan.

Digital transformation, European Commission. 2018.https://ec.europa.eu/growth/sectors/digital -econ omy/importance_en.

Fact Sheet: Key Barriers to Digital Trade. 2016. https://ustr.gov/about-us/policy-offices/press-office/fact-sheets/2016/march/fact-sheet-key-barriers-digital-trade.

Garicano, L. and Kaplan, S.N. 2001. *The Effects of Business-to-Business E-Commerce on Transaction Costs.* The Journal of Industrial Economics. No.49(4), 463–485.

Guo, S., Ding, W., Lanshina, T. 2017. *Digital Economy for Sustainable Economic Growth. The Role of the G20 and Global Governance in the Emerging Digital Economy.* International Organisations Research Journal. No.12-4 (2017), pp. 169–184. DOI: 10.17323/1996-7845-2017-04-169.

Knickrehm, M., Berthon, B., Daugherty, P. 2016. *Digital Disruption: The Growth Multiplier, Accenture.* https://www.accenture.com/_acnmedia/PDF-4/Accenture-Strategy-Digital-DisruptionGrowth-Multiplier.pdf.

Manyika J., Lund, S., Bughin J., Woetzel J., Stamenov K., Dhingra D. 2016. *Digital globalization: The new era of global flows.* McKinsey Global Institute. http://www.mckinsey.com/business-functions/digital-mckinsey/our-insights/digital-globalization-the-new-era-of-global-flows.

Measuring the Information Society Report. 2009-2017. ITU International Telecommunication Union. Geneva: Switzerland.

Mesenbourg, T.L. 2001. *Measuring the Digital Economy. U.S. Bureau of the Census.* http://www.census.gov/content/dam/Census/library/working-papers/2001/econ/digitalecon.pdf.

On Approval of the Digital Economy of the Russian Federation program. RF Government Decree dated 28. 07.2017 N 1632-p. ConsultantPlus information system.

Pravitel'stvo utverdilo plany po cifrovoj ekonomike na 520 mlrd rub. 2017. https://www.rbc.ru/technology_and_media/18/12/2017/5a37b95d9a79 47457322c9f4?from=main (rus).

Strany`-uchastnicy (EAE`S): Dolya v mirovom VVP. 2018 /https://www.economic data.ru/union. php? menu=economic-unions& un_id=27&un_ticker=EAEU&union_show=economics&ticker=EAEU-GDPShare (rus).

The EU and the digital single market. 2017. European Union. https://publications.europa.eu/en/publication-detail/-/publication/8084b7f3-6777-11e7-b2f2-01aa75ed71a1.

V RSPP obsudili perspektivy formirovaniya edinogo platezhnogo prostranstva EAES. 2018. http://asprof.ru/news/pub/367 (rus).

Inclusive Development of Society – Lumban Gaol (eds)
© *2020 Taylor & Francis Group, London, ISBN 978-1-138-33476-2*

Actual directions of scientific research in tourism economics

M.A. Morozov
Plekhanov Russian University of Economics, Moscow, Russia

N.S. Morozova
Russian New University, Moscow, Russia

A.V. Sharkova
Financial University under the Government of the Russian Federation, Moscow, Russia

T.A. Yudina
Sochi State University, Sochi, Russia

ABSTRACT: This article examines the current trends in scientific research of the tourism economy. It is noted that the study of the influence of tourism on world and national economies, as well as the study of the features of ecotourism on mega-, macro-, and meso-levels is an important issue. The traditional direction is the study of the multiplicative effect of tourism and forecasting, the results of which are used in the formation of tourism policy. It was noted that the analysis of the impact of information and communication technologies and financial and economic crises on the tourism economy became topical. Current economic research is related to the introduction of blocking technology into the tourist business, which represents a fundamentally new economy of joint use. The introduction of blocking technology in the activities of online travel agencies, metasearch systems, GDS, and airlines can lead to changes in all business processes in tourism. Prospective directions are interdisciplinary studies of tourism economics and psychology, social psychology, sociology, and other sciences; the impact of globalization on the tourism economy; mergers and acquisitions in the tourist market; and the study of risks in tourism.

1 INTRODUCTION

Tourism is an important sphere of economic activity, which is constantly growing and developing. In the first half of 2017, the growth of international tourist arrivals was 6%. The highest growth was in the Middle East (+ 9%), in Europe (+ 8%) and Africa (+ 8%) (http://media.unwto.org/press-release/2017-09-07/international-tourism-strongest-half-year-results-2010). This is the highest growth rate since 2010. In this regard, the scientific study of the economic patterns of tourism development is important and relevant.

The scientific content of the tourism economy is associated with the study of economic relations arising in the sphere of tourism in the manufacture, sale, and consumption of tourism products and services, ensuring meeting the needs of tourists. Currently, the economic importance of tourism is recognized for the world economy as a whole and for national economies. Among the first to realize the importance of tourism as an important economic activity of the country were France, Spain, Austria, Canada, Sweden, United Kingdom, and the United States, which are actively developing tourism. They began conducting the first economic studies in the field of tourism and to examine the economic contribution of tourism to the national economy.

The purpose of this article is to identify the most relevant and promising areas of research in the field of tourism economics, which will allow us to establish new trends and economic patterns of tourism development that are characteristic for the present time. This provides a basis for developing a sound economic policy for the development of tourism both internationally and at the level of individual tourist destinations.

2 LEVELS OF STUDY OF TOURISM ECONOMICS

The economy of tourism should be explored on several levels: mega-, macro, meso, and micro (Morozov & Morozova, 2012a). The meta-level corresponds to the study of the tourism economy on a global scale and is the level of the global economy of tourism. The nature of the economic tasks is related to the analysis of the role and place of tourism in the global division of labor, assessment of the contribution of tourism to world gross domestic product (GDP), etc.

Scientific study of the tourism economy at the macro-level is connected with studying the role, place, and economic value of tourism to national economies. Important study directions of the state regulation in the sphere of tourism are the problems of investment activity in the framework of national tourism, the formation of the state policy of financing of tourism projects, development and implementation of national projects and programs of development of tourism economy, etc. An example of such a program is the federal target program "Development of domestic and incoming tourism in the Russian Federation (2011–2018)." The implementation of this program is quite successful and will continue in the coming years. In August 2017, the Federal Agency for Tourism prepared the Concept of the Federal Target Program "Development of Incoming and Entry Tourism in the Russian Federation (2019–2025)" (Draft Order of the Government of the Russian Federation) . At the macro level of economic research in the field of tourism, it is of scientific interest to develop a well-founded list of indicators that will allow an objective assessment of the effectiveness of economic activity in the sphere of tourism. The performance is an indicator of the extent of achievement of the program objectives and the decisions necessary to achieve objectives. From the point of view of the tourism economy, an important indicator is the economic efficiency that measures the ratio between the obtained result and the time and resources spent on its achievement. Economic efficiency should be assessed at all levels of tourism management.

The meso-level of tourism economy associated with the study of the theory and practice of formation and development of tourist destinations. The main issues of tourism economy is to estimate the contribution of regional tourism to the gross domestic product of the region, assessment of the impact of tourism activities on the economy of the region, the development and implementation of regional tourism projects and programs, the study of factors influencing investment attractiveness of tourist destinations, etc. Federal and regional programs of development of tourism in the Russian Federation are based on the concept of cluster development of tourism. The tourist cluster is understood as a group of geographically adjoining interdependent enterprises of the tourist industry and the related organizations providing service to tourists and characterized by a community of activities and complementary to each other. However, scientific reasons for the cluster theory in the sphere of tourism require deep and all-round development. The application research m5etodov a cartographical taxonomy is of special interest when forming tourist clusters and destinations (Morozov & Morozova, 2011). This direction of scientific research is represented rather prospectively because it assumes creation of scientifically based methodology for formation of tourist clusters.

The micro-level of tourism economy involves the study of economic regularities of development and functioning of the tourist industry. The following are trends in economic studies: study of the factors influencing formation of the cost of the tourist product, the features of pricing in tourism, modern sources of financing of the enterprises in the tourism industry, and others.

In many countries of the world, tourism industry enterprises belong to small and medium-sized enterprises of national economies. In this regard, the research into new forms and methods of financing the enterprises of the tourism industry, the study of the directions for the development of public–private partnership, the analysis of the risks of entrepreneurial activity in tourism, and the impact of crises on the financial and economic performance of tourism industry enterprises are of particular scientific interest.

3 HISTORY OF ECONOMIC RESEARCH IN TOURISM

The first scientific research in the field of the economics of tourism dates back to the mid-1960s, in particular, the study on the economics of outdoor recreation (Clawson & Knetsch, 1966). In the 1970s issues of the relationship between international travel and trade were studied by Gray (1970). The scale of ongoing research in the field of the tourism economy contributed to the creation of the new specialized scientific journal *Tourism Economics*, which began publication in 1995. In 2007the International Association for Tourism Economics (IATE) was created, which brings together scientists from different countries, leading economic research in the field of tourism.

Since the end of the 20th century, the research interests of scientists in the field of the tourism economy have increased considerably. First of all, considerable scientific research is devoted to the development of a methodology for assessing the contribution of tourism to the global and national economies and the study of the multiplier effect in tourism. These studies are of particular importance from the standpoint of evidence of the importance of tourism as a driver of the economy for many countries and regions. Classical works in this direction are the studies by Archer and Fletcher (1996), Crompton et al. (2001), Frechtling (1999), and Jones and Munday (2007). This issue remains relevant to this day.

An important area of research in the field of the tourism economy is associated with the analysis of the impact of the global financial and economic crisis on tourism, which in recent decades has repeatedly occurred. These issues are highlighted in the works of Blake and Sinclair (2003) and Pambudi et al. (2009). Conceptual issues in the assessment of the role of tourism and its socioeconomic impact on society were studied in the works of Blake et al. (2008) and Wattanakuljarus and Coxhead (2008).

Developed in the 2008 international recommendations on tourism statistics UNWTO helped to provide a new factual basis for research in the field of tourism economics, which enriched and diversified the list of the problems studies. Would La forms an information base, which provides the opportunity to examine and measure the relationship of tourism with other sectors of the economy, including in comparison in terms of of the magnitude of the positive economic indicators, industry of employment, contribution of tourism to national and regional economies, and the comparison of this contribution at the intercountry and interregional level. Research conducted on forecasting of the tourism economy provides a scientific basis and improves the quality of long-term plans and program of tourism development. The first studies on the prediction of the development of tourist activities were undertaken by such scholars as Archer (1994, 1980) and Frechtling (2001).. However, the capacity of the Russian tourism questions to assess the economic contribution of tourism to the regional and national economies remains underdeveloped. One of the important problems is a scientific substantiation of the system of indicators for assessing the performance of tourism at the regional level. Currently, there is ambiguity in the understanding of a number of economic indicators related to the study of the economy of the tourist region, as well as the comparability of data in the different regional programs of tourism development. These issues were addressed in the work of Morozov and Morozova (2015).

4 A PROMISING DIRECTION OF ECONOMIC RESEARCH IN TOURISM

Priority directions in economic studies of tourism are the issues of the influence of the world financial-economic crisis on the tourism sector, which is still insufficiently investigated. It is especially important to recognize the effects of the financial and economic crises on the financial and economic activity of enterprises of tourist industry, changing demand for tourism products, attraction of investments, changes in marketing policy, and increased competition in the tourism market. In addition, currently a special significance for the Russian tourist market is study of the problems of bankruptcy of enterprises in the tourism industry, which is associated with a series of bankruptcies of large national operators that have occurred in recent decades. The work of Morozov and Morozova (2012a) is devoted to the study of the problems of bankruptcy of Russian tour operators. Current areas of research are the study of business risks and crisis management in tourism.

In the globalized tourism market competitiveness in tourism has become one of the topical areas of economic research. The works of Crouch and Ritchie (1999) and Dwyer and Kim (2003) are devoted to these issues.. The most studied form of competition is price competition, which is described by Forsyth and Dwyer (2010). Currently a topical issue for the Russian tourist market is using the methodology of dynamic pricing in tourism, where the price of tourism products is set according to the value that the tourist sees in this tourist product. The implementation of the methodology involves the use of spatial price dispersion, i.e., price changes in relation to the territory, and temporal price dispersion, i.e., price changes depending on time parameters. Price discrimination also applies, in which prices vary depending on the consumer, hence the need to study the consumer behavior of the tourist, taking into account the impact of the price policy of the tourism industry. Scientific research of changes in consumer demand and behavior depending on the price strategy of a tourist industry enterprise is one of the most important issues that require a thorough and comprehensive study.

The economy of tourism (destination), in which economic actors of the tourist market in the conditions of free competition ensure development and implementation of tourism products that meet the needs of the world tourism market, is competitive. An essential condition is the presence of effective demand. The competitiveness of the tourism industry involves its ability to adapt to the changing market situation in the tourism market and to remain relevant in this market. The success of individual enterprises in the tourism industry in competing in the globalized tourism market is not confined to national boundaries but depends on the state of affairs in the tourism industry of both the country and the world. Each country has its own concept of adjusting to competitiveness in the tourism market. In many countries, such policies shape the national tourism organization.

Most modern economists believe that the first step is to manage the competitiveness of the products. However, in tourism, from our point of view, of paramount importance is the competitiveness of the tourist destination, because it is mostly through the competitiveness of the tourism destinations that the competitiveness of the tourism product is evaluated (Morozov & Morozova, 2011, 2012a). It should be emphasized that the competitiveness of the tourist product, the enterprises of the tourism industry, and tourism destinations are interrelated as part of the whole. In many cases, the competitiveness of tourist destinations plays a pivotal role because its tourist appeal is based on tourist choice (Eskindarov, 2017).

The relevance of the study of tourist attractiveness of destinations should be noted, as tourist attractiveness is the most important factor in selecting travel destinations travel (Morozov and Morozova, 2011, 2016). The correct assessment of the attractiveness of a destination has a direct impact on the volume of tourist flows and income from tourism activity. Of scientific interest is the study of the realistically achievable competitiveness of a tourist destination. It is understood as the ability of the tourist industry of this destination to produce competitive tourist products that, on the one hand, will ensure the cost recovery for the creation of this destination and its further economic growth, and on the other hand, will promote the growth of real income of the local population and its employment for a long time.

The modern paradigm of the digital economy has a direct impact on the development of scientific research in the framework of the tourism economy. In this regard, the current generated will be the issues of studying the influence of information and communication technologies on the formation and promotion of tourism services, competitiveness at all levels of tourism management, and improving productivity in the tourism sector, which affects the economic performance of the tourism system.

In scientific economic research, the introduction of innovative technologies, which radically change business processes in the sphere of tourism, acquires special significance. Such new progressive innovations and directions of digital tourism development include blocking technologies that implement a distributed database, in which storage devices are not connected to a common server. This provides such advantages of blocking technology as openness, security, and security. Blokchein begins to actively penetrate into many areas of activity and it can be called a new economy of common use. In the short term, blocking technologies will be used by online travel agencies, metasearch systems, GDS, and airlines, which can fundamentally change many business processes in the sphere of tourism. In this regard, economic studies associated with the introduction of blocking technology in the tourism industry are very relevant.

The multidimensional nature of factors affecting tourist activity, as well as the complexity of interrelationships within tourism systems, leads to the necessity of interdisciplinary research. This is underlined by scholars such as Stabler et al. (2010), noting that prospective studies in the field of tourism economy need to expand based on psychology, social psychology, and sociology (Stabler et al., 2010). Such research will identify new factors affecting tourism demand, to evaluate ecological and environmental issues of the tourism economy. Research of the tourism economy needs to expand at the expense of interdisciplinary research, including sociological, environmental, psychological studies, economics, welfare, etc. The need for a thorough examination of such areas as experience economy, economics of participation, etc. should be emphasized.

The role of interdisciplinary scientific research related to the study of consumer behavior of tourists, depending on the influence of various factors on the tourism economy, the interrelationship of environmental problems with the development of tourism, and the impact of the tourism economy on the socio-economic status of individual regions and countries is growing.

5 CONCLUSION

The most relevant directions of scientific research in the field of tourism economics are therefore the study of the impact of financial and economic crises on the development of crisis management in the tourism sector at all levels of the tourism economy, the impact of globalization and of mergers and acquisitions on the development of tourism at the level of global and national economies, and research and risk prevention in tourism and their impact on the economic status of individual market actors and the economy as a whole.

REFERENCES

1. Archer, B. 1980. Forecasting demand, quantitative and intuitive techniques. *International Journal of Tourism Management*, March: 5–12.
2. Archer, B. 1994. Demand forecasting and estimation. In J. R. Brent Richie & C. R. Goeldner (eds.), *Travel, Tourism, and Hospitality Research: A Handbook for Managers and Researchers*. New York: John Wiley & Sons.
3. Archer, B., & Fletcher, J. 1996. The economic impact of tourism in the Seychelles. *Annals of Tourism Research*, 23(1): 32–47.
4. Blake, A. T., & Sinclair, M. T. 2003. Tourism crisis management: US response to September 11. *Annals of Tourism Research*, 30(4): 813–832.
5. Clawson, M., & Knetsch, J. L. 1966. *Economics of Outdoor Recreation*. Baltimore: Johns Hopkins University Press.
6. Crompton, J. L., S. Lee, and T. Shuster. 2001. A Guide for Undertaking Economic Impact Studies: The Springfest Festival. Journal of Travel Research, 40 (1): 79–87.
7. Crouch, G. I. & Ritchie, J. R. 1999. Tourism, Competitiveness, and Societal Prosperity. Journal of Business Research 44: 137–152.
8. Draft Order of the Government of the Russian Federation "On the Approval of the Concept of the Federal Targeted Program" Development of Domestic and Entry Tourism in the Russian Federation (2019-2025) "(prepared by Rosturism on August 14, 2017) URL: http://www.garant.ru/products/ipo/prime/doc/56624252/
9. Dwyer, L. & Kim, C. W. 2003. Destination Competitiveness: a Model and Indicators, Current Issues in Tourism, Vol. 6, No. 5: 369–413.

10. Forsyth, P. & Dwyer L. 2010. Exchange Rate Changes and the Cost Competitiveness of International Airlines: The Aviation Trade Weighted Index. Research in Transport Economics vol. 24: 12–17.

11. Frechtling D. 1999. The tourism satellite account: foundations, progress and issues. Tourism Management 20: 163–170.

12. Frechtling, D. C. 2001. *Forecasting Tourism Demand: Methods and Strategies.* Oxford: Butterworth-Heinemann.

13. Gray, H. P. 1970. *International Travel—International Trade.* Lexington, MA: D. C. Heath.

14. International tourism - strongest half-year results since 2010. Available at: http://media.unwto.org/press-release/2017-09-07/international-tourism-strongest-half-year-results-2010

15. Jones, C., & Munday, M. 2007. Exploring the environmental consequences of tourism: A satellite account approach. *Journal of Travel Research,* 46: 164.

16. Morozov, M. A., & Morozova, N. S. 2011. Assessment of competitiveness of tourist destinations on the basis of cluster approach. *International Trade and Trade Policy,* 11: 114–124.

17. Morozov, M. A., & Morozova, N. S. 2012a. The model of competitiveness estimation of tourist destinations. *International Trade and Trade Policy,* 11: 100–108.

18. Morozov, M. A., & Morozova, N. S. 2012b. On the topic of the day: Analysis of the competition and ban crust in tourism. *International Trade and Trade Policy,* 3: 61–68.

19. Morozov, M. A., & Morozova, N. S. 2015.The formation of a system of indicators to assess the performance of regions in the field of tourism. *Sochi Journal of Economy,* 2(35): 105–115.

20. Morozov, M. A., & Morozova, N. S. 2016. Attractive tourist destinations as a factor of its development. *Journal of Environmental Management and Tourism,* VII, 1(13): 105–112.

21. Pambudi, D., McCaughey, N., & Smyth, R. 2009. Computable general equilibrium estimates of the impact of the Bali bombing on the Indonesian economy. *Tourism Management,* 30.

22. Stabler, M., Papatheodorou A., & Sinclair, T. 2010. *The Economics of Tourism,* 2nd ed. London: Routledge.

23. Eskindarov, M. A. 2017. *The Development of Entrepreneurship and Business in Modern Conditions: Methodology and Organization.* General edition. Moscow. Financial University.

24. Wattanakuljarus, A., & Coxhead, I. 2008. Is tourism-based development good for the poor? A general equilibrium analysis for Thailand. *Journal of Policy Modeling,* 30(6): 929–955.

Inclusive Development of Society – Lumban Gaol (eds)
© 2020 Taylor & Francis Group, London, ISBN 978-1-138-33476-2

Mathematical tools for research on economic processes

V.B. Dzobelova, A.V. Olisaeva & M.V. Galazova
Federal State Budget Educational Institution of Higher Education (FSBEI HE), North Ossetian State University named after Kosta Levanovich Khetagurov, Vladikavkaz, North-Ossetia, Alania

N.B. Davletbaeva
Karaganda State Industrial University, Karaganda, Kazakhstan

O.A. Ishchenko-Padukova
Southern Federal University, Rostov-on-Don, Russian Federation

ABSTRACT: This article is devoted to the questions of interaction of economics and mathematics, as well as the search for ways of complementarity of these sciences. In this article, we consider some theoretical aspects of the interaction of the two disciplines of economics and mathematics. Methods of applying mathematical models in economics are also considered. The article concludes that the mathematical apparatus lets one correctly explicate and scientifically justify the statements of economic theory. The actual application of mathematics in economic research, allowing an explanation of the past, an ability to see the future, and an evaluation of the results of the research, will require significant efforts, which at the moment in the economy are not adequate.

Keywords: age of information technologies, economics, mathematical culture, mathematics, socioeconomic systems

1 INTRODUCTION

In the 21st century, exact sciences are necessary in the economic sphere of people's lives. A natural science such as mathematics rather quickly penetrates into various sciences with the help of its methods, and economics, management theory, and management are no exception. Economics and mathematics are independent sciences, possessing their own objects and subjects of research. But it should be noted that there is definitely a link between economics and mathematics, both direct and inverse. Mathematics allows people to study different methods (theoretical and practical) in order to understand the surrounding world.

There has never been a period of time when mathematics did not play an important role in the development and existence of the social sciences. Nowadays, because of the difficult political and economic situation, Russia needs good economists, managers, analysts, and organizers of economic production. Such specialists need serious training, not only physical and psychological, but also mathematical. This training would help specialists to develop logic, allow "taking a look into the future," to explore the widest range of new problems, learn to use computer technology, and also to use the obtained theoretical methods in practice. In addition, every specialist is required to constantly improve his or her qualifications in mathematical education. A modern economist must generate, not just absorb, new knowledge and abilities. The increase in requirements for applied mathematics has become the duty of economic universities.

During the years of his or her training, the future economist should clearly define for himself "what is mathematics?", as well as have an idea about the mathematical approach to the study of the real world and be able to apply it correctly. Lecturers of mathematics should provide a solid basis of knowledge about mathematical culture, help the student to master an elementary base of skills, thanks to which he or she will be able to acquire additional abilities in the future. Then well-trained economists will be able to check the results using the method of economic and mathematical analysis, suggest a forecast, prove the suggestions and hypotheses, and compile mathematical models and optimal plans for the object of their activity. It is interesting that such sciences as management and economics were the last to penetrate them. Many people still believe that these are only descriptive sciences. However, more precise calculations from economics and management are required today. And, of course, it is impossible to do this without mathematics.

2 METHODS

Mathematics has penetrated into macro- and microeconomics; it is the basis of programming. The mathematical apparatus is an acknowledged tool of management and economics; without it we would not be able to develop specific applied tasks for managing organizations or enterprises and optimize production and business, and financial regulation would be impossible without it. Mathematics teaching today should be easy, simple, and comprehensive. And students, in turn, must learn and understand the material of research or the question under study.

At all times people have been interested in mathematics, because it is an exact science, and people need accuracy and specifics in everything. The level of managers' and accountants' training depends on how accessible and comprehensive mathematics is because these professions are entirely connected with the turnover of financial resources. So, mathematics needs to be made easier at the initial stage of its study, i.e., in schools, and then when they are enrolled in higher educational institutions, students will have a wide fund of knowledge in this discipline and will have no problems with mathematics, as they do now. If mathematics comes easily, it certainly will affect the career growth of managers and accountants, and this is probably the most important goal that a person sets him- or herself when choosing these professions. Mathematics plays the most important role in the life of mankind, because without it, we would not have anything now.

There is both a direct and inversely proportional relationship between economics and mathematics: the creation of a modern mathematical apparatus allows solving as many problems as possible. In its turn, the economics itself sets tasks to mathematics, which it has to solve; economics also stimulates mathematics to search new methods of solutions. Due to this synthesis of mathematics and economics, various methods of economic research have appeared: balance method, network method, and correctional and regression analysis.

Economics has also become the basis for the appearance of some new directions in applied mathematics: programming, games theory, multivariate statistical analysis, etc. Economics has become a fertile ground for the use of mathematics; that's why the given topic is quite urgent at the present time and is profoundly studied by students and scientists (Bondarenko & Tsyplakova, 2013). .

Economics often uses numerous mathematical models and methods, considering them necessary (Dolgopolova, 2012). On the one hand, with the help of mathematics, it becomes possible to describe the most important and essential links of economic science, to obtain accessible results and conclusions. On the other hand, mathematical methods and models allow assessing general view and parameters of economic interrelations on the basis of specific observations.

3 ANALYSIS

The concept of a matrix is often used in practical economic activity. For example, the data on the output of several types of products in each quarter of a year are conveniently recorded in the form of a matrix.

The issue of the efficiency of multisectoral management is also often discussed in economics. Each branch acts at the same time as a producer of some products, on the one hand, and as a consumer of its own products or those produced by other branches, on the other. The relationship between different industries is reflected in the tables of intersectoral balance. Analysis of these tables is made with the help of the model of a multisectoral economics.

Some phenomena in economics are quite stable and regular; therefore they can be described with the help of the notion of function. For example, using the function, you can describe the dependence of demand Y on the price of goods X (Figure 1).

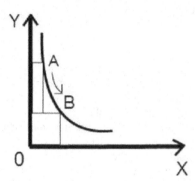

Figure 1. Dependence of demand on prices.

Figure 1 shows the inverse relationship between the price of goods X and the actual demand for given goods Y. It shows that the volume of demand decreases with an increase in price.

If we take into account the fact that most objects in economics are subject to the influence of not one but many factors, then functions of several variables should be used to study such objects. For example, the production function links the volumes of resources spent and an output (Figure 2).

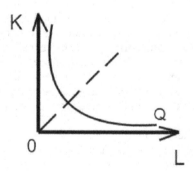

Figure 2. Production function.

Figure 2 shows a two-factor model of the production function, where L is labor, K is capital, and Q is the maximum output: $Q = F(L, K)$.

One has to deal with the concepts and methods of mathematical analysis in economic practice quite often (Rytova, 2017). For example, limit values and the derivative are used in the study of the dependence of demand on the price of goods. So, the derivative of the demand function describes a 1-point increase of price with a decrease in consumer demand. The derivative of the supply function characterizes the increase in the supply of goods from the producers with a 1-point increase of price. Marginal utility, i.e., utility from the acquisition of one additional unit of goods, determines the derivative of the utility function.

Differential calculus is used in economics to determine, for example, instantaneous water or energy consumption. But to determine the daytime output according to the function of workforce productivity, production volume, etc. the elements of integral calculus are used.

4 CONCLUSION

Thus, mathematics and economics constantly interact and mutually complement each other (Proshkin & Uvarova, 2016). Today, this relationship is not just an object of research, but is actually used in practice. Knowledge of mathematical methods and models is becoming an obligatory part of the professional knowledge for specialists in the field of economics and management.

Not only the information itself, but also precisely obtained knowledge about the socioeconomic system has become necessary for modern society. It is possible to obtain this knowledge only by applying a modern mathematical apparatus, computers, and information technologies.

Modern economic systems have sufficiently branched external and internal relationships. Therefore, in order to manage them efficiently, it is necessary to develop and fully utilize all the capabilities of the mathematical apparatus.

Mathematics and economics are in constant complementarity and interaction with each other. Taking into account the fact that the present computer age is so developed, research on these fields has moved far ahead. It is not only theoretical knowledge, but also practice (e.g., the economic practice of modern business management). Rapid development of the economics in our century, a change in the business environment, a smooth transition of economics to knowledge and interaction—all this influenced the convergence of mathematics and economics. The current state of economics is not stable enough to subordinate itself completely to mathematical culture. But very soon, one can expect this synthesis to produce such economic studies that will contribute to explaining much of the past history, assessing the present, and also foreseeing the future. The synthesis of mathematics and economics has a great and successful future, which will require the application of tremendous efforts and nonfundamental knowledge, which economics is still trying to obtain.

It seems to us that in modern conditions, due to the increasing role of applied mathematics in the economy and the emergence of new technologies of mathematical processing of information, there is an urgent need to revise the content of courses in mathematics, as well as the educational technologies used: it is important for the user to know not how the mathematical formula is obtained, but what connections it describes, and where and under what conditions in the economy it can be applied.

Thus, two completely different sciences— economics and mathematics—are characterized by close interaction. The rapid change in the business environment in the 21st century, the high pace of development of intellectual technologies, and the gradual transition from an economy based on material resources to a knowledge economy enhance this interaction. Modern socioeconomic systems are characterized by such ramified external and internal relations that determine their state and behavior that it is impossible to effectively manage them without the use of the modern mathematical apparatus and information technologies. This is confirmed by modern business management technologies used by transnational corporations.

The present application of mathematics in economic research, which allows explaining the past, seeing the future, and assessing the consequences of their actions, will require even greater effort and new fundamental knowledge, which in the economy is not yet adequate.

REFERENCES

Bondarenko, V. A., & Tsyplakova, O. N. 2013. Conditions for the formation of mathematical culture among students of economic areas. Agrarian science, creativity, growth, Vol. 1, pp. 1–286.Stavropol: AGRUS.

Dolgopolova, A. F., Gulai, T. A., & Litvin, D. B. 2012. Mathematical modeling of socio-economic systems. In *Accounting and Analytical and Financial and Economic Problems of the Development of the Region*, pp. 283–286.

Malikov, D. Z., & Ufimtseva, L. I. 2017. Use of mathematics in economics. Specifying ways of the least cost. *International Student Scientific Bulletin*, 41: 115–116.

Proshkin, P. S., & Uvarova, M. N. 2016. Mathematics in economics. In *Physics and Modern Technologies in Agrarian and Industrial Complex Proceedings of the International Youth Scientific-Practical Conference*, pp. 230–232.

Rytova, T. A. 2017. The role of mathematics in economics. *Economics*, 3(24): 65–67.

Inclusive Development of Society – Lumban Gaol (eds)
© 2020 Taylor & Francis Group, London, ISBN 978-1-138-33476-2

Indicators of digital inequality and spatial development of the ICT sector in the Russian Federation

D.A. Artemenko
Southern Federal University, Rostov-on-Don, Russian Federation

E.N. Yalunina
Ural State University of Economics, Ekaterinburg, Russian Federation

E.A. Panfilova, T.Yu. Sinyuk & T.N. Prokopets
Rostov State University of Economics RSUE (RINH), Rostov-on-Don, Russian Federation

ABSTRACT: The digital economy presents new requirements to the national instrument-evaluation system for supporting management decisions on the development of the information and communication technology or (ICT) sector. Digital inequality reflects the technological, economic, social, educational disparity of economic entities in accessing, obtaining and using information on the basis of modern IT.This article examines the contradictions in the rating dimensions of the digital economy, clarifies the methodological aspects of digital inequality, defines the features of the construction and the effectiveness of using empirical indicators of spatial unevenness in the development of the ICT sector. The article provides the rationale and the methodology for constructing the aggregate indicator ICT Development Index RF, which allows to adjust the assessment of the implementation of the state programs of the Russian Federation on the development of the digital economy, providing inclusive management with other state programs of innovation and investment orientation. From 2014 to 2016, the North-West FD is the leader among the Russian regions in the development of the ICT sector – index 77.98 p.p. At the same time, the digital backlog for the regions of this region from the leading regions was more than 21 p.p. - the North Caucasian FD.The proposed modified ICT Development Index RF is distinguished by the simplicity and flexibility of the method of construction, the conjugation of the national and regional digital profile, the use of open monitoring data of the Russian information economy, and the possibility of further modification using correlation and regression analysis.

Keywords: indicators of ICT development, digital economy , digital inequality, spatial development of ICT sector, digital profile of regions, ICT spatial development index for Russian regions

1 INTRODUCTION

In modern conditions, the digital economy gave birth to new competitive advantages and digital opportunities in all key areas of activity of economic fields. The ICT sector is a system-forming factor and an effective tool for digital modernization of national economic systems.

At the same time, digital transformations change the structure, nature and type of interactions of society, state and business. They are a source of not just new opportunities, but also new types of inequalities and barriers to the socio-economic development of national economic systems, therefore, such phenomena as digital inequality, digital barriers, digital gaps, and digital maturity become the most important object of scientific research and the subject of various econometric measurements.

The nature and specifics of the implementation of digital transformations and the digital maturity of economic entities are reflected in the works of *Westerman G., Bonnet D.* (2015), *Valdez-de-Leon* (2016), *Kane.*(2017) and others. Problems of digital inequality and digital discontinuities in the social and spatial aspects were studied in papers by *Dimagio Hargittai* (2001), *Zillien, Haufs-Brusberg* (2014), *Aissaoui* (2017) and others. The use of various empirical tools and methods for estimating the level digitalization of the activities of economic entities were considered in the works - *Sinyuk T. Yu., Panfilova E.A.* (), *Selishcheva* (2015), *Russia is online* BCG (2016), *Klochkova* (2017), *Arkhipova M. Yu., Sirotin V.P.*(2018) and others.

At the moment, there is a wide variety of global and national ratings for the development of the digital economy and the ICT sector such as the ICT Development Index (IDI), the Digital Economy and Society Index (DESI), the Network Readiness Index (NRI) Development Index, the Global Connectivity Index (GCI), the European Digital Development Index (EDDI), the e-Intensity Index BCG, the index of readiness of the Russian regions to the information society and others.

However, it is necessary to note the following:

- the ratings of the development of ICT and the digital economy have different factual and empirical research bases, reflecting various parameters such as access, speed, the willingness of users to use Internet technologies, the degree of development, the level of use of information technologies for solving various problems, etc .;
- the ratings show the impact of ICT on different sectors of the economy - tourism, finance, education, health, government, culture, etc., therefore having different levels of coverage or on the economy as a whole of the systemic impact of ICT;
- the ratings have a different component structure, as well as varying degrees of validity, indicators used, calculation methods and construction models (from simple to complex);
- the ratings have varying degrees of completeness, periodicity, reliability of source databases, because they are formed on the basis of a large amount of information, based on various sources of information (global and national statistics, reports of national institutions, studies of consulting companies, agencies, scientific publications, etc.);
- the ratings are significantly different and based on their results, having a significant diversity in the interpretation of the level of development and the impact of the digital economy and the ICT sector on economic dynamics.

All this leads to the fact that depending on the research and management tasks, various ratings fix various aspects of ICT sector development, therefore the digital inequality is reflected in different statistical structure of indicators, object expression and quantitative and qualitative certainty.

In other words, there is a contradiction in the empirical research of ICT, due to the fact that, on the one hand - the number of various ratings of the dynamics of the digital economy and digital inequality increases, and on the other hand - the systematization and comparability of the evaluation data reflecting the uneven spatial development of the digital economy and the ICT sector at various levels - global, national and regional.

In addition, the requirements to information efficiency of ratings, availability and visualization of data, their completeness, forecasting, the possibility of "matching" and comparing global, national and regional data with respect to the dynamics, variability and depth of digital transformations of national economic systems are increasing. All this forms the need for a coherent, open evaluation tool based on inclusive partnerships in the global process of managing the digital economy.

Since the development of the digital economy is uneven and not only in different sectors of the economy, but also in a spatial aspect, therefore, issues related to the systematization of databases, the construction of various indicators and ratings necessary to reflect the most important factors and the extent to which ICT affects the socio- economic development of national economies.

In this regard, it is useful to clarify the methodological aspects of the concept of digital inequality, as well as to determine the features of the construction and effectiveness of the application of empirical indicators of spatial uneven development of the ICT sector in Russia. This article is devoted to the consideration of these questions.

2 DIGITAL COORDINATES OF SPATIAL DEVELOPMENT

Digital transformations are rapidly gaining momentum in the global socio-economic space. So the number of Internet users in 2016 was 3.385 billion people, which covers 45.28% of the world population. At the same time, in 2016 in comparison with 2010, the growth of Internet users in the world amounted to 35.18%. For countries such as Iceland, South Korea, Denmark, Sweden, the proportion of the population that provides Internet users exceeds 97% (*ITU Internet users* (2018). In addition, today the number of Internet things (IoT) exceeds the world population of 8, 4 bln units, and by 2020 it is projected to exceed the 20.4 billion mark (*Shvab 2016*).

The development of the digital economy in the spatial and country aspects is uneven, as early as 1997 the UN drew the attention of the world community to the notion of "information poverty" in the context of uneven access and use of ICT, as digital inequality affects both developed and developing countries.

In the future, the need to consider the negative external influences of the development of the digital economy has attracted the attention of the scientific community more than once, and various definitions have been used such as "digital poverty", "digital divide","digital curtain","digital rupture", "digital fallout", "digital burnout "," digital inequality ", etc.

If at the initial stage of the study of digital inequality the main emphasis was placed on technological accessibility and the conditions of Internet connection, then further aspects such as cognitive and psychological readiness of users to use Internet- technology, the level of user skills for the effective use of Internet technologies, the content containing applications of digital tools, its advantages and limitations, as well as the level of information security of the use of Internet technologies.

The complexity of unambiguous definition of digital inequality lies in the fact that it has many aspects and elements of implementation such as; technological, economic, social, psychological, gender, communicative, behavioral, psychological, legal and geopolitical.

In addition, digital inequality has different levels of implementation - global, national, regional, at the level of individual companies, sectors and industries, social groups and individuals.

Various options for examining the category of digital inequality are reflected in Figure 1.

Figure 1.   Various aspects of the consideration of the category digital inequality.

Source: compiled by the authors

189

According to the authors, digital inequality reflects the technological, economic, social, educational disparity of economic entities in accessing, obtaining and using information on the basis of modern IT.

As emphasized by *Zillien N., Haufs-Brusberg M.* (2014), unlike traditional types of socio-economic inequality, the digital inequality does not have "direct" estimates (for example, as inequality in income levels of the population). The digital inequality, in the opinion of the authors, characterizes a number of "indirect" and mediated parameters and development factors, the combinations of which have a nonlinear dependence and the effectiveness of the action, which makes it difficult to unambiguously evaluate measures and directions for overcoming the digital inequality.

To a certain extent, this explains the multiplicity of the rating tools used in assessing the effectiveness of the development of the digital economy, both globally and nationally, since the position of the national economy in the international rating acts as a kind of competitive "marker" for its sustainable social and economic development.

The most popular rating of the dynamics of the digital economy is the international rating of the ICT Development Index (IDI), the International Telecommunication Union (ITU), which reflects the readiness for the information society and the level of development of the national ICT sector.

The five leaders of this rating are South Korea, Iceland, Sweden, Denmark and the United Kingdom. According to the ICT Development Index (IDI) in 2016, Russia ranked 43rd with an index value of 6.91. In 2017, Russia moved from 43 places to 45th place, with the value of the index of development of information technology - 7.5.

Dynamics of changes in the ICT Development Index (IDI) for the Russian Federation for 2016-2017 (Figure 2).

The cross-country ICT Development Index comparisons show that the development of the ICT sector in Russia is still inadequate, there is a backlog of world leaders, and also reflect the unrealized potential of existing infrastructures and technologies. Although this index does not have a regional "sweep," it only confirms the need to develop a strategic and evaluation tool based on open data used in comparing the development of the ICT sector.

From 2010-2017 in the Russian Federation, the most important national documents and government programs on the main areas of development of the economy were adopted - "Information Society (2011-2020)" (*State program,* 2010), "The State program" (2017), as well as the national statistical monitoring of the development of the information society in Russia (*Monitoring 2018*). At the same time, it should be noted that statistical observation and performance evaluation by state development programs by the economy figure is realized independently of each other and from associated subprograms and projects on innovative development of the country (*Sinyuk* 2016).

Figure 2. ICT development index (IDI) RF 2016-2017.
Source: Rating of Russia. ICT

This forms the need to develop an "end-to-end" ICT development index of the Russian Federation that could be developed at the national and regional level, and also focus "point" management influences on the main "input" parameters of the ICT sector development associated with the world indicators of the digital economy development. The next section of the article is devoted to the method of building the ICT Development Index RF

3   METHODS OF TESTING AND ANALYSIS OF RESULTS

The calculation is based on the principle of modifying the index of the ICT Development Index (IDI). By analogy with the ICT Development Index (IDI), the index in the author's interpretations is a composite index that combines 12 indicators into one benchmark measure. It is used to monitor and compare developments in information and communication technology (ICT) over time. The IDI is divided into the following three sub-indices - Access sub-index, Use sub-index, Skills sub-index.

Subindices were calculated by summing the weighted values of the indicators included in the corresponding subgroup. Weights and normalization of the data were carried out in accordance with the method of The ICT Development Index (IDI).

With the exception of Skills sub-index, since four new indicators were selected for this sub-index, which required a redistribution of the weighting coefficients. In addition, the author's interpretation has changed the structure of the indicators, this is justified by the fact that there is no reliable statistically significant data for all regions of the Russian Federation from 2014 to 2017 and for all indicators of monitoring the development of the information society in the Russian Federation. The initial array of data is the figures from the official statistical monitoring by the subjects of the Russian Federation. (*Monitoring* 2018).

The index is calculated using the formula:

$$
\begin{aligned}
ICT \; Development \; Index \; RF \; (IDI) \;=\; & 0,4 Accesse \; sub - index' \\
& + \; 0,4 Use \; sub - index' \;+\; 0,2 Skills \; sub - index
\end{aligned}
\tag{1}
$$

Where:

Access sub-index (5 indicators). The indicators included in this sub-index were given equal weights
(one-fifth weight each):

- Telephone density of fixed communication (including payphones) per 100 population (1.3.1)
- Penetration of mobile radiotelephone (cellular) communication for 100 people (1.3.2)
- Number of fixed broadband Internet subscribers per 100 population (1.3.8) Share of households
- with a personal computer in the total number of households (2.6.2)
- Specific gravity of households with access to the Internet from a home computer, in the total
- number of households (2.6.5)

Use sub-index (3 indicators). The indicators included in this sub-index were given equal weights
(one-third weight each):

- Proportion of households with access to the Internet in the total number of households (2.6.6)
- Number of fixed broadband Internet subscribers per 100 population (1.3.8)
- The proportion of the population that is active in the total population (2.6.8)

Skills sub-index (4 indicators). The indicators included in this sub-index were given equal weights
(one-quarter weight each):

- Share of the employed population aged 25-64, who has a higher education in the total number
- of employed population of the corresponding age group (1.1.1)
- Number of students enrolled in educational programs of higher education - undergraduate,
- specialist, magistracy programs per 10,000 people (1.1.2)
- Proportion of students in general education institutions (1.1.4)
- Share of employed in the ICT sector in the total number of employed (1.5.1)

Before running the statistical analyses, the indicators to be included in the index were defined based on data availability. Before running the PCA, Bartlett's test of sphericity was performed to find out whether the indicators initially chosen are correlated. For the IDI, the distance to a reference measure was used as the normalization method. The reference measure is the ideal value that could be reached for each variable (similar to a goalpost). In all of the indicators chosen, this will be 100. After the data had been normalized, they were rescaled to identical ranges, from 1-10. This was necessary in order to compare the values of the indicators and the sub-indices. In choosing the weights, the results of the Principal Components Analysis (PCA) were taken into consideration, PCA identifies the relative importance of the indicators selected in each subgroup. The weights were computed by performing the following steps:

- The component loadings were squared and divided by the share of variance explained by the component.
- The results were then multiplied by the ratio of the variance explained by the component and total variance.
- The derived weights were rescaled to sum up to 100 (to increase comparability).

The three steps were performed for the access, use and skills sub-indices. The respective weights derived from the PCA helped in identifying the relative importance of each indicator.

Sub-indices were computed by summing up the weighted values of the indicators included in the respective subgroup.

ICT use sub-indices were given 40 per cent weight each, and the skills sub-index (because it is based on proxy indicators) 20 per cent weight. The final index value was then computed by summing up the weighted sub-indices.

While modifying the ICT Development Index RF, the following changes were made in the author's interpretation:

- in accordance with the developed methodology of the ICT Development Index, a selection of similar indicators has been made, according to which statistical monitoring is carried out in the context of the subjects of the Russian Federation;
- retaining the principles of logical construction, the indicators for the Skills sub-index block were replaced. It should be noted that in this block not only a meaningful change has been made, but also a structural change. From the position of the authors, the contextual content of the block in the realities of adaptation to Russian conditions will be more fully reflected in the given four indicators.
- structural change in the number of indicators required a change in the weights of the indicators that form the Skills sub-index. Instead of a weight factor of 0.33 points, a new weighting factor of 0.25 points was introduced.

The development of ICT Development Index RF made it possible to determine the regional dynamics and spatial development of the ICT sector in different regions, to build regional profiles on the level of ICT sector development, and to determine the degree of uneven regional development of the ICT sector in Russia.

From 2014 to 2016, the North-West FD (N-WFD) is the leader among the Russian regions in the development of the ICT sector with an index of 77.98 in 2014, and 77.95 pp in 2015, in

2016 - 77.16 percentage points, which is above the national average by 8.1 percentage points. However, in 2017, leadership was transferred to the Central FD (CFD), where the value of the ICT development index was -77.35 percentage points, or 6.19 percentage points higher than the national average.

The change in the Russian regional leader in terms of the development of the ICT sector in Russia was predetermined by a downward trend in ICT development indicators in the North-West Federal District for three years from 2015-2017.

During the period under review, the lowest values of the ICT RF index were observed for the North Caucasian FD (N-CFD). At the same time, the digital backlog for the regions of this region from the leading regions was more than 21 pp, and on average in the Russian Federation by 15 pp. This indicates a significant degree of spatial unevenness in the development of the ICT sector between the Russian regions (Figure 3).

In addition, when building the ICT Development Index RF, it is possible to reflect the "digital profile" of each region of the Russian Federation at the first and second aggregation levels of the composite index. This allows identify the most problematic developments in the ICT sector in each region and the main directions of the point management impact on improving the development of the ICT sector in a particular region. As an example, the "Digital Profile of the Southern FD" is presented in 2014 and 2017 (Figure 4).

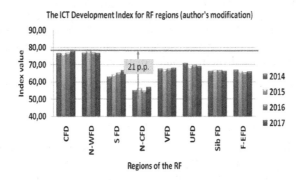

Figure 3.   The ICT development index for RF regions in 2014 - 2017 (author's modification).
Source: compiled by the authors

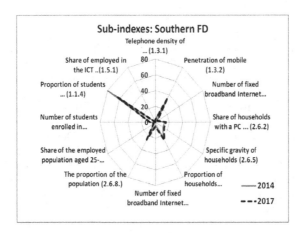

Figure 4.   "Digital Profile" of the Southern FD in 2014 -2017.
Source: compiled by the authors

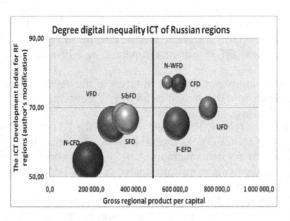

Figure 5.   Degree digital inequality ICT of Russian regions.
Source: compiled by the authors

When comparing the level of ICT Development Index RF and the indicator of gross regional product per capita, it is possible to determine the degree of digital inequality in the regions, and also to determine the threshold investment step for the development of the ICT sector in the amount of 500 thousand rubles GRP per capital (Figure 5)

The proposed modified Development Index RF in the author's interpretation is distinguished by the simplicity and flexibility of the method of construction, the conjugation of the national and regional digital profile, the use of open monitoring data of the digital economy of the Russian Federation, and the possibility of further modification using correlation and regression analysis.

4   CONCLUSION

The digital economy is a source of not only new opportunities, but also new types of inequalities and barriers to the socio-economic development of national economies, such as digital inequality.

The complexity of unambiguous definition of digital inequality lies in the fact that it has many aspects and elements of implementation such as - technological, economic, social, psychological, gender, communicative, behavioral, psychological, legal and geopolitical aspects.

Digital inequality has different levels of implementation for instance; global, national, regional, at the level of individual companies, sectors and industries, social groups and individuals.

Digital inequality reflects the technological, economic, social, educational disparity of economic entities in accessing, obtaining and using information on the basis of modern IT. Digital inequality characterizes the set of "indirect" and mediated parameters and development factors, the combinations of which have non-linear dependence and effectiveness of action, therefore the number of different global and national estimated development ratings increases with the figure of the economy and ICT sector covering various aspects and characteristics of digital inequality.

The article describes the methodology of building ICT Development Index RF in the author's interpretation. The proposed ICT Development Index RF index is distinguished by the simplicity of the method of construction, the conjugation of the national and regional digital profile, the use of open monitoring data of the digital economy of the Russian Federation, and the possibility of further modification using correlation and regression analysis.

For the systematization and comparability of estimates reflecting the uneven spatial development of the digital economy and the ICT sector at various levels as global, national and regional, there is a need to use an open evaluation tool based on inclusive partnerships in the global process of managing the digital economy.

REFERENCES

Aissaoui, N. (2017) Is Inequality Harmful for Broadband Diffusion and Economic Growth?. Asian Economic and Financial Review, vol. 7, issue 8, p.p. 799-808.

Arkhipova, M.Yu., *Sirotin, V.P.* (2018) Determinants of digital development of RF subjects//Statistics in the digital economy: training and use. Materials of the international scientific-practical conference. P. 29–31.

Dimagio, P. Hargittai, E. (2001) From the "Digital Divide" to the "Digital Inequality": a study of the use of the Internet as penetration increases//Working papers from Princeton University No 47.2001.

ITU Internet users by region and country, 2010-2016. - [Electronic resource]. - Access mode: https://www. itu.int/en/ITUD/Statistics/Pages/stat/treemap.aspx (circulation date 03.08. 2018).

Kane, G. (2017) Don't Forget the Basics in Digital Transformation. MIT Sloan Management Review, 2017.- No. 2.- P. 15–23.

Klochkova, E.N. (2017) Toolkit for assessing the development of the information society in the context of globalization: methodical approaches and reasons for differentiation: monograph. - M: Prospectus.

Monitoring the development of the information society RF -[Electronic resource]. - Access mode: http:// www.gks.ru/wps (circulation date 03.08. 2018).

Rating of Russia. ICT Development Index (IDI) Access mode: https://www.itu.int/net4/ITU-D/idi/2017 (circulation date 03.08. 2018)

Russia is online. Report 2016 *The Boston Consulting Group (BCG).* [Electronic resource]: Access mode http://russiaonline.info/(reference date 05.08. 2018)

Selishcheva, T.A. (2015) Digital inequality of Russian regions and problems of sustainable development. International Economic Symposium - 2015 materials of International scientific conferences dedicated to the 75th anniversary of the Faculty of Economics of the St. Petersburg State University.P. 23–34.

Shvab, K. (2016) Chetvertaya promyshlennaya revolutsia.- M.- Esmo.-138p.

Sinyuk, T.Yu. (2016) Polifactorial monitoring of state management of regional socio-economic systems. Abstract of the dis. ... candidate of economic sciences/Ros. acad. b. hoz-va and state. service under the President of the Russian Federation. Rostov-on-Don.

Sinyuk, T.Yu., Panfilova, E.A. (2013) Cognitive boundaries of the formation of the regional policy of the Russian economic space//Russian economic online magazine. № 1. P. 16.

State program (2017) "Digital Economy of the Russian Federation". Order of the Government of the Russian Federation of July 28, 2017 No. 1632-r.

State program (2010) "Information Society (2011-2020)". Government of the Russian Federation Order No. 1815-r dated October 20, 2010.

Valdez-de-Leon, O. (2016) A Digital Maturity Model for Telecommunications Service Providers. Technology Innovation Management Review.- №6.- P. 19–32.

Westerman, G., Bonnet, D. (2014) The Four Levels Of Digital Maturity. MIT Sloan Management Review, 2014.- №8.- P. 32-46.

Zillien, N, Haufs-Brusberg, M (2014) Wissenskluft und Digital Divide. Nomos Verlagsgesellschaft (Baden-Baden) 121 ISBN 978-3-8329-7857-0.

Inclusive Development of Society – Lumban Gaol (eds)

Definition of directivity of trainees' professional activity in supplementary education programs

A.A. Pirumyan
Southern Federal University, Academy of Psychology and Educational Sciences, Rostov-on-Don, Russian Federation

A.N. Algaev
Chechen State University, Grozny, Chechen Republic, Russian Federation

J.A. Markaryan
Don State Technical University, Rostov-on-Don, Russian Federation

P.A. Novozhilov
Ural State University of Economics, Continuing Education Institute, Yekaterinburg, Russian Federation

I.T. Rustamova
Russian University of Transport, Moscow, Russian Federation

ABSTRACT: The particularities of learners' professional directivity in the course of pursuing knowledge in psychology without basic psychological education are discussed in this article. The findings of the research on professional activity directivity of people in the course of obtaining a qualification as a psychologist are presented. The research methods consisted of the authors' questionnaire made up of 16 questions developed by A. A. Pirumyan in collaboration with I. V. Abakumova. The respondents were the listeners of the professional retraining programs at Institute of Psychology, Management and Business of Southern Federal University of Rostov-on-Don. The number of respondents was 61, who were between the ages of 24 and 58. The four trainees' groups characterized by their professional directivity were identified in the course of the examination: (1) the persons focused on career realization (or initiation); (2) the persons focused on further psychological activity; (3) the persons focused on improvement of self-control system; and (4) the persons focused on a poly-optional career.

1 INTRODUCTION

In modern conditions the issue of supplementary professional education is particularly important. Higher education provides the opportunity to gain knowledge, abilities, skills, and qualities necessary for professional and social promotion, personal development, and improvement of all spheres of activities (Shmelyova, 2010.). Changes of economic, social, and technological conditions in modern society lead to people thinking of changing their profession, choosing another directivity of professional activity, and getting further supplementary education and qualification (Bazhenova, 2012). More and more people are dissatisfied with their current career and make a decision to change the direction of their professional activity.

One of the main tools for such changes are courses and programs of professional retraining (A. V. Darinsky, S. I. Zmeev, V. A. Bordovsky, T. G. Brazhe, S. G. Vershlovsky, V. V. Krayevsky, etc.). One of the ways of realizing continuous education throughout life is completing programs of professional retraining.

At the present stage of social and economic transformation in Russia great attention is paid to development of personal general abilities, growth of professional consciousness, and motivation to get supplementary education and a new job. This subject has been studied in detail by E. A. Klimov, N. S. Glukhanyuk, I. V. Dubrovina, D. N. Zabrodin, T. V. Kudryavtsev, Y. K. Strelkov, V. D. Shadrikov, R. V. Gabdreev, etc. Today the research into professional orientation and motives of adults' educational activity is the focus of attention. They turn out to be an important link in the regulation of a new future professional activity (Hrapenko, 2015).

Motivation is the main driving force in personal behavior and activity and also in the course of future expert formation (Chan et al., 2015). Therefore, the issue of trainees' motives is of a special importance (Rean, 2002).

We consider features of professional directivity of adults, their educational motives, and motivation of professional activity during professional retraining in psychology.

2 RESEARCH METHODOLOGY

According to A. V. Kiryakova, professional directivity is considered to be a selective positive attitude to profession and an integrated dynamic property of a personality characterized by dominating extramental attitude to any profession predetermining both preparation for the forthcoming activity, and success of its implementation" (Kiryakova, 2008). The demand for psychological knowledge is in many respects caused by personal needs for self-knowledge, self-development, optimization of both people's own lives and the lives of others and their relationship with them (Khoroshavina & Stymkovsky, 2013) As the scope of tasks solved at a particular age has specific features, different competences may appear at the forefront that require further development, including psychological knowledge and abilities (Klimov, 2007).

There is also a problem of studying features of obtaining a qualification as a psychologist by adults who do not have basic psychological education. We consider driving motives of those who aim to gain knowledge in the field of psychology, and we investigate how their educational motivation corresponds to their further professional development (Dementy, 2009). The answer to this question can be given by carrying out some special psychological research.

This research attempts to reveal features of the trainees' division into several groups based on the directivity of their professional activity. The participants of the professional retraining programs in psychology studying on the basis of the first higher education (not psychological) are considered in the research.

The authors' questionnaire consisting of 16 questions was developed in order to identify the reasons encouraging the respondents to be active in the program of professional retraining, as well as to understand the level of the interest in obtaining a qualification as a psychologist, psychological knowledge, and aspects of further professional activity.

The practical significance of the research lies in the fact that the data collected applying the author's questionnaire make it possible to apply various approaches to the contents and provide the learning materials to people with a different directivity of professional activity. Therefore, the increasing quality of education trainees' satisfaction with training contributes to further progress of students and improves attendance at classes (Gorges & Kandler, 2012). The perennial problem of low educational motivation of professional retraining program trainees may be connected with the fact that they are not satisfied (and even do not understand) those basic requirements that determine the choice of profession and further professional directivity (Trofimova, 2014). Understanding of these basic requirements is an important tool that can be used for optimization of educational processes, raising the trainees' interest and their educational motivation.

The theoretical significance of the study lies in generalization and expansion of ideas about motivating trainees of professional retraining programs in psychology.

When completing the retaining program, adult trainees gain a large volume of new knowledge from unfamiliar spheres and develop professional psychological culture within a short time (Pintrich, 2014). Unlike the situation with gaining the first higher education, trainees studying in the programs of professional retraining have experience of professional activity.

Their choice of the field of retraining is therefore conscious and independent, and they have strong personal motives (Abakumova et al., 2016).

The features of professional self-determination and methods of management of this process, particular qualities of personal transformation in the course of planning a career, and aspects of professional personal self-determination have been considered in the studies of such scientists as E. A. Klimov, S. N. Chistyakova, S. S. Ilyin, G. A. Schröder, L. M. Mitino, and M. A.Baturin. But there has been little research into the issue of psychological features of professional retraining, although it has become increasingly important in the recent years (Perch, 2013).

One of the aspects relevant to this problem is getting insight into educational and professional motivation as well as investigation of the reasons that have brought trainees to the program of professional retraining and the motives inducing them to acquire a new profession (Tusheva, 2011).

It is extremely important to define those features of students' motivation, which may promote the maximum development of the personality in the best way, as the attitude to a profession and motives for its choice are the factors contributing to success of vocational education (Yong-Ming & Yueh-Min, 2015).

3 METHODS OF TESTING AND ANALYSIS OF RESULTS

The objective of this research is definition and study of different directivities of professional activity and also the motivation of people in the course of getting a new qualification as a psychologist.

The subjects of the research were 61 trainees of programs of professional retraining at the Institute of Psychology, Management and Business of Southern Federal University between the ages of 24 and 58, with the average age being 31.2 ± 8.6. Women made up 84% of the respondents and men, 16%.

The topics of the research were the features of professional directivity of persons acquiring a qualification as a psychologist through professional retraining programs.

As a result of the theoretical analysis of the scientific literature on the research problem, we have suggested that persons getting psychological education as an additional qualification may have different directions of professional activity.

To clarify the reasons that induced the respondents to undergo the professional training program as well as to determine the level of interest in the development of the profession of psychologist and the purpose of obtaining psychological knowledge and aspects of professional activity, the authors' questionnaire consisting of 16 questions was worked out by A. A. Pirumyan in co-authorship with I. V. Abakumova.

Questions on the authors' questionnaire were divided into three categories:

1. Personal data
2. Motivation for the professional choice
3. Attitude to educational and professional activity

Processing of personal data was carried out with the help of content analysis. According to the results of the survey the following determining factors were identified:

1. Designation of other professional areas
2. Indication of the professional activity of the psychologist
3. The use of psychological knowledge "for themselves"
4. Lack of a specific practical purpose in getting the education

In accordance with the selected factors in the respondents' responses, keywords and phrases were picked out that allowed noting the presence or absence of any factors and classifying them into the appropriate category.

3.1 Results of the inquiry

Having applied the authors' questionnaire, the four categories of the trainees getting a qualification as a psychologist in the programs of professional retraining have been identified. We highlight the most important questions, the answers to which make it possible to refer respondents to a particular category:

1. Persons focused on promoting their career through receiving or improving their knowledge in the field of psychology. These are people who do not work in the field of psychology but see the purpose of training to obtain and improve psychological knowledge as well as the use of this knowledge for the development of their current career. Thus, the aim of these students in increasing their psychological competence tends to be professional development and professional growth in other spheres.
2. Persons focused on further psychological activity. These people made the decision to change their professional directivity completely and focused on the further use of the obtained knowledge in the direct practice of psychological counseling.
3. Persons focused on improvement of their self-control system. The priority for these people is to solve their own psychological issues. These people use the professional retraining program tend first and foremost to raise awareness of the ways, techniques, and tools of self-knowledge and the organization of more effective interaction with other people.
4. Persons focused on a poly-optional career. These are people who get professional psychological education among other forms of improving their own education. They are not completely sure of the choice of any profession and therefore look for various options. Thus, such students "take up" educational diplomas, but they are unlikely to apply the acquired knowledge and to promote their further career as psychologists.

In Figure 1 the percentage ratio of different sample groups of the respondents is shown: the first group, notionally called "career realization," made up 16%; the second group, with the conditional name "mastering the profession of psychologist," made up 54%; the third group, conditionally called "self-cognition," made up 18%; and the fourth group, "poly-optional career"), 12%.

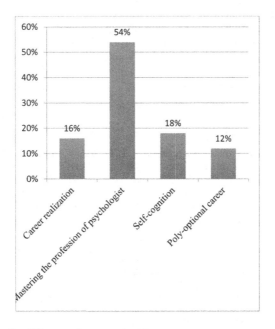

Figure 1.   Percentage ratio of the sample respondents' groups.

4 CONCLUSIONS

Taking into account the results of the research, it is possible to draw the following conclusions.

1. In the current context of extreme and radical changes we have observed job displacement on a vast scale. Hence there has been an increasing demand for new knowledge to create new jobs. There is no doubt that in conditions of constant uncertainty of life any forms of further professional education will become more popular.
2. Receiving further psychological education has essential differences from the other types of education. First, it implies personal requirements of trainees to develop their own personalities, and also to develop special psychological consciousness and form professional psychological thinking.
   An important feature is also a large variety of activities of a professional psychologist such as consulting, psychotherapeutic, diagnostic, educational, and applied spheres. Besides, unlike all other types of professional retraining, a considerable number of people come to educational programs to understand themselves and to solve their own life problems. That means psychological education is both a way of developing a profession and the opportunity to solve personal problems or to re-comprehend their life by raising their self-awareness.
3. The major elements in completing the programs of professional retraining are professional directivity as integration of conscious and unconscious ideas about the new profession; specific features of its development; and the motives inducing the person to be trained.
4. The empirical research revealed four groups of trainees of further education who differ in their professional directivity: (1) persons focused on upgrading (or initiating) a career; (2) persons focused on further psychological activity; (3) persons focused on improvement of systems of self-control; and (4) persons focused on a poly-optional career.

The data obtained in the study can be applied both to trainees of the programs of professional retraining for raising the trainees' awareness of the training process and solving problems of low motivation, and to the professors and managers of professional retraining programs as well as administration of higher education institutions on the basis of which these programs are organized.

The identification of several distinct groups of trainees offers the opportunity to provide individual psychological consulting with the purpose of raising trainees' awareness of their motives, professional directivity, and for training improvement.

The insight into heterogeneity of the trainees and their various professional directivities directed at differentiation of the training process is able to provide an essential gain in satisfaction with training and the growth of educational motivation promoting more harmonious and quicker development of a new profession.

REFERENCES

Abakumova, I. V., Ermakov, P. N., & Kagermazova, L.T. 2016. *Technologies of the Directed Broadcast of Meanings in Practice of Educational Process.* Moscow: CREDO.

Bazhenova, V. S. 2012.System of additional professional education. *The Higher Education in Russia,* 6: 83–90.

Chan, D. K. C., Yang, S.X., Hamamura, T., et al. 2015. *Motivation and Emotion,* 39(6): 908–925.

Dementy, L. I. 2009. Problems of professional identity and a marginality in a situation of conscious change of a profession (on the example of receiving the second higher psychological higher education). *Izvestiya ALTGU,* 2: 48–53.

Gorges, J., Kandler, C. 2012. Adults' learning motivation: Expectancy of success, value, and the role of affective memories. *Learning and Individual Differences,* 22(5): 610–617.

Hrapenko, I. B. 2015.Formation of professional thinking of the psychologist in the conditions of continuous multilevel preparation. *KPZh,* 5-2: 373–378.

Khoroshavina, G. D., & Stymkovsky, V. I. 2013. Implementation of additional professional programs as condition of obtaining new professional competences by students. *Social and Economic Phenomena and Processes,* 12(58): 257–259.

Kiryakova, A. V. 2008. *Aksiologiya of Education: Basic Researches in Pedagogics*. Moscow: House of Pedagogics, IPK Regional Public Institution Public Educational Institution.

Klimov, E. A. 2007. *A Psychologist: Introduction to a Profession*. Moscow: Academy.

Pintrich, P. R. 2015. A conceptual framework for assessing motivation and self-regulated learning in college students. *Educational Psychology Review*, 16(4): 385–407.

Rean, A. A. 2002. *Psikhologiya i pedagogica*. St. Petersburg.

Shmelyova, I. A. 2010. Introduction to a Profession. In Psychology. St. Petersburg.

Trofimova, L. N. 2014. A problem of motivation of educational activity at students of professional retraining. *Modern Problems of Science and Education*, 6:701.

Tusheva, E. S. 2011. The strategic and significant directions of a research of professional retraining of experts in the field of correctional pedagogics. *Magister Dixit*, 3: 25.

Yong-Ming, H., & Yueh-Min, H. 2015. A scaffolding strategy to develop handheld sensor-based vocabulary games for improving students' learning motivation and performance. *Educational Technology Researchand Development*, 63(5): 691–708.

Inclusive Development of Society – Lumban Gaol (eds)
© 2020 Taylor & Francis Group, London, ISBN 978-1-138-33476-2

Streamlining the use of hotel business financing sources

S.G. Kilinkarova
Department of Economics, Management and Finance, Pyatigorsk State University, Pyatigorsk, Russia

R.V. Shkhagoshev
Department of State and Municipal Management, South-Russian Institute of Management, Branch of the Russian Academy of National Economy and Public Administration under the President of the Russian Federation, Rostov-on-Don, Russia

T.G. Pogorelova
Accounting and Audit, Southern Federal University, Rostov-on-Don, Russia

A.U. Polenova
English Language Department for Humanities, Southern Federal University, Rostov-on-Don, Russia

A.O. Kuzmina
Management, State and Municipal Management, K. G. Razumovsky Moscow State University of Technologies and Management (the First Cossack University), Moscow, Russia

ABSTRACT: Any change to the structure of the backbone of hotel business—its capital—has to be supervised skillfully, as it permeates all results of activity, from return on assets up to solvency margin. It is the efficient allocation of financing sources that can ensure a business its independence and good standing, which is the guarantee of sustainable development and full-fledged functioning in conditions of an increasing number of bankruptcies. This article considers trends, scopes, and sources of financing for a certain standard subject of hotel business within one country. Changes in the scope and structure of financing are analyzed. The authors have determined that hotel business organizations do not perform record-keeping except for tax purposes; financial accounting and reporting are done by them only to the extent that is required for calculating and reporting taxes, there being no grounds for saying if they are valid. Moreover, it has been found that private hotels do not use long-term crediting facilities as financing sources for their business. In this article, the main stages of developing the formation policy for a hotel's own financial resources are suggested, as are the criteria of the process of streamlining the internal and external sources for the hotel businesses under study to form their own financial resources.

Keywords: bank loans, financing sources, financing structure, hotel business

1 INTRODUCTION

In recent years, the Russian Federation (RF) has hosted major world events and musical, sports, scientific, and economic competitions, with more of them planned for the future. The Universiade in Kazan, the Olympic Games in Sochi, the Eurovision Song Contest in Moscow, the Economic forum in Vladivostok, alongside quite a few other events have boosted the development of the hotel industry in the RF. The 2018 FIFA World Cup has enabled an entire number of Russia's cities to renovate their historical districts and build new traffic. Tourist routes are launched and hotels and mini-hotels attract customers. The country's largest megalopolises move into a higher gear of hospitality, which certainly improves the rating of these cities in the eyes of both Russian tourists and foreign guests.

In the RF, the main hotel business development problems are the following:

- Low level of service, compared with prestigious world resorts, associated with underestimation of qualified personnel by the employers, negligence of service, weak integration of modern technologies, and other factors
- The prolonged financial crisis compromising any sphere of economic activity and reducing citizens' purchasing power
- The investments into tourism industry being focused in resorts of the Krasnodar Territory, the Crimea, Moscow, Kazan, and Saint Petersburg
- The lack of an approved and high-quality certification and licensing system for tourism services
- Quite a high overall level of corruption in the country
- Shortage of personnel
- Not quite favorable conditions for attracting additional sources of investment into facilities and resources
- Underdeveloped mechanisms for facilitating a higher workload for hotels at the state level and the lack of advertising Russian tourism products abroad

The purpose of this study was to provide the hotel business entities with financial independence and sustainability based on the effective distribution of funding sources.

The objectives of the study are the following: to identify the essence of financing sources for the subjects of the hotel business; to analyze their role in commercial activities; and to characterize the management of own and borrowed capital by the current subject of the hotel business and suggest ways to optimize financing sources.

Solving these problems will allow creating conditions for

- Ensuring easy credit terms for hotel business subjects
- A favorable tax regime for investors
- Taking into account the industry-specific features of the way composition of costs for services is formed
- Broadening the scope of the financing sources the hotel business attracts.

Nevertheless, in spite of the aforementioned factors, in the RF, the tourism industry is one of the priorities and the most promising one. Russia features immense capacities, including in its attractiveness for guests from other countries owing to the richness of its nature, diversity of climatic areas, and great history and traditions.

2 THE NOTION, ESSENCE, AND CHARACTERISTICS OF HOTEL BUSINESS FINANCING SOURCES

In choosing the sources of financing the activity of hotel business organizations, five principal problems have to be solved:

- Identifying the needs of short- and long-term capital
- Finding out possible changes in the composition of assets and capital in order to determine their optimum composition and structure
- Ensuring continuous solvency and therefore good standing
- Using the own and loan funds at the best value
- Reducing the costs for financing the business activity

The structure of financial resources can affect market capitalization of hotels. The generally accepted economic theory does not give a significant role to this or to the proportion of the own and the loan capital (Kalinina, 2015), yet the actual data regarding the capital structure enable the hotel business owners to tailor such a proportion of sources under which an enterprise could maintain stable growth rates and confidently develop among its competitors and, which is of no less importance, in conditions of market instability.

Usually, the structure of financial resources of a hotel encompasses several sources of financing simultaneously. In terms of the resource streamlining strategy, this is indicative of a conscious and well-considered financial policy of the hotel business. Meanwhile, from the standpoint of the current financial flows, the situation may both be planned and quite frequently spontaneous. This can potentially cause worse financial results of activity of the hotel, lower return on the own capital, and a number of other negative factors.

The sources of formation of an organization's own financial resources are subdivided into external and internal ones (Bhattacharya &).

Among the internal sources, the undistributed profits of the hotel, depreciation charges on the own fixed assets being used, and other internal financial resources formation sources can be named.

External sources refer to attracting additional equity or stock capital and the enterprise's receiving nonrepayable financial aid.

Own capital features a number of absolute advantages, yet it also has some disadvantages, as shown in Table 1.

Thus, an enterprise that uses internal sources of financing investment has the highest financial sustainability, but thereby limits the pace of its development.

On the other hand, there is loan capital—the one that is provided for the hotel business by third parties. It incorporates both long-term and short-term funds.

Just like the own sources, the loan ones have a number of positive and negative aspects, as shown in Table 2.

The disadvantage of attracting borrowed sources of financing is the complexity and duration of the procedure for raising funds; the need to provide guarantees of financial sustainability; increased risk of insolvency and bankruptcy; decrease in profits due to the need to

Table 1. Advantages and disadvantages of own financing sources.

Advantages	Disadvantages
1. It is easy to attract: no permission of another subject is required. 2. It contributes to the best profit- generating effect, as there are no expenses for the loan interest. 3. It ensures financial stability of the enterprise, guaranteeing its solvency.	1. Attracting this resource is somewhat limited. 2. The resource is a high-cost one. 3. There is no growth of profitability ratios of the own capital that is provided at the expense of loan funds.

Source: Compiled by the authors.

Table 2. Advantages and disadvantages of loan financing sources.

Advantages	Disadvantages
1. Vast opportunities for attracting financing 2. Growth of potential of the enterprise 3. Increase of the profitability indicator of the own capital 4. The low cost of resources as compared to the own ones	1. A high risk of the good standing getting destabilized and the enterprise losing solvency 2. Decrease of the generated profits indicator due to paying the loan interest or lease rate 3. A strong correlation between the cost of loan funds and the market condition 4. A quite complicated procedure of attracting that requires additional expenses, such as guarantee.

Source: Compiled by the authors.

repay on borrowed sources; and the possibility of losing ownership and management of the company.

Currently, in spite of the challenging situation in the economy, the crediting processes keep gaining momentum (Prohorovs &, 2015). In particular, during the post-crisis period, hotel business was hard pressed to take on loans for funds added to current assets in order to have the opportunity to continue functioning in the market.

Under a market economy, most frequently, the hotels use the five-element system of the own financing:

- Self-financing
- The opportunity to attract resources via the capital market
- Bank loans
- Financing at the expense of the budget
- Mutual financing of organizations

Regardless of the size of a hotel business, it needs sources of funds in order to successfully finance its activity—both for the current and prospective transactions. Financial results of activity are directly associated with the structure of sources of financing, the speed of attracting them, and the efficiency of using them.

3 ANALYSIS OF THE ACTIVITY FINANCING SOURCES OF A HOTEL BUSINESS ORGANIZATION

The organization under study is a micro-enterprise, just like the numerous similar businesses of the tourism industry. It was registered in 2006 with the inspectorate of the Federal Tax Service for Rostov Region.

The authorized capital of the hotel amounts to 10,000 rubles, which is the compulsory minimum required by the law, and its sole founder is an individual.

In 2016, an income of 6.09 million rubles was obtained that was determined by a cash method according to the Russian accounting standards; this is 33.12% higher than the indicator for the similar period of time of the previous year. The revenue of the hotel increased by 32.15% in 2016—up to 6.23 million rubles. This is confirmed by the company records. In 2016, the sales revenue of the private hotel increased by 32.15% and was 6.23 million rubles as compared to 4.72 million rubles a year earlier. According to the company records, its sales revenue grew up to 5.32 million rubles in 2015. The company's sales volume increased by 16.25% up to 5.48 million rubles in 2015.

The main financial indicators of commercial activity of the hotel in 2015–2017 are presented in Table 3.

The data of Table 3 show that in 2017 the sales revenue of the hotel under analysis decreased as compared to the previous period by 1.029 million rubles or 16.51%. However, the size of tax indicator equalized to the profit tax increased by 9.86%, which amounted to 14,000 rubles in absolute terms, making 156,000 rubles.

In 2016, the organization saw a considerable growth of sales revenue—by 1.516 million rubles—against a slight increase of the uniform tax under the simplified taxation system, by

Table 3. Economic indicators of the 2015–2017 dynamics of the hotel under study.

Indicator	2015	2016	Deviation 2015–2016 (+, -)	2017	Deviation 2016–2017 (+, -)	2016-2017 growth rate, %
Revenues on sales, million rubles	4.715	6.231	1.516	5.202	−1.029	−16.51
Tax according to the simplified taxation system, thousand rubles	141	142	1	156	14	9.86

1,000 rubles. However, in 2017, the sales revenue indicator fell considerably, with the company's tax burden getting higher at the same time.

The organization under analysis has objects of fixed assets on the balance sheet that are listed in Table 4.

According to Table 4, the depreciation accrued on fixed assets of the hotel under analysis amounts to more than 60% of the initial fixed assets value as of the beginning of 2018, bearing in mind that the land plots value is not subject to wear.

It has to be pointed out as well that the land plots were not revaluated and they can cost much more in actual prices, but in this case the taxable base will increase too.

Thus, it can be concluded that the hotel under study, just like the prevailing part of similar hotel business subjects, performs record-keeping for tax purposes only, calculating the unified tax according to the simplified taxation system and paying the land tax, property tax, transport tax, compulsory insurance contributions, and personal income tax as a tax agent. Financial accounting and reporting are done only to the extent that is required for calculating and reporting taxes, there being no grounds for saying if they are valid.

4 MANAGEMENT OF THE OWN AND LOAN CAPITAL BY THE HOTEL BUSINESS SUBJECT

The structure of capital of the hotel under study is not optimum. From year to year, the own capital occupies a minor share in the financing sources structure while short-term liabilities prevail and take up more than a half of the total of the organization's capital.

The composition of the organization's property formation sources does not feature broad indicators; it is given in Table 5.

The data of Table 5 show that the total quantity of property formation sources of the hotel is smaller in 2016 as compared to 2015 to a greater extent due to decrease of undistributed profits and also due to the essential repayment of short-term loan liabilities, which brought about the change of the property formation sources structure.

So, while at the beginning of 2016 the organization had the undistributed profits indicator prevailing (67.2%), by 2017 the situation changed drastically and it is accounts payable (57.66%) that started to occupy the principal percentage of property formation sources, having grown several scores of times and amounted to 3.737 million rubles. With regard to

Table 4. Composition of the fixed assets of the hotel under analysis as of January 1, 2018.

Name	Initial value, rubles	Depreciation accrued, rubles	Depreciated book value, rubles
Building of the hotel	2,840,000	1,798,666.67	1,041,333.33
Hangar	255,000	205,425.81	49,574.19
Block container 3*2,0*2,5	94,439.95	58,237.97	36,201.98
Eastern fence	23,301.46	10,874.01	12,427.45
Land plot of 1,803 sq. m.	570,154		570,154
Land plot of 5,543 sq. m.	185,510.97		185,510.97
Car	460,000	337,333.33	122,666.67
Heating system	140,000	95,666.67	44,333.33
Metal fencing	305,000	249,083.33	55,916.67
Letter V dismountable hangar	733,950	721,717.5	12,232.5
Letter G dismountable hangar	733,950	709,485	24,465
Letter D dismountable hangar	2,857,341	1,333,425.8	1,523,915.2
Total	9,198,647.38	5,519,916.09	3,678,731.29

Table 5.   Composition and structure of property formation sources of the hotel under study in 2015–2017.

Indicators	2015 Amount, thousand rubles	% of the total	2016 Amount, thousand rubles	% of the total	2017 Amount, thousand rubles	% of the total
Authorized capital	10	0.15	10	0.15	10	0.11
Undistributed profits	4,610	67.2	2,052	31.66	2,936	33.73
Short-term loan funds	2,100	30.61	682	10.52	0	0
Accounts payable	140	2.04	3,737	57.66	5,759	66.16
Total	6,860	100	6,481	100	8,705	100

Source: Compiled by the authors.

this, the undistributed profits indicator fell more than twofold, making 2.052 million rubles and accounting for only 31.66% in the structure of property formation sources.

In 2017, the trend having begun in the previous year continued: the indicator of accounts payable of the hotel keeps increasing, prevailing in the structure of property formation sources (66.16% of the total amount of the sources) and reaching 5.759 million rubles in absolute value. The share of the undistributed profits indicator of the hotel under analysis increased, making up 33.73%, which is 2.936 million rubles in absolute value.

The authorized capital indicator of this hotel business subject remained unchanged in absolute value, 10,000 rubles, throughout the analyzed period of time and occupied a minor percentage in the overall structure of property formation sources— 2015–2016— 0.15%, and as of the beginning of 2018 its share amounted to 0.11% due to the growth of other sources.

For three years, the hotel business studied had consistently been reducing its dependence on bank crediting, and by the beginning of 2018, the short-term loan funds had been repaid in full. Meanwhile, as of the beginning of 2016, this indicator accounted for 30.61% of the total quantity of the sources of financing.

In general, the total of property formation sources of the hotel was increasing. As for favorable aspects, the increase of the own capital in 2017 that took place due to the growth of undistributed profits can be considered one.

A fact that can be regarded as a negative one is the sharp growth of the accounts payable balance in relation to currency both in dynamics and in structure. This leads to higher risks of activity and stepped-up financial imbalance (Frolova & Pogorelova, 2017).

The distinctive feature of financing sources of the hotel under study is the fact that in the period analyzed it did not attract any long-term liabilities whatsoever.

The lack of formation of both added and reserve capital is observed. In organizations where creation of reserve capital is not regulated, it is not created at all. The hotel under study does not reevaluate its fixed assets, which explains the absence of added capital.

Table 6 shows the composition and structure of the sources of financing of the hotel business under study.

The main factor of increase of the own capital of the hotel being analyzed was the growth of accrued depreciation charges and undistributed profits. The resulting behavior of the own capital elements has remained the same for 2015–2017.

Based on the calculations performed, it can be concluded that as of the beginning of 2018, the greatest part in the financing sources structure of the hotel belongs to the loan source represented by accounts payable that increased several scores of times within the period of time

Table 6. Composition and structure of financing sources of the hotel under study in 2015–2017.

Indicators	2015 Amount, thousand rubles	% of the total	2016 Amount, thousand rubles	% of the total	2017 Amount, thousand rubles	% of the total
Authorized capital	10	0.09	10	0.09	10	0.07
Undistributed profits	4,610	42.67	2,052	18.3	2,936	20.64
Depreciation	3,943	36.5	4,733	42.21	5,520	38.8
Short-term loan funds	2,100	19.44	682	6.08	0	0
Accounts payable	140	1.3	3,737	33.32	5,759	40.49
Total	10,803	100	11,214	100	14,225	100

Source: Compiled by the authors.

from 2015 through 2017. A minor role in the financing sources formation is played by the authorized capital that retains its value throughout the analyzed time span.

The hotel under study should expand its production capacities in order to ensure efficient profits distribution and achieve financial balance at the expense of the internal sources. As a rule, reduction of scope of the own capital is a consequence of inefficient, unprofitable activity of the enterprise.

Summing up the foregoing, it can be stated that currently the financial provision of the hotel under consideration is in an unsatisfactory condition; the own funds are not sufficient for providing even the non-current assets, long-term loan sources of financing are not used, the authorized capital is actually nominal only, with the added and reserve capital lacking completely. In order to develop successfully, the enterprise has to streamline its sources of financing.

5 DIRECTIONS FOR STREAMLINING THE SOURCES OF FINANCING FOR THE HOTEL BUSINESS UNDER STUDY

For the hotel business under study, the process of optimization of the internal and external sources of formation of the own financial resources should be based on the following criteria:

• The minimum total cost of attracting the own financial resources has to be ensured.
• The initial founders of the enterprise have to keep control of the enterprise management.

Management of the own capital of hotels has to involve determining the optimum proportion of the own and loan financial resources too.

Developing the financing sources at the expense of the own resources reduces only some financial risks in business, yet meanwhile it cuts down the speed of business size increment considerably, the revenue first of all. On the contrary, given the correct financial strategy and high-quality financial management, attracting additional loan capital can boost the hotel business owners' incomes on their invested capital. The reason for this consists in the fact that the increase of financial resources leads to the pro rata increase in sales volume and frequently that of the net profit, provided the management is skillful. This is especially relevant for smaller and medium-sized companies.

The optimum level of corporate giving (CG) of a hotel company is positively related to the general market demand and the competitive advantage of CG and is negatively related to the induced cost of services rendering practice. Moreover, a positive or neutral relationship between CG and a hospitality firm performance depend on the ability of CG to create a competitive advantage of differentiation of the brand and customers' loyalty for increasing the profits (Chen et al., 2017).

However, the overloaded with loan funds structure of capital puts excessively high requirements for its profitability, as the probability of nonpayments increases and risks for investors go up. On top of that, having noticed the high percentage of loan funds, customers and suppliers of the company may start looking for more reliable partners, which will result in decline of the revenue. On the other hand, a too small share of loan capital means making insufficient use of a financing source that is potentially cheaper than the own capital (Orlova & Petrushina, 2017). Such a structure leads to higher expenses for capital and excessive requirements for profitability of the future investments. The optimum proportion of financing sources represents such a combination of the own and loan sources under which the optimum balance of levels is ensured, i.e., the market value of the enterprise is maximum. When streamlining the capital, each part of it has to be taken into account. (Lutsenko, 2016).

Based on these particularities and the financial analysis conducted, it can be concluded that if the management of the hotel uses the loan capital, they will obtain a higher potential and an opportunity of increment of return on the own capital, but the company's financial stability will be compromised. However, in case the management decides to use the own capital as the source of financing, conversely, good standing of the enterprise will be the highest but the opportunities of the profits growth will be limited (Frolovaet al.).

As a result of the research conducted, it has been found that the financing sources mechanism of hotel business subjects has to be improved, with two areas of focus singled out within it:

- Recommendations on improving the organization's self-financing strategy
- Measures for improving the organization's loan financing.

For the purposes of improving the policy of self-financing, the enterprise should look for ways to increase the net profits and for this indicator to achieve 40% within the financing sources structure (as of the beginning of 2018, it was 20.64% in the hotel under study). In order to gear up profits, the quality control of the services rendered has to be enhanced, with special attention to be paid to advertisement and presentation of new products. A relevant system of profits and losses accounting has to be implemented for pinpointing the internal resources of sources of profits. (Karapetyants et al., 2017).

In order to manage accounts receivable, the enterprise should conduct an active policy of recovering the accounts receivable and shape and group the information about debtors, payment deadlines, any overdue accounts receivable, as well as about certain debtors the delay of settlements with which results in the current solvency issues of the hotel. Keeping a calendar schedule of proceeds on receivables is a good idea for the convenient analysis of accounts receivable.

For managing accounts payable, active measures for reducing them should be taken. Just like in the accounts receivable monitoring, the hotels should keep a payment calendar schedule for their accounts payable. Given the mutual agreement of the parties, the existing accounts payable could be paid off by counter obligations obtained on the basis of rendering tourism industry services.

6 CONCLUSION

1. Among the internal sources of formation of a hotel's own financial resources, the leading place is occupied by the profits remaining at the disposal of the hotel; it shapes the greater part of the own financial resources. Depreciation allowances play a certain role among the internal sources too, although they do not increase the amount of the hotel's own capital. Other internal sources do not have a noticeable part to play in formation of the own financial resources.

2. Among the external sources for forming the hotel's own financial resources, the leading place belongs to attracting additional equity or stock capital. One of the external sources of formation of the own financial resources can be the nonrepayable financial aid provided to the hotels. Commercial loans, options, pledge transactions, factoring transactions, leasing, etc. are the instruments of financing the hotel activity also.

3. Currently, hotel business financing is in a unsatisfactory condition because of a shortage of the own funds for self-financing, the lack of sufficient state financial support, a high cost and risk-taking of innovations, the long-term nature of payback in innovation projects, and conservative investors prevailing instead of the aggressive ones. For further successful development, Russian hotel businesses under study have to complete two tasks: the first one is to streamline sources of financing for development of new projects, and the second one is to learn how to select such innovation projects as to yield actual rewards even under crisis conditions.

4. The analysis of an existing hotel has shown that financial accounting and reporting are done only to the extent that is required for calculating and reporting taxes, there being no grounds for saying if they are valid. As of the beginning of 2018, the greatest percentage in the hotel financing sources structure is accounted for by the loan source represented by accounts payable. Within the period of time from 2015 up to 2017, it has increased scores of times. The authorized capital plays a minor part in forming the sources of financing, keeping its importance throughout the entire period of time under analysis.

5. The negative fact is the sharp growth of the accounts payable balance in relation to currency, both in dynamics and in structure. This leads to higher risks of activity and stepped-up financial imbalance. For most hotels, the distinctive feature of financing sources is the absence of long-term liabilities.

6. For the purpose of streamlining the hotels' financing structure, it is first and foremost recommended to
   - Look for ways of enhancing the net profits in order to achieve 40% in the structure of financing sources in this indicator
   - Pursue an active policy of recovering the accounts receivable, shape and group the information about debtors, payment deadlines, any overdue accounts receivable, as well as about certain debtors the delay of settlements with which results in the enterprise's current solvency issues
   - Take active measures for reducing the short-term accounts payable, etc.

7. The research conducted allows wording the conclusion that the hotel managers have to consider all feasible options of obtaining long-term loans for improvement of their business.

REFERENCES

Bhattacharya, S., & Londhe, B. R. Dr. 2014. Micro entrepreneurship: Sources of finance & related constraints. *Economics and Finance*, 11:775–783.

Chen, Ming-Hsiang, Lin, Chien-Pang, Li Tian, & Yang,. 2017. A theoretical link between corporate giving and hospitality firm performance. International Journal of Hospitality Management, 66 (September): 130–134.

Frolova, I. V., Matytsyna, T. V., Pogorelova, T. G., & Likhatskaya, E. A. 2017. Harmonization of the tax portfolio of an organization by means of situational matrix modeling.In *Managing Service, Education and Knowledge Management in the Knowledge Economic Era. Proceedings of the Annual International Conference on Management and Technology in Knowledge, Service, Tourism and Hospitality (SERVE 2016), 4th*, pp. 69–74.

Frolova, I. V., Matytsyna, T. V., Pogorelova, T. G., Polenova, A. U., & Lebedeva, N. U. 2018. Improving the efficiency of hotel business through the use of tax alternatives. In *Financial and Economic Tools Used in the World Hospitality Industry: Proceedings of the 5th International Conference on Management and Technology in Knowledge,* Service, Tourism & Hospitality 2017 (SERVE 2017).

Frolova, I. V., & Pogorelova, T. G. 2017. Financial analysis of the business entity as a factor of economic growth. In I. V. Shevchenko (ed.), *Economic Development of Russia: Traps, Forks and Rethinking of Growth. Proceedings of the International Scientific and Practical Conference*, pp. 231–234.

Kalinina, D. 2015. Money is not enough. There is an option not to borrow from the bank. *Financial Director*, 1: 11–13.

Karapetyants,I., Kostuhin, Y., Tolstykh, T., Shkarupeta, E., & Krasnikova, A. 2017. *Establishment of Research Competencies in the Context of Russian* Digitalization. *Proceedings of the 30th International Business Information Management Association Conference (IBIMA)*, November 8–9, 2017 Madrid, Spain, pp. 845–854.

Lutsenko, S. V. 2016. The dilemma of choosing a source of finance. *Society and Economics*,9:22–32.

Orlova, T. S., & Pervushina, V. Yu. 2017. Formation of the capital structure of the enterprise. *Problems of Humanitarian and Socio-economic Sciences*, 6(11): 42–49.

Prohorovs, A., & Beizitere, I. 2015. Trends, sources and amounts of financing for micro-enterprises in Latvia. *Social and Behavioral Sciences*,: 404–410.

Inclusive Development of Society – Lumban Gaol (eds)
© 2020 Taylor & Francis Group, London, ISBN 978-1-138-33476-2

Budget instruments for smoothing socioeconomic polarization in the development of the subsidized region

M.V. Alikaeva, L.O. Aslanova, M.B. Ksanaeva & L.S. Chechenova
Kabardino-Balkarian State University named after H. M. Berbekov, Nalchik, Russian Federation

D.A. Karashaeva
Ministry of Finance of Kabardino-Balkarian Republic, Nalchik, Russian Federation

ABSTRACT: This article discusses the tools of smoothing the socioeconomic` polarization in the development of the subsidized region. This issue acquires an urgency because overcoming territorial disparities in the development of the regions is one of the key ways of a breakthrough development of the economy of the region as a whole. That actualizes the need for detailed consideration of the problem of the polarization of territories in the development of regions and the most accessible instruments to influence them. The main goal of the article is to investigate the relationship between the influence of budget management tools and the evening-out of polarization using the example of one of the regions within the North Caucasian Federal District (NCFD): the Kabardino-Balkarian Republic (KBR). Such a research focus led to the formulation and solution of the following tasks: to justify a system of indicators that allow determining the level of socioeconomic polarization; to assess the state of the economy of municipalities of the KBR in terms of the level of socioeconomic polarization; to study the indicators of the polarization gap in dynamics and in aggregate for the municipalities of the KBR; and to reveal the general tendencies and regularities characterizing the polarization situation for the Republic as a whole for the period from 2011 to 2017.

The use of the following research methods in social and economic phenomena—abstract-logical; statistical; expert assessments; the method of graphical interpretation; absolute, relative, and average values; and graphical and tabular analysis—allows the authors to provide reasoned estimates and conclusions. The proposed indicators of the polarization disruption can be used both for assessing the level of development and identifying municipalities and constituent entities of the Russian Federation that claim to be leaders, which in the future can become a pole of growth and have the potential to positively influence the breakthrough development of the national economy.

Keywords: subsidized region, subventions, subsidies, polarization, disruption of polarization, municipalities, budget expenditures, gratuitous receipts, shipment of goods, work and services, average monthly nominal salary of employees of organizations, asymmetry, growth pole.

1 INTRODUCTION

problems of uneven development of regions and methods of impact have been repeatedly touched upon by many Russian and foreign scientists. An essential role in the formation of this study was provided by the works of I. Wallerstein, J. Friedman, P. James, and F. Perru. In the Russian scientific school, polarization and economic spatial development are described in detail in the works of B. L. Lavrovsky, P. Pozdnyakova, and E. A. Shiltsin. Research of A. A. Zinoviev and S. N. Saaya discloses statistical and organizational mechanisms for smoothing polarization in regional development. The smoothing of polarization in the aspect of subsidized regions was previously highlighted in the research and publications of M. V. Alikaeva and D. A. Karashaeva. The significance of the share of subsidized regions in the economy of Russia (Lavrovsky & Shiltsin, 2016), including the regions of the North Caucasian Federal District (NCFD) that occupy a prominent place, is important, which in itself is an indicator of certain problems and disproportionality in the economic development for such regions. There are a number of features inherent in all these problematic regions: increased seismicity characteristic for these regions of the crossed and mountainous terrain (James & Martin, 1988); the budget sector remains a significant contributor (public administration, education, and health care form up to 40–50% of gross regional product [GRP]) in the structure of the economy of most subsidized regions; strengthening of public administration and construction in the real sector over the past decade over the insufficient private capital activity; and the continuing low level of economic and tax potentials over the relative increase in GRP and the considerable financial assistance of the center (Milchakov, 2017).

All these factors lead to the functioning of the region's economy in the "center-periphery" format (Wallerstein, 2011). In this regard, the issue of mobilizing own resources and increasing the use of domestic sources (Hirshman, 1966) for the development of subsidized regions becomes even more important. Overcoming territorial disproportions in the development of regions is one of the key ways to overcome many crisis moments in the economy of the region as a whole (Pozdnyakova et al., 2016). This necessitates a detailed consideration of the problem of the polarization of territories in the development of regions and the most accessible instruments of influence on them.

2 RESEARCH METHODOLOGY

In accordance with the system approach, a model was developed for diagnosing the effect of budgetary mechanisms on indicators, characterizing the polarization in the region. The importance of budgetary mechanisms in the development of a region in terms of smoothing out polarization processes is most illustrative, if we consider the most direct, directly affecting indicators. Such indicators are the following:

1. The indicator "expenditures carried out by the budgets of municipal districts" is formed as all MO expenditures from all types of revenues (including taxes, fines, fees, other payments going to the district budget, income from the sale, lease of property, municipal transfers, interbudgetary transfers, and other receipts). It reflects the financial viability of the district as an administrative unit, while at the same time characterizing, to some extent, along with autonomy, the inclusion in the system of government authorities of the management with all the attendant levers of influence within the region as a system of interacting co-subordinate elements (Alikayeva et al., 2018).
2. Secondary with respect to the first indicator, which follows from the economic point of view, is "gratuitous receipts from a higher budget to the budgets of districts." It reflects interbudgetary transfers in the form of subsidies, grants, subventions, and other gratuitous payments from the higher budget (the republican budget, the budget of the subject). This indicator more clearly characterizes the degree of participation of the centralized administration of the region in financial policy and, in general, in the economic life of the district (Alikayeva & Karashayeva, 2012).

However, as practice shows, both so-called "budget" indicators need an amendment to the universal weight. Therefore, in calculations, these figures are taken into account with the distribution on the population. The results of the impact of the chosen factors will clearly illustrate the welfare indicators calculated for a certain territory. Statistical indicators are representative enough for these purposes:

1. "Shipment of own production goods, performed works and services (GWS) on its own," thousand rubles (calculated by statistical agencies for municipal entities), as an analogue of gross domestic product (GDP, which is calculated for the region as a whole), but with an amendment to the population (Lavrovsky & Shiltsin, 2016)

2. "Average monthly nominal salary of organizations' employees"; the value is also calculated by statistical bodies according to data officially submitted by organizations.

The following indicators make it possible to assess the result of the influence of the budgetary factor on the polarization level: (1) *polarization gap*, defined as the ratio of the indicator of a particular municipal formation to the average value of the indicator in the republic; (2) *the magnitude of polarization*, defined as the ratio of the maximum value of the indicator of the municipal formation to the minimum value of the indicator of municipal formation (Karashayeva, 2013).

The magnitude of the polarization discontinuity is expressed in fractional terms and allows us to classify the region as "crisis" (values less than 0.5), "weak" (from 0.5 to 1), "safe" (from 1 to 1.5), and "strong" (over 1.5) (Saaya, 2008).

It is proposed to consider the polarization situation in the following foreshortenings:

1. Ratio of polarization discontinuities in one indicator for all regions of the region as a whole relative to each other, to understand the general trend of its dynamics and to characterize the degree of polarization of the region on this basis
2. Ratio of the polarization discontinuities of various indicators within each municipal region separately, to assess the possible interrelationships, regularities, and determinants of the phenomena under consideration

3 METHODS OF TESTING AND ANALYSIS OF RESULTS

Considering the indicators of the polarization disruption in their dynamics and in aggregate across all municipal formations, it is possible to single out general trends and patterns that characterize the polarization situation for the republic as a whole. The breakdown of the polarization according to the expenditures of the budget per thousand people (Figure 1), when considering the aggregate for all municipalities in dynamics over the years, is characterized by the identification of a certain wave-like trend and at the same time tends to increase. The trend lies in such a way that the regions show a wave-like change in the index of the polarization discontinuity, but they are located at the same time with respect to each other. Leaders retain their primacy, and outsiders remain among the laggards. This indicates some regularities in the budget policy, which the management subjects are purposefully or reactively adhered to. The overall spread of the gap varies from 0.7 to 1.6 with an extreme in 2014 (3.7 Maisky district), which indicates an increase in costs compared to the rest of the municipalities. There is an increase in the range of values of the polarization discontinuity in 2014. Peaks corresponding to the total increase in expenditures per thousand people took place in 2014 and 2017. At the same time, the leading district continued to occupy the leading positions in the break of polarization (Chereksky, Zolsky, Prokhladnensky, Nalchik). Identifying features characteristic of this indicator that distinguish the given polarization

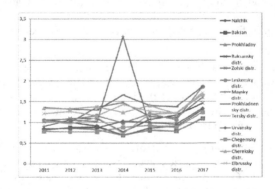

Figure 1. Breakdown of the polarization according to the expenditures of the budget per thousand of the population.

index from the others, it can be noted that the points of the polarization discontinuity graphs on each time interval of the period under consideration are rather crowded; that is, the difference in the absolute values of the polarization discontinuity lies in a narrow corridor, which indicates the presence of a relatively small span of polarization.

In this case, the range of values is divided into "weak" and "safe," but the difference between them in conventional units is not so great. The only time period when the density of points is dissipated is in 2014, when a number of MOs showed peak falls or take-offs. In the years to come, this gap was overcome.

Turning to the characteristics of the next indicator, the following main features of the breakdown of polarization in terms of budget expenditures due to donations to the budget per thousand of population can be noted. The disruption of polarization in terms of budget expenditures due to donations to the budget per thousand population (Figure 2) has a greater range between the extreme values.

The corridor is from 0.65 (Prokhladny district) to 1.55 (Chereksky district) during the period under review with an expansion of the range, that is, an increase in the range of polarization for a number of municipalities in 2015. However, until 2014 it is not possible to identify a single trend in the behavior of the graphs. After 2014 and until the end of the period under consideration, it is obvious that downward trends in most areas (except Urvansky, Zolsky, and Maisky) are synchronized and the scope of polarization is reduced; that is, the differences between the areas in this indicator become more smoothed.

The disruption of the polarization in the shipment of the GWS (Figure 3) per thousand population is a figure with a significant sweep of polarization, as the graphs show the distribution along the entire coordinate plane.

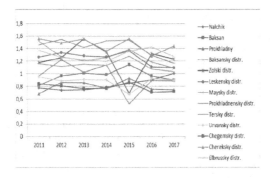

Figure 2.   The breakdown of the polarization of budget expenditures due to gratuitous revenues to the budget per thousand people.

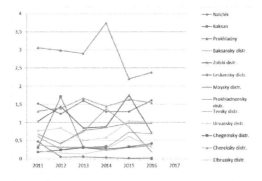

Figure 3.   Disruption of polarization in shipping GWS per thousand population.

Analysis of the dynamics of polarization in the shipment of goods, works, services (GWS) (Figure 3), an indicator chosen as an analogue of GDP per capita in application to municipalities, indicates the absence of a pronounced trend, and, therefore, it is already obvious that this indicator is weakly dependent, and perhaps completely independent of the influence of budgetary factors. Individual extremes correspond to the rise or fall of GDP in the corresponding area at specific times. The highest values throughout the period are demonstrated by the industrialized Prokladny and the lowest, Chereksky district. In the range closest to the value of 1, and thus approaching the medium-republican, there are points of growth of the Nalchik, Zolsky, and Urvan municipal districts.

The break in the polarization in terms of the average monthly wage (Figure 4) is the most static during the period under review.

The curves in Figure 4 are almost parallel to the abscissa and do not have extrema. The highest rates are in Nalchik, with the highest wage in the region, which is significantly different from the average for the republic, and the discontinuity of polarization is 1.4. Consideration of the polarization characteristics separately for the district allowed to draw conclusions regarding the interrelationships of factors, the manageability of budgetary tools and the development potential of each region.

3.1 Urban district of Nalchik

This municipal entity, being the administrative, social, economic, and cultural center of the region, focuses on one territory the basic capacities of the region in all spheres of activity: industrial, transport, recreational, and others (Figure 5).

3.2 Urban district of Baksan

This district, being in previous years an administrative center and an integral part of the Baksan municipal district, is currently allocated to the city district of the republican subordination with a separate budget with the status of a municipal formation. The presence of a number of industries—textile, agricultural, industrial —allows us to demonstrate high indicators relative to the adjacent territories (Figure 6).

3.3 City district of Prokhladny

The city of Prokhladny is the leading city in the economic life of the region after the capital. This is the territory of the concentration of transport hubs: railway; automobile; and a number of various industries, including food, alcohol, technical, light industry, and others (Figure 7). Obviously, the stably high rates of the polarization discontinuity curve for the average wage (above 1 throughout the period, and therefore higher than the regional average) are directly related to the prosperous production sphere.

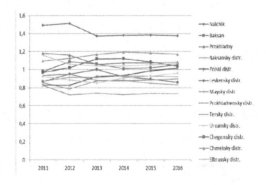

Figure 4.   Breakdown of the polarization in terms of the indicator is the average monthly nominal wage of employees of organizations.

3.4 Baksansky district

In this municipal region, the basis of the economy is composed of agriculture and processing industries (Figure 8).

3.5 Zolsky district

Zolsky municipal district is a territory whose geographic location determined the specialization and direction of economic development: agriculture, gardening, and tourism (Figure 9).

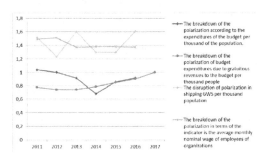

Figure 5.   The discontinuity of polarization in the aggregate of indicators for the city of Nalchik.

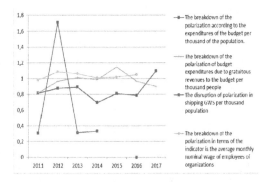

Figure 6.   The discontinuity of polarization with respect to the aggregate of indices is Baksan.

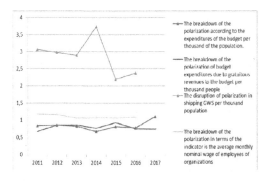

Figure 7.   The discontinuity of the polarization with respect to the set of indices is Prokhladny.

217

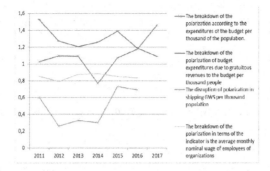

Figure 8.   The discontinuity of polarization in terms of the totality of indices: Baksansky district.

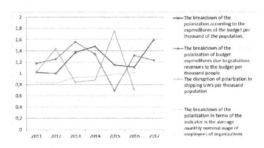

Figure 9.   The discontinuity of polarization in terms of the totality of the indices: Zolsky region.

3.6   Leskensky district

The given municipal formation is characterized by a low diversification of manufactures, specialized mainly in agriculture (plant growing). The indicators of the polarization disruption in shipping GWS are the lowest among all municipalities throughout the entire period under review. The indicators do not rise above 0.2 (Figure 10).

3.7   Maisky district

In this district the priority is in the development of the food industry, processing industry, and agricultural complex. At the same time, the district is not among the leaders, as illustrated by the graph of the polarization disruption in the shipment of the GWS (Figure 11).

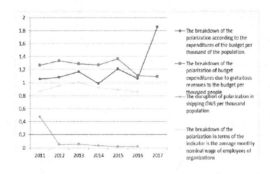

Figure 10.   The discontinuity of polarization in terms of the set of indicators: Leskensky district.

218

3.8 Prokhladnensky district

The economy of the region is mainly of an agricultural nature, where the enterprises of the food industry operate. The analysis of the indicators of the polarization of macroeconomic indicators reflects the situation shown in Figure 12.

3.9 Tersky district

In the Tersky region, the economy is represented by agriculture and food and processing enterprises, and a diamond manufacturing plant operates (Figure 13).

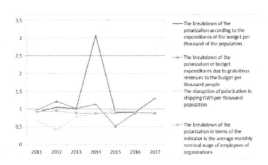

Figure 11.   The discontinuity of polarization in terms of the set of indicators: Maisky district.

3.10 Chegemsky district

In this region, located in the mountainous part of the republic, the economy is represented by tourism, agriculture, and processing enterprises.
   Analysis of the polarization indicators demonstrates the following situation (Figure 14).

3.11 Chereksky district

This municipal area has, among its distinctive features, its geographical position, the high-lands, a significant part of which is occupied by the territory of the reserve. In many ways, this is due to the specific nature of economic activity: agriculture and tourism. This is one of the

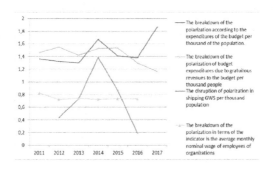

Figure 12.   Discontinuity of polarization in terms of the set of indicators: Prohladnensky district.

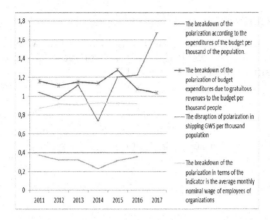

Figure 13.   Polarization discontinuity in terms of the totality of indicators: Tersky district.

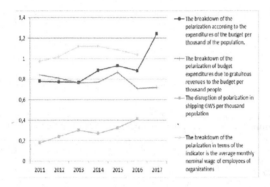

Figure 14.   Discontinuity of polarization in terms of the totality of the indicators: Chegemsky district.

few territories where all indicators are in the category of "safe," and in terms of the polarization of the shipment of GWSs per thousand people in certain periods, "strong" (Figure 15).

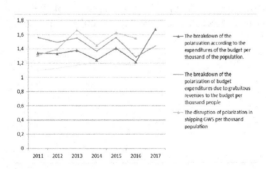

Figure 15.   Polarization discontinuity in the aggregate of indices: Chereksky region.

3.12 *Elbrussky district*

The main part of the territory is occupied by a mountain massif. In the past, the mining industry was developed in the region. Currently, the priority in economic activity in this territory is tourism and recreational activities (Figure 16).

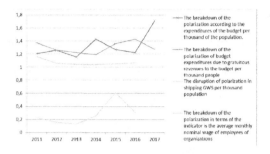

Figure 16. Polarization discontinuity in terms of the set of indicators: Elbrussky region.

4 CONCLUSIONS

The configuration of the polarization situation in the region objectively appears in the results of the study. However, when formulating conclusions on the results of observations, it is necessary to take into account a certain error in the calculations, which results from the specifics of subsidized regions with a high unemployment rate, or when the self-employed population does not submit information on incomes or on the produced GWSs to statistical bodies. Summarizing the conclusions obtained by calculation, one can reach certain regularities, which are not the problems formulated at the beginning of the study. Nevertheless, the following regularities occur regardless of the completeness of the statistical data: (1) The polarization breakdown for gratuitous receipts and for budgetary expenditures has general trends in most of the cases examined. (2) Indicators characterizing the level of wage polarization vary little; they are inert or practically unchanged. This indicates a weak determination of objective factors or low representativeness of statistical information.

Thus, it can be concluded that the policy of evening-out the polarization of financial instruments within the KBR region is functioning, but the low values of the indicators in question suggest that measures to support municipal areas are insufficient to form growth points that can claim to be a leader, to become a pole of growth and to have the potential to positively influence the development of the economy in the region. The distortions and asymmetry that develop reactively under the influence of unregulated factors takes place. Perfection of the regional policy on this issue can be conducted in various directions. An effective instrument in such a situation can be monitoring of the main macroeconomic indicators of the region from the point of view of polarization processes.

ACKNOWLEDGMENTS

The work was supported by the RFBR project No. 18-010-00885 titled "Creation of a model for evening-out the polarization of the development of municipal entities of the region as a condition for transition to an innovative economy."

REFERENCES

Alikaeva, M. V., Ashinova, I. V., Ksanaeva, M. B., & Karashaeva, D. A. 2018. Axiomatics of various ap-proaches to the study of polarization of social and economic development of regional economies. *Economics and Entrepreneurship*, 8:381.

Alikaeva, M. V., & Karashaeva, D. A. 2012. Use of tools for smoothing polarization in the economic policy of the CBD. *Modern Problems of Science and Education*, 4: 193.

Districts and cities of Kabardino-Balkaria. 2016. Federal State Statistics Service (Rosstat). Statistical collection. Nalchik.

Friedmann, J. 1966. *Regional Development Policy: A Case Study of Venezuela*. Cambridge, MA: MIT Press.

Hirshman, A. 1961. *The Strategy of Economic Development*, 2nd ed. New Haven, CT: Yale University Press.

James, P., & Martin, D. 1988. *All Possible Worlds: The History of the Geographer*. Moscow: Progress.

Jarimov, A. A. 1995. *The Region in a Single Market Space*. Rostov-on-Don.

Karashayeva, D. A. 2013. The mechanism of smoothing socio-economic polarization in the conditions of the subsidized region (on the materials of the Kabardino-Balkaria Republic). dissertation for the competition uch. PhD degree, Vladikavkaz.

Lavrovsky, B. L., & Shiltsin, E. A. 2016. Estimation and forecast of the spatial configuration of the gross product of the regions of Russia. *Economy of the Region*, 12(2:383–395.

Milchakov, M. V. 2017. Highly subsidized regions of Russia: The conditions for the formation of budgets and mechanisms for state support. *Financial Journal*, 1:22.

Perru, F. 2010. Economic space: Theory and applications. *Spatial Economics*, 2.

Pertsov, L. 2012. Analysis of the financial state of municipal entities and proposals for strengthening the revenue base of local budgets. *Municipal Power*, 1:51–55.

Pozdnyakova, P., Lavrovsky, B., & Masakov, V. 2016. The policy of regional equalization in Russia (basic approaches and principles). *Issues of Economics*, 10:74.

Saaya, S. N. 2008. Organizational-economic mechanism for smoothing polarization in regional de-velopment. Abstract of thesis for the competition uch. PhD degree, Samara State Economic University, Novosibirsk.

Wallerstein, I. 2011. *The Modern World-System I: Capitalist Agriculture and the Origins of the European World-Economy in the Sixteenth Century*, with a new prologue. Berkeley: University of California Press.

Zinoviev, A. A. 2012. Development of regional management by smoothing spatial polarization on the basis of the implementation of infrastructure projects. Author's abstract of the dissertation for the competition uch. PhD degree, Un-t control. "TISBI," Kazan.

Inclusive Development of Society – Lumban Gaol (eds)
© 2020 Taylor & Francis Group, London, ISBN 978-1-138-33476-2

The development of environmental accounting in the digital economy

O.G. Vandina
Armavir Pedagogical University, Armavir, The Russian Federation

S.A. Mohammed & G.A. Artemenko
Southern Federal University, Rostov-on-Don, The Russian Federation

N.N. Khakhonova & E.A. Panfilova
Rostov State University of Economics RSUE (RINH), Rostov-on-Don, The Russian Federation

ABSTRACT: This article shows the administrative specifics of the use of IT in the accounting system of a company and defines the advantages and the drawbacks of informatization of accounting in the context of the modern stage of development of digital economy. We conducted a content-analysis of the scientific literature on environmental accounting that permitted us to reveal methodological differences related to the understanding of the idea of environmental accounting, its objects, goals, and tools, which reflect the interdisciplinary nature of environmental accounting and the need for the development of a "triadic" system of markers applicable to financial, social, and environmental results of the activity of a company.

Keywords: digital economy, environmental accounting, goals of environmental accounting, objects of environmental accounting, sustainable development, tools of environmental accounting

1 INTRODUCTION

Global development of the "digital economy" creates new conditions and possibilities for the realization of accounting management with the use of information and communication technologies in the sphere of accounting and financial reporting to provide competitive and "sustainable development" of business units (Eremeeva, 2015).

Within this context there is an increased demand for national accounting systems capable of using efficient tools of environmental accounting both on macro- and microeconomic levels on the basis of the conception of "sustainable development" (Malinovskaya, 2013).

On one hand the speed of processing accounting and administrative information has increased, but on the other hand, the conceptual, instrumental, and analytic basis of accounting management is poorly adjusted to the integration of economic, environmental, and social components of interaction between a company and its stakeholders (Clarke, 2001).

These factors determine the development of a new race and a new type of digital competition of ecosystems of companies. The use of IT programs at all the stages of goods and services lifecycles amplifies the intensity and information integration of companies with their stakeholders, and it creates both competitive strengths when forming the chains of the costs of companies and certain restrictions for the sustainable development of companies.

The national accounting systems is crucial in the pursuit of sustainable development, as it is the main source of information about the economy and is widely used to assess economic indicators, but while mining revenues are recorded in national accounts, the simultaneous depletion of mineral reserves is not (Lange, 2014).

The difficulties with forming complex environmental and economic accounting into national systems of accounting both on macroeconomic and microeconomic levels are determined by the following factors:

- The specifics of functioning of national accounting systems
- The plurality of subjects of both external and internal environmental accounting
- A wide variety of financial, managerial, and accounting methods and tools of realization of environmental accounting
- Different milestones for external and internal environmental accounting
- The absence of a universal conception of environmental accounting within the system of other kinds of accounting management
- Uneven development of the main structural elements of environmental accounting (financial accounting of environmental activity of environment audit, environmental management, environmental reporting, environmental statistics, tax regulators of environmental activity).

Various aspects of environmental accounting are addressed in the works of Eremeeva (2015), Malinovskaya (2013), Abdulmanova and Moskaleva (2011), Volodin and Gazaryan (2016), Ilyicheva (2010), Sergeeva (2011), Gubaydulina and Ishmeeva (2015), Khmelev and Suglobov (2011), Rubanova (2005), Clarke (2001), Lange (2014), Akdogan and Hicyorulmaz (2014), Van (2012), and other authors.

From this point of view there is a need to study the specifics of the impact of the digital economy on the processes of environmentalization of accounting, specify the conceptual moments of environmental accounting of the business activity of companies, and reveal the advantages and disadvantages of the informatization of accounting in the context of the modern stage of the development of digital economy. These questions are studied in this article.

2 DIGITAL COORDINATES OF THE DEVELOPMENT OF ACCOUNTING MANAGEMENT

The notion of the "digital economy" was created by American information scientist Nicholas Negroponte in 1995. The digital economy reflects a certain part of economic affairs intermediated by the Internet, mobile communications, and the development of information and communication technologies (ICT) (Negroponte, 1995). So, digital economics creates a new technological paradigm of economic production (Ivanov & Malinetsky, 2017). In 2016 10% of the world gross domestic product (GDP) was accumulated with the help of the blockchain information technology (World Bank).

In turn the wide use of information models of economic reality creates a new context of efficient management of the changes and competitive strengths of organizations, transforming the inner and outer elements of the sustainable development of the eco-surroundings of companies and national economic systems.

The digital economy changes the qualitative and quantitative parameters of administrative and accounting information and transforms the functionality of the accounting reports; the new digital assets of a company are formed (Frolova et al., 2015).

According to K. Schwab (2016), since 2015 the key factors of the development of a new stage of the "digital economy" have been gaining strength as the fourth industrial revolution, known as "Industry 4.0" or the fourth technological paradigm. The digital economy transforms the entire sphere of social relations, production, logistics, information flows, the market, and consumption (Karapetyants et al., 2017a, 2017b; Tolstykh et al., 2017).

According to English professor T. Clark (2001) the modern accounting system in organizations falls behind the informational needs of business and is still on the level of the era of early capitalism within the first industrial revolution "Industry 1.0." It prefers standard accounting with stakeholders; it is oriented to the accounting of financial assets and financial risks of business; and it slowly develops the "triadic" system of indicators that can be applied for financial, social, and environmental results of the activity of organizations.

These are the key aspect of the digital economy:

- First, the tempo of development of digital economy increases; unlike during the previous revolutions, the digital economy changes not linearly, but in exponential tempo, and it requires the use of advanced digital technologies for the efficient management of organizations.
- Second, fast changes in the sphere of economic management and business eco-systems cause the modification of organizational, informational, and administrative borders of companies; this process broadens their networking cooperation with society and the state ["digital economy changes not only 'what' and 'how' we do, but also 'who' we are" (Schwab, 2016)].
- Third, the transformation of the external and inner system parameters of the development and functioning of business units, spheres, and national economies (the integration of virtual and real subsystems of economic management of business units) takes place.

So, on the basis of the digital technologies of "Industry 4.0," including blockchain (IB), the number of mediators decreases, crypto-currencies and new financial tools and services are created, the types of transactions change, taxation in real time takes place, the speed of document flow increases, the number of current assets grows, and also the transparency of transactions and the accounting information of organizations increases.

For the financial sphere the sustainability of its development is defined by the fact that the activization of the process of its "digitization" leads to the exactly opposite results, on the one hand: the sphere of virtual decentralization financial flows expands. On the other hand, there is a need for the integratedness of the control over the usage of financial resources to estimate the financial results of the activity of business units and to provide the sustainability of the national financial architecture as a whole (Global financial and foreign exchange system in 2030).

The expected mega-trends of "Industry 4.0" by 2025 are shown in Figure 1.

At the same time the expected mega-trends of "Industry 4.0" by 2025 not only create the never-seen-before possibilities for the changes of the financial and social landscape, but also create a "wave" of potential dangers (Schwab, 2016), determining the key factors and new business risks (Kleiner, 2018) (dehumanization of administration; failures and "cracks" of digital technologies; imperfection of the artificial intelligence; eco-terrorism; the growth of the

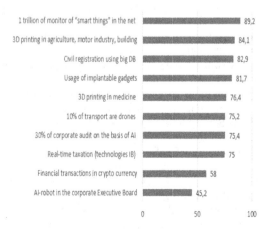

Figure 1. Expected mega-trends of "Industry 4.0" by 2025 (% of the number of answers of the questioned respondents).

Source: Report "Deep change: technological turning points and their impact on society."

costs of providing the information' logistic, material, and environmental safety of corporate governance, etc.).

3 CONCEPTUAL ASPECTS OF ENVIRONMENTAL ACCOUNTING

The main premises and conditions of environmentalization of accounting management are linked to the need to provide safe environmental development of national economic systems within international environmental safety, and are also connected with the stiffening of legal and environmental requirements, which influence the sustainable growth of business units, in the sphere of environment preservation.

Today the interaction between the spheres pf economy and ecology is contradictory and lately the intensity of the "burden" of economy toward the environment has been increasing, which leads to the growth of abatement costs (Abdulmanova & Moskaleva, 2011).

On the one hand, the economy actively uses the natural resources, including recreation services of ecosystems, which decrease every year due to soil degradation, desert invasion, and air and water pollution. The speed of the reduction of the volumes and areas of natural ecosystems is about 0.5–1% a year and, according to experts' measurements, such a speed of "shrinking" can cause a complete phase-out of natural ecosystems by 2030 (Girusova, 2002). On the other hand the transformations taking place in ecosystems limit the realization of competitive and environmental potential of business units both on national and global levels of economic development. For example, changes in environment influence the increase in the costs in different economic sectors: in agriculture, materials, and logistic sectors of the economy because the more affordable and high-quality resources have already been used.

Thus, the processes of environmentalization of economic life of business units become actual due to the fact that the problems of environmental development on all the levels of interaction by business units become sources of instability and factors of environmental risks, connected with the depletion of natural resources, environmental pollution, and climatic changes.

Environmental accounting is a relatively new element of the system of accounting course areas and a legitimate result of the process of differentiation of the information basis of accounting. According to a valid remark by Y. V. Sokolov, "The information structure of accounting is very dynamic, there are no and there can't be any boundaries between its parts. They are always vague. And what was called accounting in the beginning of the XX c., in the last third of this century started changing rapidly" (Sokolov & Sokolov, 2009).

At present, the scientists who research the problematics of environmental accounting can't come to a single decision about understanding the essence, subjects, goals, elements, methods, and tools of environmental accounting. Functional and status features of environmental accounting within the system of other branches of accounting are still not clearly determined.

According to N. N. Rubanova (2005), the subjects of environmental accounting are natural assets, environmental funds, environmental reserves, the results of activity of business units, and environmental costs (environmental compliance costs). Other authors such as O. N. Volodin and N.M. Gazaryan (2016) state that the subjects of environmental accounting are physical and cost indicators of the state of the environment and also unfavorable events and environmental risks and the information model of environmental compliance of a company as a whole. According to E. A. Sergeeva (2011), the most important subjects of environmental accounting are the parameters of the environmental protection to model the enterprise value.

According to E. V. Ilyicheva (2010), S. A. Khmelev and A. E. Suglobov (2011), and N. N. Rubanova (2005) the main subjects of environmental accounting are environmental costs and liabilities of business units. From the point of view of A. S. Ishmeeva (2015), among the subjects of environmental accounting is the information about environmental and ecological processes. We need to mention that the subjects of environmental activity of business units have extended beyond the cost side and can cover not only the results, but also the consequences of environmental activity of business units, which is why the subjects of environmental accounting include external readings, which reflect anthropogenic load of the enterprise on the environment.

There are also different opinions concerning the tools of environmental accounting. For example, E. A. Sergeeva (2011) thinks that the tools of environmental accounting are built upon the methods of inner management and financial and strategical accounting. According to O. N. Volodin (2016), the tools are not only the procedures of logging, collecting, and accounting of information but also the processes of monitoring the subjects of environmental accounting. Among the methods of environmental accounting E. V. Ilyicheva (2010) mentions the processes of collecting, logging, assimilating, evaluating, analyzing, controlling, and the methods of planning and projecting for environmental accounting. Other authors within environmental accounting apply not only traditional methods of management accounting but also the methods of environment audit (Volodin, 2015).

Target parameters of environmental accounting are understood by some scientists within the general system of enterprise management (Ilyicheva, 2010), as for the express purposes of environmental accounting, they can be viewed as used for purposes of external environmental reporting (Sergeeva, 2011), in order to control environmental compliance of business, rational use of natural resources, environmental safety, and for environmental reporting and its audit (Volodin, 2015; Volodin & Gazaryan, 2016), in order to determine the environmental potential of business (Rubanova, 2005).

The content analysis of scientific literature on environmental accounting lets us reveal the methodological difficulties related to understanding of the essence of environmental accounting; its subjects, goals, tools applied, and complex methods of environmental accounting; and the problems connected with the absence of a common definition of environmental costs and expenditures and the absence of the ways of integrating environmental information into the system of making effective managerial decisions within environmental management.

The integrated nature of environmental accounting is depicted in Figure 2.

It is important to mention that the essential features of environmental accounting can't be completely, pure and simple, reduced to only bookkeeping, financial, and management accounting because environmental accounting of an informationally integrated accounting system includes elements of different kinds of accounting (bookkeeping, financial, management, and tax), concerns environmental activity of business units, and covers external and internal subjects of environmental accounting for purposes of environmental management and control.

In the context of the digital economy a new type of digital competition appears among the ecosystems of companies. The usage of IT programs at all the stages of realization of product and service lifecycles increases the intensity and informational integration of companies with their stakeholders, and it includes both competitive strengths during the formation of chains of the costs of companies and certain limitations for the sustainable development of companies.

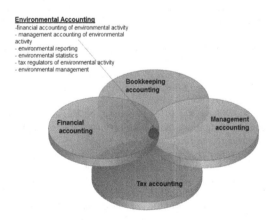

Figure 2.   Integrated nature of environmental accounting.
Source: Compiled by the authors.

4 CONCLUSION

The digital economy influences the whole system of accounting management in a contradictory and ambiguous way. We can see, on the one hand, the speed of processing of accounting and managerial information of companies, but on the other hand, the conceptual and tool and analytic base of accounting management is still poorly adjusted to the integration of economic, environmental, and social components of interactions between a company and its stakeholders.

Accounting management does not fully reflect the entirety of informational needs of business in the context of the digital economy, still being on the level of the era of the first industrial revolution; it still prefers standardization of accounting with its stakeholders, emphasis on financial risks of a company, and orientation toward the accounting of the financial assets of the business, due to a slow development of a "triadic" system of readings, applicable both for social and environmental results of the activity of companies. The digital economy creates not only never-seen-before possibilities for changes of financial and social landscape, but also a "wave" of potential dangers and new business risks (dehumanization of administration; failures and "cracks" of digital technologies; imperfection of the artificial intelligence; eco-terrorism; the growth of the costs of providing the information, logistic, material, and environmental safety of corporate governance, etc.).

The content analysis of the scientific literature on environmental accounting we conducted allowed us to reveal the methodological difficulties related to the understanding of the essence of environmental accounting—its subjects, goals, tools applied, and complex methods of environmental accounting—and the problems connected with the absence of a common definition of environmental costs and expenditures and the absence of the ways of integrating environmental information into the system of making effective managerial decisions within environmental management.

Environmental accounting has an interdisciplinary nature covering financial, managerial, bookkeeping and accounting of environmental activity of an enterprise and also environmental accounting, tax regulators of environmental activity, elements of environmental management, and control and audit for purposes of sustainable and competitive development of companies.

REFERENCES

Abdulmanova, A. Sh., & Moskaleva, A. Z. 2011. Accounting problems in the environmental accounting system. In *Proceedings of the Institute of Management Systems SGEU*, 1:107–110.
Akdogan, H., & Hicyorulmaz, E. 2014. The importance of the sustainability of environmental accounting. *International Conference on Economic Sciences and Business Administration*, 1(1):14–25.
Clarke, T. 2001. Balancing the bottom line: Financial, social and environmental performance. *Journal of General Management*, 26 (4):1627.
"Deep change: Technology." Report, International Expert Council, September 2015. Available at: https://www.weforum.org/agenda/2016/01/the-fourth-industrial-revolution-what-it-means-and-how-to-respond (accessed March 7, 2018).
Eremeeva, O. S. 2015. Ecological information expands the boundaries of traditional accounting. *Ecological Bulletin of Russia*, 9:50–59.
Frolova, I. V., Panfilova, E. A., Matytsyna, T. V., Lebedeva, N. Yu., & Likhatskaya, E. A. 2015. *The 4th International Congress & Interdisciplinary Behavior and Social Science*, Bali.
Girusova, E. V. (ed.). 2002. *Ecology and Environmental Economics*, S.60. Moscow: V. N. Lopatin.
Global financial and foreign exchange system in 2030. Available at: http://www3.weforum.org/docs/WEF_Global_Future_Council_Financial_Monetary_Systems_report_2018.pdf (accessed March 7, 2018).
Gubaidullina, I. N., & Ishmeeva, A. S. 2015. Environmental accounting as a factor in the economic security of an economic entity. *Modern Problems of Science and Education*, 11: 657.
Ilyicheva, E. V. 2010. Environmental accounting in the conditions of implementing an Environmental Balance Policy. Dissertation, Doctor Econ Sciences. 08.00.12. Eagle.

Ivanov, V. V., & Malinetsky, G. G. 2017. Digital economy: Myths, reality, perspective. RAS. M., Available at: www.ras.ru (accessed March 7, 2018).

Karapetyants, I., Kostuhin, Y., Tolstykh, T., Shkarupeta, E., & Syshsikova, E. 2017a. Transformation of logistical processes in digital economy. In *Proceedings of the 30th International Business Information Management Association Conference (IBIMA)*, November 8–9, 2017, Madrid, Spain, pp. 838–844.

Karapetyants, I., Kostuhin, Y., Tolstykh, T., Shkarupeta, E., & Krasnikova A. 2017b. Establishment of research competencies in the context of Russian digitalization. In *Proceedings of the 30th International Business Information Management Association Conference (IBIMA)*, November 8–9, 2017, Madrid, Spain, pp. 845–854.

Khmelev, S. A., & Suglobov, A. E. 2011. Methodological aspects of environmental-oriented accounting and auditing in order to ensure the economic security of industrial enterprises. *Vector Science of Togliatti State University*, 3:95–101.

Kleiner, G. B. 2018. Systematic accounting of the consequences of digitalization of society and security problems. *Scientific Works of the Free Economic Society of Russia*, 210:63–73.

Lange, G.-M. 2014. Environmental accounting. In *Handbook of Sustainable Development*, pp. 319–335. Cheltenham: Edward Elgar Publishing.

Malinovskaya, N. V. 2013. The development of environmental auditing in Russia. *International Accounting*, 43:29–36.

Negroponte, N. 1995. *Being Digital*. New York: Knopf.

Rubanova, N. N. 2005. Ecological accounting at the enterprises of the building materials industry. dissertation Candidate Econ Sciences: 08.00.12. Stavropol.

Schwab, K. 2016. *The Fourth Industrial Revolution*, 138c. K. Schwab, "Eksmo."

Sergeeva, E. A. 2011. On environmental-oriented reporting in the analytical assessment of the sustainability of enterprises. *Bulletin of the Taganrog Institute of Management and Economics*, 1:50–53.

Sokolov, Ya. V., & Sokolov, V. Ya. 2009. *Accounting History*, 287c. Moscow: Mag.

Tolstykh, T. O., Shkarupeta, E. V., Shishkin, I. A., Dudareva, O. V., & Golub, N. N. 2017. Evaluation of the digitalization potential of region's economy. In E. Popkova (ed.), *The Impact of Information on Modern Humans (HOSMC)*, pp. 736–743. Advances in Intelligent Systems and Computing, Vol. 622. Cham, Switzerland: Springer International.

Van, H. 2012. Environmental accounting: A new challenge for the accounting system. *Public Finance Quarterly*, 57(4): 437–452.

Volodin, O. 2015. Ecological accounting at the enterprise. *Modern Society and Science: Socio-economic Problems in the Studies of Teachers of the University: Sat. Scientific st. on the Basis of the International Scientific-Practical Conference*, Volgograd, pp. 73–79.

Volodin, O. N., & Gazaryan, N. M. 2016. Environmental accounting as a factor in improving the competitiveness of the company. *Economy and Entrepreneurship*. 11:852–860.

World Bank. Report "Digital dividends" on the digital economy. Available at: http://www.worldbank.org/en/publication/wdr2016 (accessed March 8, 2018).

Inclusive Development of Society – Lumban Gaol (eds)
© 2020 Taylor & Francis Group, London, ISBN 978-1-138-33476-2

Application of foresight for assessment of topicality of the Master's thesis

S. Sagintayeva, R. Zhanbayev, A. Abildina & G. Nurzhaubayeva
Almaty University of Power Engineering and Telecommunication, Almaty, Kazakhstan

G. Revalde
Riga Technical University, Riga, Latvia

ABSTRACT: In this work we propose an algorithm for the assessment of the relevance and significance to potential commercial attractiveness of the student's Master thesis. The proposed algorithm is based on modernized foresight technologies and applies principles of fuzzy logic. The developed algorithm is the first step to create a system for evaluation of potentially attractive research topics for knowledge transfer at the university and state level.

1   INTRODUCTION

Nowadays there are rising concerns about the quality of the higher education due to the fast development of technologies and unpredictable future needs. There are a large number of educational technologies; however most of them do not consider that learning should be directed to the future, should teach to predict and be ahead of modern advances in science and technology. It is obvious that the main goal of educational activity is the formation of a personality that expresses the full potential of a person. This leads to an increase in the importance of the paradigm of higher education, which treats students as active, responsible and full-fledged subjects of educational activity, along with faculty and employers (Lebedeva, 2008).

The current situation in Kazakhstan essentially raises the most interesting problem from the field of institutional economics, which can be formulated as the next question: "Is there a cheap method for mobilizing intellectual resources in breakthrough scientific fields, on the condition of very low communication connectivity in the original scientific and technical environment?"

It can be emphasized that the communication connectivity of the scientific and technical environment is indeed very low, which is directly confirmed by the almost complete absence of references to the works of Kazakhstani scientists in scientific articles published by Kazakhstani authors. The intellectual environment in Kazakhstan is fragmented and overcoming such a state of affairs, at a minimum, requires the reorientation of a significant number of researchers to new scientific directions. Traditional methods currently used in Kazakhstan (through material incentives) are very costly.

There is a problem of an adequate choice of specific scientific and technical tasks that are really relevant for the economy of Kazakhstan. A review of the dissertation works carried out in Kazakhstan in recent years shows that in all these works, the "relevance" section seems to be quite reasonably stated, however the low level of public-private partnership in the scientific and technical field suggests otherwise. We have seen that in Kazakhstan their results in many of cases do not find real practical application and do not contribute to the commercialization. At present, this quite huge resource is mainly spent on the fulfillment of formal requirements imposed by the governments of university administration.

Foresight is known as a tool for assessing the long-term perspective of science, technology, economics and society, allowing determination of the strategic directions of research and new technologies contributing to the greatest socio-economic effect. Foresight is often used by scientific centers and governments in different countries of the world to assess large projects and R&D strategies (e.g. Iden, 2017, Lee, 2018). There are not so many works investigating possibility to apply foresight methodology to modernize the university level research strategies.

The purpose of our research is to consider the possibility of using foresight, as a technology, in the educational process and evaluate possibilities of modernizing existing foresight technologies, designed to identify the most promising areas of scientific and technical activities, aimed at ensuring the correction of the progress of master's and later PhD theses for increasing their economic efficiency, including the commercialization of scientific-technical activities.

In this paper we propose an algorithm based on modernization of existing foresight methodologies aimed for evaluation of topicality of Master's theses (identifying the most promising areas of scientific activity). This is the first step for solving the problem if and how the Master's (and in later stage PhD) theses can be used as a cheap resource for the formulating policy of innovative development of university's and national level.

2 METHOD

Foresight, as parallel to the evolving direction of futures studies, was formed as an applied research method for solving practical problems. Foresight focuses on the development of practical measures to approximate selected strategic guidelines. Therefore, with the development of foresight, the work to determine the prospects for the development of the forecasting object for particularly important projects, as a rule, does not end, but is continuous, so that when significant changes occur in the conditions of the project implementation or during its implementation, the timely adjustments were made to the forecast. This makes the foresight an effective strategic management tool.

About 33 foresight methods are known (see e.g. Popper, 2008). They can be divided according types of techniques – qualitative, quantitative, semi-quantitative. Usually, in each of the foresight projects, a combination of different techniques is used. If state foresight, financed by government agencies, uses certain methods, the business sector with developing foresight projects uses other methods, whereas the research sector uses all of the above methods more equally.

The most frequently applied method is the Delphi method, expert analysis, literature review, scenarios, and then critical technologies, brainstorming and SWOT.

It should be noted that the set of techniques used often depends on who is the sponsor of the project. The development of foresight as a method of research has helped in a wide range of studies, like expert surveys by the Delphi method, first applied in strategic military planning at RAND Corporation in 1953 in the US. The Delphi method is known as semi-quantitative method which applies mathematical principles to manipulate data derived from subjectivity, rational judgments, probabilities, values and viewpoints of experts, commentators or similar sources (Mozuni, 2017). The Delphi method, as a poll in several iterations of the same group of experts, allows constructing a forecast based on the principle of scientific consensus regarding the perspectives and priorities of scientific and technological development. Development and modifications of the Delphi and other methods are described in detail by Mozuni and Jonas (Mozuni, 2017).

We propose to implement a modified Delphi method with a computational algorithm of analysis using fuzzy logic to get the most relevant themes of Master's thesis (Mozuni, 2009).

3 EXPERT SELECTION AND COLLECTING SOURCE DATA

The first step in the implementation of the goal of our research is to create the foundations for increasing the communication connectivity of the scientific and educational space. There is a classic problem of selection and motivation of experts for foresight exercises. Foresight methods are aimed at solving this problem through the use of statistical methods and a broad base of expert assessments. To attract a sufficient number of qualified experts can be a costly approach. In our case by application of foresight to the evaluation of master's thesis, the question of the necessity of attracting experts of high qualification (and verification of this qualification) is partially solved using university resources. This paper shows that there is the possibility of obtaining expert assessments in the mode of self-organization (at least, if we talk about the assessment of doctoral and Master's theses). Namely, the defense of the thesis requires the participation of two actors - the actual dissertator and his supervisor.

However, in the existing form, foresight methods solve the problem of assessing the quality of such scientific materials as master's theses only partially, since the question remains about attracting a significant number of experts whose task, moreover, is to work with a large data array. The question of the motivation of experts and the assessment of the quality of the examination itself remains open in this case. There are no guarantees that a particular expert will not score points by a superficial review of work or from specific subjective considerations. Namely, the system of expert assessments can be both, a control tool and a management tool. With a simple example is can be shown that it is possible to use the procedure for grading students by university teachers to evaluate their own business and professional qualities.

Suppose that it is possible to get "true" grades (the scale is not yet concretely specified) of a certain array of scientific and technical works (further for definiteness, master's theses will be considered). In this case, it is immediately possible to assess the competence/objectivity of the expert himself; to do this, it is enough to compare the marks he gave with the "true" ones. Apparently, a certain ambiguity arises here, because possible deviations will be due to several factors acting simultaneously:

- the expert in good faith is mistaken (factor of insufficient competence);
- the judgments of the expert are influenced by subjective factors (for example, an underestimation of the assessment of the work, the author of which is unpleasant to the expert);
- The expert treats this work in bad faith (the factor of insufficient motivation, for example, estimates are made on the basis of superficial judgments).

The purpose of the algorithm being developed, therefore, should be including the selection of the results for the indicated factors, as well as obtaining estimates as close as possible to the true ones, based on data in which the subjectivity factor is present. In this case, it is possible to provide not only high-quality expertise without the involvement of third-party resources but also to use the very fact of its conduct to increase the level of scientific and technical activity.

In general case, the evaluation is carried out from data on the implementation of n of the same type of scientific work. In our case, we use the Master's thesis. For the convenience, we assume that for the conducting expert assessments, the most relevant information relating to the dissertation will be converted into a compact form. It is also assumed that there will be a possibility of rating for persons who are not specialists in a particular narrow field of science. (This provides the expansion of the expert base and the possibility of using statistical methods.)

Each of the N experts will receive materials reflecting the work of n (minimum 5) Master's theses. The theses are distributed accidentally. Respondents are encouraged to rank these theses, assigning from each of them a number from 1 to 5 (1 is the best of the given set, 5 is the worst, the assignment of the same value to many works is not allowed), the nature of the data is illustrated in Table 1. The fundamental difference from the existing approaches (used, for example, by the National Scientific Councils of the Republic of Kazakhstan when conducting an international examination) is that the respondents do not set the absolute values of the points, but conduct a comparison of several dissertations with each other.

For additional control, the respondent is also offered to assign one of the numbers from 1 to 5 of his (her) supervised Master's thesis. The ranking is carried out according to one or several comparison criteria. The ranking is compulsory and according to the conditions of conducting the assessment, procedure respondents must be announced in advance that the conclusion will be made from this assessment, including concerning their competence and honesty. This will serve as an additional incentive for making an objective conclusion. At the

Table 1. An example of a table containing the raw data.

	Number of thesis							
	1	2	3	n	n+1	n+2	n+3	n+4
Name 1		3		1		2		
Name 2	1		5				4	
Name N		5			2			3

same time, it is the factor of mandatory ranking that excludes the possibility of forming a conclusion like "we are all competent and highly qualified specialists".

4 RESULTS AND DISCUSSION

In the first (basic) approximation, the procedure for parallel assessment of the level of master's theses and the integrity/competence of scientific leaders acting as experts is based on the following counting method. Each of N survey participants issues 5 ratings, ranking 5 received materials, that is, all 5N ratings are displayed. If the materials are distributed evenly, this means that each participant receives 5 ratings, i.e. the score obtained varies from a minimum of 5 to a maximum of 25. Number 5 is chosen from psychological considerations. A smaller number gives insufficient statistics; further, its increase raises considerable difficulties with ranking. Indeed, psychologically, from five works, it is easy to identify the best and the worst, and then repeat the procedure concerning the two remaining jobs (two stages of ranking). Implementation of such a ranking in three stages may already lead to difficulties due to the need for the parallel accounting of a sufficiently large volume of material.

So, in the first approximation, the estimation procedure is expressed by the formula

$$I_n = \sum_{K=1}^{s} w_{nk} \tag{1}$$

where w_{nk} is one of the numbers in the natural series from 1 to s (for the reasons given above, it is expedient to choose this number equal to five), which the expert assigns to n^{th} work in the process of ranking, In is the estimate corresponding to the first approximation.

In the next step, the mapping K is constructed using the typical method of fuzzy logic (Semenenko and Chernaev, 2014):

$$\{I_n\} \xrightarrow{K} \{Q_m\} \tag{2}$$

which translates the obtained score into a ball scale.

In the simplest case, this mapping corresponds to splitting the entire possible interval of change in the I_n estimates into several subintervals, each of which corresponds to a particular administrative action. Simplifying, assessments should be transferred to a coarser scale, for example, "excellent", "good", "satisfactory", "bad", to eliminate the influence of errors. Besides, it is this mapping that corresponds to the procedure for using neural networks to solve the problem, as will be clear from the sequel.

The disadvantage of formula (1) is the inability to take into account the integrity/competence of a particular expert. Rewrite formula (1) in the form with another upper limit of summing, corresponding to the total number of participants N in the procedure of mutual evaluation.

$$I_{n0} = \sum_{k=1}^{N} w_{nk} \tag{3}$$

Formally, formula (3) looks the same as (1), but it contains elements of the matrix $N \times (N-1)$, which assume a zero value if an expert with the number n did not evaluate the operation with the number k. This form of estimation formula allows you to directly enter weight coefficients S_k, reflecting the integrity/competence of a particular expert. We have

$$I_n = \sum_{k=1}^{N} S_k w_{nk} \tag{4}$$

Thus, the problem is reduced to the definition of coefficients reflecting the characteristics of a particular expert. Its solution allows, on the one hand, characterizing a particular scientific supervisor (in the case of master's theses), on the other hand, it allows for obtaining relevant assessments of the quality of work performed.

The simplest method for determining weight coefficients is based on a direct comparison of the judgment made by a particular expertise with a collective opinion. Such a comparison is carried out as follows.

Obtaining the set of estimates $\{I_{n0}\}_{i=1}^{i=N}$ naturally, allows us to order N evaluated works in the ranking of the first approximation by obtaining a sequence of integers $\{q_{n0}\}_{i=1}^{i=N}$. Each 5 papers ranked by each of the experts in this sequence occupy certain positions: by striking from the entire sequence $\{q_{n0}\}_{i=1}^{i=N}$ the numbers of those works which are not included in the set of works ranked by a separate expert, we obtain the ranking necessary for comparison (a particular case when two or more jobs occupy the same position in the ranking $\{q_{n0}\}_{i=1}^{i=5}$, are considered separately below). The result of this ranking is denoted by the sequence $\{r_{mk}\}_{i=1}^{i=5}$, respectively by $\{r_{mk0}\}_{i=1}^{i=5}$ we denote the ranking generated by a separate expert.

The result of the procedure just described can be represented as a substitution (5)

$$\begin{matrix} 1 & 2 & 3 & 4 & 5 \\ 3 & 2 & 1 & 5 & 4 \end{matrix} \tag{5}$$

In the upper line of the substitution, there is a ranking obtained on the basis of a collective opinion, in the lower one - based on the judgments of the individual expert. The numerical distance between the rankings is defined as the sum of the squares of the differences between the positions occupied by the particular work in the upper and lower rows (5). In Example (5), the distance corresponding to work with the number "3" is 2, which is shown by a curly bracket. In the general case, the distance is given by the formula

$$Q_k = \sqrt{\sum_{m=1}^{5} (r_{mk0} - r_{mk})^2} \tag{6}$$

Or, for the case of ordering the form (5)

$$Q_k = \sqrt{\sum_{m=1}^{5} (m - r_{mk})^2} \tag{7}$$

Weighted coefficients S_k, respectively, are given by the formula:

$$S_k = \frac{Q_k}{\sum_{i=1}^{N} Q_i} \tag{8}$$

Formula (8) provides the possibility of an iteration procedure, which allows calculating the final values of weight coefficients. At the ith step of this procedure, the search for the ranking $\{I_n^i\}$, is carried out, for which the values of the weight coefficients $\{S_n^{i-1}\}$ obtained in the i step are used. By ranking $\{I_n^i\}$, the coefficients $\{S_n^i\}$, are determined, the procedure is repeated until the required accuracy is reached.

The experts' self-assessment is used for additional verification of integrity/competence, as well as the adequacy of the proposed procedure. It is primeval enough to implement it utilizing qualitative comparison of the expert's honesty conducted through the above-described procedure for calculating the coefficients $\{S_n^i\}$, with the data obtained when comparing the expert's self-assessment and ranking $\{I_n^i\}$.

In future, the algorithm showed before can be applied for constructing an artificial intelligence system designed to obtain expert assessments using modernized foresight methods using an example when the mapping (2), built on the principles of fuzzy logic, is binary. Binary mapping is also of direct practical interest for Kazakhstani universities regarding improving the quality of education in the magistracy. As mentioned before, as even a cursory review of Master's theses in technical disciplines shows that the overwhelming majorities of them are hopelessly outdated and have no practical significance. It is evident that attempts to force the student to work on a dissertation, a topic that cannot be interesting, lead to the fact that work on the dissertation will be reduced only to the fulfillment of formal requirements. Consequently, the problem arises on the selection of topics of dissertation papers on the criterion of relevance, which implies the possibility of using a binary assessment ("relevant - irrelevant"). This task is also of interest from the point of view of improving the economic efficiency of universities, by blocking further work on topics that do not promise a clear economic return. Besides, this example is very convenient for working out the proposed assessment methodology, since it significantly expands many experts, since it is possible to make an adequate judgment about the commercial relevance of a particular work without being a narrow specialist in a particular field.

It can be seen that for binary splitting suggests that the described method is, in fact, a numerical implementation of an analog of a Hopfield-type neural network with a matrix of weight coefficients w_{nk} (Hopfield, 1985). The analogy is not direct since the specified matrix is not symmetric, but this suggests that the corresponding method can be transformed to the prototype of the artificial intelligence, and for the case when the partition corresponding to specific administrative actions is not binary. In general, the output is preserved, with the difference that it is necessary to proceed to the use of K-valued logic (Hopfield, 1985). We will continue to develop the system implementing the binary splitting. Next steps will be the experimental validation of the proposed algorithm in real conditions.

5 CONCLUSIONS

In our research, we propose that the labor and time spent on the preparation of Master's and doctoral (Ph.D.) thesis are a resource that can be used for the innovative development of countries' economy. At present, their results rarely contribute to the development of public-private partnerships in terms of commercialization of the results of scientific and technical activities. Our work is devoted to creating the system and method to evaluate the Master's and PhD theses as a resource that employing modernized foresight can be applied to the innovative development. In this work, we report an algorithm, created using modernized foresight approach based on the principles of fuzzy logic, to obtain the topical and most relevant Master's thesis by expert estimates for the potential knowledge transfer. In this paper, we demonstrate the elaborated theoretical model and algorithm for expert assessment processing, including the mechanism for evaluating the expert integrity/competence. The work will be continued introducing the binary splitting and experimental validation in real conditions.

ACKNOWLEDGMENTS

This study was funded and supported by the Committee of the science of the Ministry of Education and Science of the Republic of Kazakhstan Project AP05132160 "Development and implementation in the educational process of foresight-oriented methods of educational work of doctoral and master's students."

REFERENCES

Hopfield J. & Tank D. 1985. "'Neural' Computation of Decisions in Optimization Problems", *Biol. Cybern* 52: 141–152.

Iden J., Methlie L.B., Christensen G.E. 2017. The nature of strategic foresight research: a systematic literature review. *Technological Forecasting & Social Change* 116: 87–97.

Lebedeva S.A. 2008. The philosophy of the social sciences and humanities: study guide, ed. Moscow, Academic Project: 733 (in Russian).

Lee J. &Yang J.S. 2018. Government R&D investment decision-making in the energy sector: LCOE foresight model reveals what regression analysis cannot. *Energy Strategy Reviews* 21: 1–15.

Mozuni M. & Jonas W. 2017. An Introduction to the Morphological Delphi Method for Design: A Tool for Future-Oriented Design Research, *She Ji: The Journal of Design, Economics, and Innovation* 3(4): 303–318.

Popper, R. 2008. Foresight Methodology, in Georghiou, L., Cassingena, J., Keenan, M., Miles, I. and Popper, R. (eds.), The Handbook of Technology Foresight, Edward Elgar, Cheltenham: 44–88.

Semenenko M.G.& Chernaev S.I. 2014. User Functions in Excel 2013: Development of Fuzzy Logic Applications. *Successes of Modern Natural Science* 3 (in Russian).

Inclusive Development of Society – Lumban Gaol (eds)
© 2020 Taylor & Francis Group, London, ISBN 978-1-138-33476-2

Application of XBRL as an electronic format for financial communications in the Russian Federation

T.V. Matytsyna, A.S. Surov, A.A. Galkin & M.M. Kazhu
Southern Federal University, Rostov-on-Don, Russian Federation

O.A. Anichkina
Federal State Budget Educational Institution of Higher Education "K.G. Razumovsky Moscow State University of Technologies and Management (the First Cossack University)", Moscow, Russian Federation

ABSTRACT: This article considers the application of the XBRL format as a digital platform for the organization of financial communications in the Russian Federation. The main subjects of the pilot project for the implementation of the XBRL format in the Russian Federation were identified. The article reveals the main advantages and limitations of the use of the XBRL format and the complexity and difficulty of the transition of financial communications for the Russian economy.

Keywords: advantages and limitations of using XBRL format, digital economy, language, reporting, technology, XBRL

1   INTRODUCTION

The processes of harmonization of financial communications based on International Financial Reporting Standards (IFRS) are built not only by convergence of national accounting systems, but also through the use of technical platforms such as the XBRL format (eXtensible Business Reporting Language). According to A. Markelevich (2015), "XBRL – language for the electronic communication of business and financial data which is revolutionizing business reporting around the world. It is a tool to bridge potential language barriers and unify financial reporting."

The concept of integrated reporting, combining financial, social, and environmental reporting of the company, is the basis for improving the process of financial communications between the supervisory authorities, national statistical services, company management, and their stakeholders.

The XBRL format allows you to process digital and text arrays of information in retrospect and compare the indicators of companies. The principle of combining the financial and nonfinancial performance of economic entities into a single system and the transition to the XBRL format is the most important technological step in the formation of integrated reporting.

At the same time, integrated reporting and IFRS reporting have a variability in the structure and capabilities to provide reporting information. The individuality of the implementation of the taxonomy of IFRS-based reporting based on the XBRL format requires additional actions, costs, and use of special software, which increases the company's costs when providing integrated reporting.

The XBRL format allows presenting business data in a single digital format, such as financial, supervisory, and statistical indicators of the functioning of economic entities, which makes it possible to simplify the financial communications of companies, supervisory authorities, and national statistical services.

The possibilities and limitations of the application of the XBRL format in national accounting systems are the subject of heated discussions and were considered in the works of R. Debreceny and G. L. Gray (2001), M. Alles and M. Piechocki (2012), A. Abdullah (2009), M. Enachi (2013), A. Markelevich (2015), F. Radu (2016), and other authors.

The prospects for the transition and the specifics of the application of the XBRL format for Russian enterprises are reflected in the works of such authors as A. B. Polozov (2010) L. E. Kaspin (2013), T. V. Morozova et al. (2018), and others.

In the Russian Federation, a pilot project on reporting in the new electronic format XBRL was implemented in July 2016.

The initiator of the introduction of the XBRL format in the Russian Federation is the Central Bank of Russia, which, from January 1, 2018 has obliged the use of this format for the following organizations: insurance organizations and mutual insurance societies; nonstate pension funds; professional participants of the securities market (PPSU), trade organizers and the central counterparty; and the joint-stock investment funds and management companies of investment funds.

This article is devoted to a discussion of certain outcomes of the pilot project implementation, the specifics of its implementation, as well as consideration of the advantages and limitations of the use of the digital format XBRL for Russian enterprises.

2 XBRL FORMAT OF FINANCIAL COMMUNICATIONS IN THE DIGITAL ECONOMY

The XBRL format as the electronic basis for of financial communications in the digital economy is a diverse set of scientific papers, which focus on various aspects of using XBRL in national accounting systems. In his comparative study on the use of the XBRL format in the United States and the United Kingdom, A. Abdullah (2009)indicated that there are variations in the adoption of the XBRL between the two countries. According to the authors, XBRL seems to have taken a deeper root in the United States for the relatively stronger institutional coercive pressures.

M. Alles and M Piechocki (2012) argued that XBRL can improve corporate governance and the management process in a company. According to the authors, this is achieved because XBRL can be used to democratize data for different groups of users, so XBRL allows us to consider corporate data from different points of view, which leads to the creation of new knowledge.

Three factors are noted in the R. Debreceny and G. L. Gray (2001) study of the use of Web correction of financial reporting on the Internet. Second, accounting recognition (problem recognition) should be reliably analyzed. Third, standard mechanisms should be applied in a consistent manner. According to the authors, in order to win business competition, companies must switch from obsolete PDF and HTML formats to the new XBRL digital format, which allows "more automated analysis and polling of basic information on several platforms" and forms a "new communication channel for interested parties."

M. Enachi (2013) notes the advantages of switching to the XBRL format for Romanian enterprises, associated with the fact that information comes from various regulations.

A. Markelevich (2015) emphasizes that XBRL is oriented to translate financial statements to foreign investors, among others, who can rely on information in XBRL-tagged financial reports to make investment decisions without having to translate financial statements from the local language.

The study "XBRL in the digital age"– by F. Radu (2016) states that it is possible to use it. According to the author, the XBRL format is applicable not only for external verification of financial statements, but also for the company's internal control system.

The processes of digitalization of financial communications have led to the appearance of the XBRL format. XBRL has a number of characteristics as a result of which its implementation can be successful and will bring significant benefits (Kaspin, 2013; Morozova, 2018):

- Elimination of redundancy and duplication of reporting data by building a unified system for collecting and processing reports based on IFRS
- Increase of reliability and quality of reporting data by unification and automation of data collection processes
- Increase of transparency and openness of financial information
- Reduction of the burden on accountable organizations; by automating the process of collecting reporting data and their automatic validation, the compilers of reporting will free up resources, which will allow them to concentrate their efforts on data analysis
- In the future, the unification of formats for interagency and international electronic data interchange

3 APPLICATION OF XBRL IN THE RUSSIAN FEDERATION

In the Russian Federation, the project on reporting in the new electronic format XBRL was launched in July 2016.

XBRL format allows presenting business data in a single digital format, such as financial, supervisory, and statistical indicators of the functioning of economic entities, which makes it possible to simplify the financial communications of companies, supervisory authorities, and national statistical services.

Starting from January 1, 2018, reporting in the new electronic format XBRL became mandatory for the following participants of the financial market:

- For insurance organizations and mutual insurance societies (Bank of Russia Ordinance No. 4584-U of October 25, 2017)
- The nonstate pension funds (Bank of Russia Directive No. 4623-U of November 27, 2017)
- The professional participants of the securities market (PPSU), trade organizers, clearing organizations, and persons performing the functions of a central counterparty (Bank of Russia Directive No. 4621-U of November 27, 2017)
- From April 5, 2018 for joint-stock investment funds, management companies of investment funds (IF), unit investment funds, and nonstate pension funds (Bank of Russia Ordinance No. 4715-U of 08.02.2018)

Stages of the implementation of the XBRL project in the Russian Federation from 2018-2022 are shown in Figure 1.

The transition of credit rating agencies and insurance brokers to reporting in the XBRL format is scheduled for 2019; microfinance organizations, specialized depositaries, credit

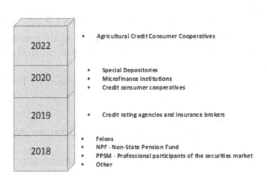

Figure 1. Stages of the implementation of the XBRL project in the Russian Federation from 2018-2022.

Source: Herald XBRL (2018).

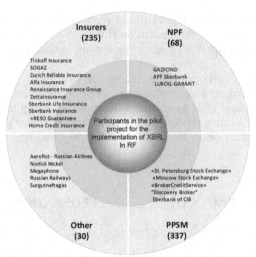

Insurers
(235)

NPF
(68)

Tinkoff Insurance
SOGAZ
Zurich Reliable Insurance
Alfa Insurance
Renaissance Insurance Group
ZettaInsurance
Sberbank Life Insurance
Sberbank Insurance
«RESO Guarantee»
Home Credit Insurance

GAZFOND
APF Sberbank
LUKOIL-GARANT

Participants in the pilot
project for the
implementation of XBRL
In RF

Aeroflot - Russian Airlines
Norilsk Nickel
Megaphone
Russian Railways
Surgutneftegas

«St. Petersburg Stock Exchange»
«Moscow Stock Exchange»
«BrokerCreditService»
"Discovery Broker"
Sberbank of CIB

Other
(30)

PPSM
(337)

NPF - Non-State Pension Fund
PPSM - Professional participants of the securities market

Figure 2. Participants in the pilot project for the implementation of XBRL in the Russian Federation, 2018.

Source: Herald XBRL (2018).

consumer cooperatives, and housing funded cooperatives for 2020; and agricultural credit consumer cooperatives and pawnshops for 2022.

More than 700 organizations are involved in the pilot project for the implementation of the XBRL format in the Russian Federation: nonstate pension funds occupy 10%; professional participants of the securities market, 49%; participants in the insurance market, 35%; and others, 6% (Figure 2).

XBRL is the international standard for the publication, sharing, and financial analysis of reporting data, which simplifies the preparation and publication of financial documents. The basic idea is to collect data once and convert them into several formats using automatic processing.

Reporting, presented in doc or pdf format, does not allow it to be processed, to make several inquiries at the same time, and to receive certain information and as a result is perceived as a display of certain symbols that do not have their characteristics and history. The XBRL format, in turn, gives each unit of information individual tags (tags) and provides translation (translation) of accounting concepts in a computer-friendly set of financial data.

The XBRL standard creates a link between information technology (XML) and standards (GAAP), as the taxonomy of XBRL is the translation of GAAP into XML documents. Work based on XML provides the financial community a method of preparation based on standards, publishing in various formats, and the correct extraction and automatic exchange of financial statements of companies.

Advantages of implementation of XBRL as a digital format of financial communications can be identified in the following areas:

1. *Uniform digital model of requirements*
   * Business control and relationships in the transferred taxonomy remain on the side of the organization.
   * Reporting contains references to regulatory requirements and clarifications in taxonomy.
   * Visualization of requirements is part of a taxonomy.
   * Supervisors fill in and transfer figures from the account, so there is no need to form.

2. *Reduction of redundancy and duplication of requirements*
   - More than 30,000 indicators in the supervisory requirements for nonfinancial and financial reporting have been converted into 12,000 indicators in taxonomy.
   - Ninety-four forms in the supervisory requirements of the professional participants of the securities market are duplicated to 58 forms.

3. *International integration*
   - The Bank of Russia taxonomy is built on the basis of the IFRS taxonomy and includes the IFRS taxonomy indicators for the international XBRL 2.1 specification.

4. *The single metalanguage of the Central Bank of Russia*
   - A complete cycle of the infrastructure for receiving, processing, storing, and analyzing the XBRL reporting in the Bank of Russia has been generated and tested.
   - In-depth analytics on the basis of a single taxonomy data model

5. *Cross-sectoral oversight*
   - A single model of digital data for all markets
   - Single cloud of digital directories
   - Cross-sectoral monitoring of financial and non-financial reporting indicators

6. *Transparency in the process of making changes to requirements*
   - Changes to the taxonomy are made twice a year.
   - The publication of a preliminary taxonomy version for the market takes place 3 months before the entry into force.

7. *IT cost reduction*
   - With the expansion of the project, the unification of the technology approach will allow supervisors to reduce the cost of owning IT reporting systems.

8. *Data quality*
   - More than 1,700 business-control in taxonomy are used for automated validation of reporting on the side of supervised and the Bank of Russia.
   - Methodologically equal indicators are filled once.

The advantages of implementation XBRL as a digital format of financial communications in the Russian Federation are shown in Figure 3.

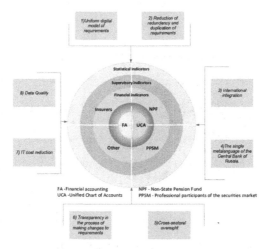

Figure 3.   Advantages of implementation XBRL as a digital format of financial communications in the Russian Federation.

Source: Compiled by the authors.

The use of the XBRL format has its own implementation constraints. First of all, there are large financial costs for special software and a small number of technically ready users in Russia able to read and process documents in this format.

The Bank of Russia has prepared a free data converter that allows the following:

1. The XBRL converter receives, on the one hand, any XBRL taxonomy (in this case, the taxonomy of the Bank of Russia) and, on the other hand, an external file of the specified format with the data and indicators of the reporting subject.
2. The XBRL converter automatically fills in the taxonomy with data, evaluates the data for compliance with business taxonomy checks, and reports errors if necessary.
3. The result of the XBRL converter is an XBRL format file, which can be signed and sent to the controller.

Among the pilot enterprises, the software of Bank of Russia Converter uses 37%; Fujitsu Interstage Xwand, 15%; Own solution, 11%; HomnetXBRL, 7%; Orticon XBRL Processor Module, 7%; and other software products, 4% (Figure 4).

It is planned that in 2021 the "Converter" of the Bank of Russia will cease to exist and organizations will prepare XBRL reporting files based on their own third-party IT solutions integrated into their IT infrastructure, which will also allow them to use these products for deeper internal analysis of their own internal information.

Now, the main difficulty in preparing reports in XBRL format is not the conversion itself but the preparation of primary data and indicators that will be converted to the XBRL format. The process of forming XBRL reporting is reduced to filling in the reporting forms, but more detailed in comparison with the usual ones.

XBRL is the optimal format in terms of control. However, market participants will have to build new business processes to record their operations and automate them deeper. However, as noted at the 17th meeting of the European office of XBRL International "XBRL Europe," the consolidated financial statements in XBRL format are 11%, and in XML, 89% (Official website of XBRL, 2018).

The decision on the expediency of the transition of banks in the Russian Federation to the XBRL format will be made after the transition of noncredit financial organizations to the XBRL format. If the results of the pilot project are positive, the development of the XBRL taxonomy for non-financial organizations will become the basic platform for the development of taxonomy for Russian banks.

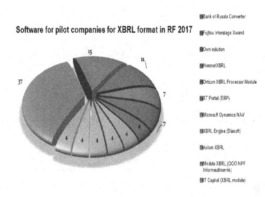

Figure 4.   Software for pilot companies for XBRL format in the Russian Federation 2018.
Source: Herald XBRL (2018).

4 CONCLUSION

In international practice, the XBRL format is used to exchange business information between companies and supervisors, financial market regulators and statistical services, etc.

The initiator of the introduction of the XBRL format in the Russian Federation is the Central Bank of Russia, which, from January 1, 2018, has obliged the use of this format for the following organizations: insurance organizations and mutual insurance societies; nonstate pension funds; professional participants of the securities market (PPSU), the trade organizers, the central counterparty; And joint-stock investment funds and management companies of investment funds.

The processes of harmonization of financial communications based on IFRS are built not only by convergence of national accounting systems, but also through the use of technical platforms such as the XBRL format.

More than 30,000 indicators in the supervisory requirements for nonfinancial and financial reporting have been converted into 12,000 indicators in taxonomy, as well as 94 forms in the supervisory requirements of the professional participants of the securities market.

The Bank of Russia taxonomy is based the IFRS taxonomy and includes the IFRS taxonomy indicators for the international XBRL 2.1 specification.

The use of the XBRL format has its own implementation constraints. First, this is a large financial cost for special software and there is a small number of technically ready users and specialists in this field who are able to read and process documents of this format.

The decision on the expediency of the transition of credit institutions to the XBRL format will be made after the transition of noncredit financial organizations to the XBRL format. If the results of the pilot project are positive, the development of the XBRL taxonomy for nonfinancial organizations will become the basic platform for the development of taxonomy for Russian banks.

REFERENCES

Abdullah, A. 2009. Institutionalisation of XBRL in the USA and UK. *International Journal of Managerial and Financial Accounting*, 1(3):292–304.

Alles, M., & Piechocki, M. 2012. Will XBRL improve corporate governance? *International Journal of Accounting Information Systems*, 13(2):91–108.

Bank of Russia Directive No. 4623-U of November 27, 2017. Available at: http://cbr.ru/analytics/Default.aspx?PrtID=na_vr&docid=461

Bank of Russia Ordinance No. 4584-U of October 25, 2017. Available at: http://cbr.ru/analytics/Default.aspx?PrtID=na_vr&docid=455

Bank of Russia Ordinance No. 4715-U of August 2, 2018. Available at: http://cbr.ru/analytics/Default.aspx?PrtID=na_vr&docid=572

Debreceny, R., & Gray, G. L. 2001. The production and use of semantically rich accounting reports on the Internet: XML and XBRL. *International Journal of Accounting Information Systems*, 2(1):47–74.

Enachi, M. 2013. XBRL and financial Reporting Transparency. *BRAND: Broad Research in Accounting, Negotiation, and Distribution*, 4(1):10–19.

Herald XBRL. 2018. Bank of Russia No. 12018. Available at: http://www.cbr.ru/publ/XBRL/XBRL_2018-01.pdf

Kaspin, L. E. 2013. Possibilities of using XBRL in the formation of integrated reporting. *Innovative Development of the Economy*, 1(13):148–149.

Markelevich, A. 2015. The Israeli XBRL adoption experience. *Accounting Perspectives*, 14(2):117–133.

Morozova, T. V., Safonova, E. G., & Kalacheva, O. N. 2018. Assessment of the impact on taxonomy of IFRS-reporting format XBRL. *Azimuth of Scientific Research: Economics and Management*, 7 (23),237–241.

Official website of XBRL. 2018. Available at: https://xbrl.ru/events

Polozov, A. B. 2010. XBRL: electronic language IFRS-reporting. Corporate *Financial Reporting. International Standards*. 4(42):82–96.

Radu, F. 2016. XBRL: The business Language in the Digital age. *Ovidius University Annals, Economic Sciences Series*, XVI(2):589–594.

Inclusive Development of Society – Lumban Gaol (eds)
© *2020 Taylor & Francis Group, London, ISBN 978-1-138-33476-2*

Alternative culture as part of *genius regionis*

M. Trousil & B. Klímová
University of Hradec Kralove, Hradec Kralove, Czech Republic

ABSTRACT: A qualitative case study explores the interconnection of the natural and cultural uniqueness of the selected Podkrkonoší region in the Czech Republic. An alternative culture is shown as part of *genius regionis* and emphasizes its importance in the context of the development of post-Communist society. The study describes it as a phenomenon ideologically connected to the underground before 1989 and with alternative culture and the local sandstone subsoil. The methods include a review of available relevant literature in the acknowledged databases such as Web of Science or Scopus and Internet sources, observations, and informal and unstructured interviews conducted in the period between 2011 and September 2017. The findings show that the art itself combines alternative oriented personalities, and the local artistic spirit can act as a magnet for the inhabitants of other regions.

1 INTRODUCTION

The authors of this qualitative case study point to the phenomenon of alternative culture in the (sub)region of Podkrkonoší (the foothills of the Krkonoše Mountains, the highest mountains in the Czech Republic) with overlapping to neighboring regions. In the studied area, from the geopolitical point of view of the small post-Communist state in Central Europe, a specific cultural alternative to the current mainstream culture, which is a part of the local *genius regionis*, has been identified and its specificity is mainly connected to the sandstone and following the value base of the underground before 1989.

The alternative culture following the anti-Communist underground is more interesting nowadays. Some Czech philosophers (Bělohradský et al., 2010) speak of the onset of the so-called neo-normalization, which is manifested, among other things, by increasing state control of cultural and social life and resistance to the so-called nonadaptive minorities and migrants. This was characteristic of the period of the so-called normalization (from the 1970s to the 1980s after the political release in the period of the Prague Spring that ended with the occupation of Czechoslovakia). The so-called neo-normalization is reflected, for example, in the field of cultural events by the preference of authorized artists from the period of the Communist regime, the so-called "normalizing stars" in the spirit of "ostalgia" (Cooke, 2003). In the political field, then, the generations rise to standardization of the elite students (e.g., the current prime minister of the Czech government is registered in the Slovak Republic as a former Communist state security agent) and the use of the standardization language.

Alternative culture can be understood as a dynamic concept, embracing one or more subcultural styles, with a historical and spatial context (center × periphery, with different levels of economic and (sub)cultural capital). Although there is a consensus that alternative culture is culture outside of the mainstream, it must be remembered that it may sooner or later be absorbed by the mainstream (Heath & Potter, 2005; Rozsak, 2015). This was done with the counterculture of the 1960s in the subsequent period of "culture of narcissism" (Lasch, 1991), "bobos" (Brooks, 2001), and the commodification of cultural authenticity in response to the process of McDonaldization (Ritzer, 2014). The subject of commodification and "sacrifice" of the tourism industry today is not only authentic cultural "products," but also places and regions themselves (Su, 2011). In this study, alternative culture forms a contrast against popular culture.

Furthermore, the study of local culture in a local/regional context can be considered as an important part of the effort to understand social phenomena and processes that affect the level of social cohesion, cultural openness, and the potential of local development. Identification of *genius regionis* should be an important part of destination marketing in tourism, but also identity formation of local people.

Genius regionis, a concept based on the extension of *genius loci*, contains many subjective, difficult to measure factors. Therefore, there is little literature that deals with it, and the term *genius loci* is more familiar in terms of architecture and art (Norberg-Schultz, 1980). The term *genius regionis* itself is not given much attention in the literature. The Czech geographer Jaroslav Vencálek can be described as the pioneer of its conceptualization: "*Genius regionis* expresses the effect of a large number of extraordinary features and meanings that characterize the region, these characteristics and meanings being completely unrepeatable in the region" (Vencálek 1998: 123).

The author also states: "If we are to approach our understanding of *genius regionis* [. . .], we must try to understand, in addition to expressing the essence of those peculiarities, how these peculiarities and significance function in the landscape" (Vencálek 1998: 123). *Genius loci* and *genius regionis* are, in his view, very subjective, but they can be talked about by identifying the extraordinary characteristics of a given place/region.

The study thus shows a unique interconnection of one of the natural features of the region (sandstone subsoil) with the local alternative culture, which is part of the *genius regionis* of the Podkrkonoší region in the Czech Republic.

2 METHODS

The methods include a review of available relevant literature in the acknowledged databases such as Web of Science or Scopus and Internet sources, observations, and informal and unstructured interviews conducted in the period between 2011 and September 2017. Participants in the interviews were 21 people who were actively involved in organizing alternative cultural events in the region. They all ran them in their free time and (according to their words) without financial enrichment. They were predominantly people with secondary, artistic education, aged 30–58 years. The observations were performed by the authors by attending several cultural events during the year (on average 10 events per year). These events usually included various festivals (e.g., anniversary of the establishment of a village or a town), fairs, or fun fairs. The mainstream events featured mainly music production that can be commonly heard on Czech radio.

The key research question was as follows:

How is the interconnection of nature and alternative culture manifested in the local genius regionis?

3 RESULTS

3.1 Genius regionis *of the Podkrkonoší region*

From the point of view of administrative division, the region does not have a given border but is part of the Czech Republic's marketing/tourist regions, defined by the CzechTourism State Agency. The region is situated in the tourist region of the Krkonošemountains and Podkrkonoší region (CzechTourism, 2017). Since 2000 there has been a voluntary union of municipalities, towns, and companies of Podzvičinsko (derived from Zvičina Hill, 671 m), which is one of the carriers of destination marketing of the Podkrkonoší region. The absence of an administrative region makes it possible to use a definition based on the volition of local institutional actors and the regional identity that makes up the membership of the organization. In fact, *genius regionis* is very stable in the countryside, but this is not necessarily guaranteed by the regional administration. *Genius regionis* exists independently of the purely rational perception of the reality of the territory (Vencálek, 1998).

The area from a higher territorial division is part of the Hradec Králové region (northeast of the Czech Republic), which is the fifth smallest of the 14 regions of the Czech Republic (Czech Statistical Office, 2017) with a population of 550,800 (as of January 1, 2017). It is an agro-industrial region with a significant share of tourism, an aging population, and relatively low unemployment.

The population of the Podkrkonoší region was about 50,000 in 2010. There are 7 municipalities/towns with a total population of 32,900 (as of January 1, 2017). These are Dvůr Králové nad Labem (15,800), Hořice in Podkrkonoší (8,600), Lázně Bělohrad (3,700), Bílá Třemešná (1,300), Pecka (1,300), Mostek (1,200), and Miletín (900). In other municipalities, the average population was only 315 (Czech Statistical Office, 2017).

The boundaries of the area blend with the boundaries of the cadastral territories of municipalities involved in an association, but not all smaller municipalities do so. At the same time, there are relations with neighboring regions (short-term commuting for work, culture, education, etc.). The strongest connections (one-day commuting for work, education, culture, etc.), given by the transport infrastructure, are located 20 km away from the center of the region, with the regional capital of Hradec Králové (92,900 inhabitants as of January 1, 2017) and the district town Jičín (16,000 inhabitants) (Czech Statistical Office, 2017). Jičín is a political-administrative center and the capital of the Bohemian Paradise tourist region, which is famous for its sandstone rock towns (UNESCO Geopark) and sandstone subsoil connected with the monitored Podkrkonoší region.

The specific peculiarity and historical significance of the region is the interconnection of the sandstone subsoil with the local economy, culture, and art. It is not only the use of natural resources in the stone industry (for example, the export of sandstone treated or its use in local construction, as reflected, for example, in the sandstone basement of older houses) but in its artistic processing. In the region itself and its immediate vicinity, at least five major real estate (sandstone) cultural monuments can be found. The first one is the Baroque Hospital Kuks (beginning of the 18th century), with sculptures of the most significant sculptor of the Czech Baroque—Matyáš Bernard Braun (1684–1738). Another is the Baroque Walenstein lodge in Jičín (17th century), the reservoir with sandstone dam Les Království (technical monument and one of the oldest Czech dams from the beginning of the 20th century), and in Hořice the monumental Neo-Renaissance cemetery portal (beginning of the 19th century) and the Neo-Gothic Town Hall (end of the 19th century).

Connection of art and stone seems to be partly reflected in the work of local natives. These include Karel Zeman (1910–1989), whose film *The Road to Prehistory* (1955) received international recognition. A significant Czech writer of children's literature on the subject of Stone Age, Eduard Štorch (1878–1956) was born in the region. In Dvůr Králové nad Labem there is a permanent exhibition of paintings by Zdeněk Burian (1905–1981), who was the most prominent Czech painter of prehistoric motifs and books that were also published abroad.

The local sandstone is the cause of the fact that there are dozens of abandoned quarries in the region. In Hořice in Podkrkonoší, the Secondary School of Stonemasonry and Sculpture was founded in 1884. It was the first of its kind in Austria-Hungary (Tomíčková, 2011) and is currently the only one of its kind in the Czech Republic. Among its graduates there were significant Czech artists, such as Jan Štursa (1880–1925), recognized abroad. The town since 1966 has organized in the abandoned quarry of St. Joseph an international meeting of sculptors, the so-called Sculpture Symposium (Biennial), whose helpers now include people from the organizers of other "alternative" events. After the Velvet Revolution, the school graduates are, in most cases, the organizers of alternative cultural events.

An important regional specificity is also the local tradition of growing fruit trees. This is evidenced by the existence of the Research Institute of Breeding and Fruit Production in Holovousy and the Holovousy raspberry apple ceremony (e.g., a contest for the best apple dessert). Destination marketing itself characterizes the area as a landscape of fruit and stone.

The extraordinary features and meanings of the region, the so-called *genius regionis*, consist of the existence of a sandstone subsoil previously connected with the local industry and today with culture and art. According to the authors' information, as well as in the interviews, there is no high frequency of alternative cultural events in any other region of the Czech Republic (except for large cities of more than 100,000 inhabitants) and not even connected directly with stone/sandstone.

3.2 *Alternative culture and* genius regionis

From the point of view of external expressions, in the Podkrkonoší region, the most obvious alternative culture can be seen as a mix of reggae, hippies, New Age, and, to a lesser extent, punk. In addition to the music, the genres of reggae, ska, big beat, psychedelic rock, punk, world music, gypsy jazz, and funk are based on fine arts (especially sculpture). Ideologically and partly personally, it continues the counterculture of the 1960s and the pre-revolutionary underground, which is reflected mainly in the principles of functioning, organization, cultural actions, values, but also in the personal connection to it. From the interviews and available materials, direct underground connection was not recorded in the Secondary Scholl of Stone-masonry and Sculpture school in the 1970s. In the 1980s, however, punk fans appeared at school, especially around its graduate Štěpán Málek, today's frontman of the nationally significant punk band. In the 1980s, the founder of the nation-famous punk band Support Lesbiens also studied at this school.

Prior to 1989, according to the interviewees, alternative culture was held mainly in private flats, leased by students of the school who did not want to live in a dormitory. After 1989, graduates of the school participated in the organization of alternative cultural events.

The prerevolutionary underground is today connected with four major figures of the local culture. One of them is a local native, a participant in prerevolutionary illegal actions and a publisher of the then underground magazine *Sklepník*. Another two came from the neighboring region Bohemian Paradise (renowned with sandstone rock towns) after 1989 and built a tea and oriental goods store in Hořice. The father of these men was a political prisoner and their uncle was a respected artist. The elder brother of the brothers organized an illegal underground jam session (with the attendance of several dozen people and neighboring regions) in one of the caves of the sandstone rock town of Bohemian Paradise in the 1980s. The second brother became a cofounder and sponsor of the Cultural Reggae Vibez, 14 film festivals (from 1998 to 2012), and an 8-year (2006–2014) irregular summer tea room in the city park with rock, punk, jazz, oriental, and classical music concerts, or theater and film performances.

There is a certain similarity with the situation before 1989. Similarly, as the majority of the society closed in its living rooms and second homes (cottages or weekend houses), the regime's opponents closed into subcultural communities. As Václav Havel (Vaněk, 2010) stated: "In no other Communist country, music, literature, film, or art, except in the late 1960s, were suppressed as in our country. [...] Still, there were musicians who, despite persecution, went patiently for their purpose, which was surprisingly nothing more than to freely play and sing what they liked."

The most significant manifestation of the alternative against the current mainstream culture is the 16-year-old Cultural Reggae Vibez Music Festival in the abandoned sandstone quarry of St. Joseph, with a visit of about 2,000 people. Though it began as a small musical event, artists from the genre of reggae, ska, and world music from around the world perform there today (Jamaica, France, Great Britain, Morocco, etc.). For the years 1995 until 2014, the Big Beat and Punk Festival Garden Festival was held at the quarry, the main organizer of which is one of the members of the prerevolutionary underground movement.

Already since 1969 (with the regime-influenced interruptions) in Hořice in Podkrkonoší, there is also the so-called Welcoming of Spring, connected with the landscape parade and evening music program (big beat, rock, punk, etc.). Its current main organizer is the representative of the local prerevolutionary underground movement along with one of the reggae festival organizers. By September 2017, 21 graduate students of the school had already had 21, the so-called musical stories (3–4 times a year) in local pubs, which can be genuinely written especially for big beats, psychedelic, punk, or reggae. From 1998 to 2012, 14 film festivals took place in the city.

All these events were co-organized and attended by the graduates of the stonemasonry and sculpture school.

3.3 *Relationship to place/region, working with space*

The fundamental specificity of the monitored events of alternative culture is the creative use of space, which is still not used in artistic terms, and its overall character is emphasized.

The interviewees have a strong relationship to a place and a region, especially through participating in local cultural events even if they live far or away from it temporarily or for a long time. Some of them moved permanently to the region after their studies at the local Secondary School of Stonemasonry and Sculpture. The "Podkrkonoší" region means for them a picturesque foothill landscape, sandstone, associated stone carvings, and an alternative culture that forms part of their regional identity and whose appearance and frequency of events are not similar to other smaller towns in the Czech Republic.

The potential of the region/place is used for organizing cultural events. In Hořice alone, the use of the aforementioned, in the forest abandoned, sandstone quarry, fascinates visitors and musicians of the reggae festival. The organizers doubt that the festival could be held elsewhere in order to preserve its uniqueness. In the same quarry, also the international sculpture symposium and the Garden Festival are held every year.

In addition to the Wallenstein loggia, where the space is adapted to the needs of the program and it itself becomes a theater scene, it is necessary to mention the Baroque monument Hospital Kuks, which is connected with the phenomenon of the so-called 17th- and 18th-century Baroque landscapes. The area was used in 2016 and 2017 to organize theater and musical performances, including the "offer" of a certain spiritual overlap in connection with the so-called New Age spirituality (e.g., tea rooms, sales of blowers, and esoteric literature).

Some cultural events were held on private land, respectively, in the gardens of family houses, similar to what happened in the prerevolutionary underground. In the garden, courtyard, and in the barn of the family house (old farmhouse), a small Majdafest festival has been held since 2016.

4 CONCLUSION

Genius regionis of the studied area consists of a unique interconnection of the local sandstone subsoil with an extraordinary amount of artistic expressions and alternative cultural events whose organizers are directly or indirectly associated with stone processing. This interconnection is unparalleled in any region of the Czech Republic. The graduates of the Secondary School of Stonemasonry and Sculpture in Hořice in Podrkonoší or otherwise artistically oriented persons belong to most organizers of alternative cultural events. Their links to the Czech anti-Communist cultural underground determine to a great extent the character of cultural events and are the source of contacts with the bearers of alternative cultures outside the art-oriented artistic scene.

Czechoslovakia, respectively the Czech Republic, underwent two totalitarian regimes (Nazism, Communism) in the 20th century, which led to the destruction of a significant part of the local cultural elite (besides the execution, e.g., the liquidation of the Jews and the postwar expulsion of the Germans). This led, together with state control of citizens, to weakening cultural diversity and creating illegal alternatives to permitted culture. The arrival of liberal democracy and the market economy, along with free choice, also meant the commercialization of culture in conjunction with the elements of "ideologically harmless" (a "normalization" art that does not contain criticism of existing social and political circumstances), neo-normalizing entertainment that does not need to be thought. Thus, the more significant seem to be the existing alternatives that can be perceived as carriers of new cultural diversity and initiators of cultural life in peripheral regions—the regions outside the big cities.

In further exploration of the topic, it will be advisable to focus attention on the absence of reflection of this fact in the marketing of the destination, the attitudes of the local population to this phenomenon, and the monitoring of changes in the nature of alternative cultural events with emphasis on the risk of commercialization and the maintenance of generational continuity.

REFERENCES

Bělohradský, et al. 2010. *Kritika depolitizovaného rozumu*. [Criticism of depolitized reason.] Všeň: Grimmus.

Brooks, D. 2001. *Bobos in Paradise: The New Upper Class and How They Got There*. New York: Simon & Schuster.

Cooke, P. 2003. Ostalgie's not what it used to be. *German Politics and Society*, 22(4):134–150.

Czech Statistical Office. 2017. Available at: from https://www.czso.cz/csu/czso/pocet-obyvatel-v-obcich-k-112017 (accessed September 15, 2017).

CzechTourism. 2017. Available at:http://www.czechtourism.cz/nase-sluzby-pro-vas/spoluprace-s-regiony/mapa-turistickych-regionu-a-oblasti,-kontakty/(accessed September 15, 2017).

Heath, J., & Potter, A. 2005. *The Rebel Sell*. Toronto: HarperCollins.

Lasch, C. 1991. *Culture of Narcissism: American Life in an Age of Diminishing Expectation*. New York: W. W. Norton.

Norberg-Schultz, C. 1980. *Genius Loci: Towards a Phenome nology of Architecture*. New York: Rizzoli.

Ritzer, G. 2014. *McDonalddization of Society*. Thousand Oaks: SAGE.

Roszak, T. 2015. *Zrod kontrakultury: úvahy o technokratické společnosti a mládeži v opozici*. [Birth of contra culture: Considerations of technocratic society and youth in opposition.] Prague: Malvern.

Su, X. 2011. Commodification and the selling of ethnic music to tourism. *Geoforum*, 42(4):496–505.

Tomíčková, O. 2011. *Zmizelé Čechy – Hořice*. [The dis-appeared Bohemia Hořice]. Litomyšl: Paseka.

Vaněk, M. 2010. *Byl to jenom rock'n'roll?: Hudební al-ternativa v komunistickém Československu, 1956–1989*. [It was just rock'n'roll?: A musical alternative in the communist Czechoslovakia, 1956–1989.] Prague: Academia.

Vencálek, J. 1998. *Protisměry územní identity*. [Countermeasures of territorial identity.] Český Těšín: OLZA.

Financial security monitoring in terms of threats and vulnerabilities associated with people's financial literacy level

E.N. Alifanova, Yu.S. Evlakhova, E.S. Zakharchenko & A.V. Ilyin
Rostov State University of Economics, Rostov-on-Don, Russia

ABSTRACT: This article is a review of national financial security monitoring in terms of new threats arising from expansion of financial facilities and the low level of people's financial literacy. The Strategy of Increasing Financial Literacy in the Russian Federation for 2017–2023 was put into effect by the Order of the Russian Government. Thus, the article proposes to supplement the financial security indicators with those related to the people's financial behavior. The general scientific methods used were the analysis and synthesis methods of analogy and deduction. The empirical methods used in the study were observation, comparative and statistical analysis, and evaluation. The method of tabular interpretation was also used.

1 INTRODUCTION

Financial security monitoring is an important stage in the provision of financial security because it makes possible the determination of the effects of the measures taken, on the one hand, and identification of deviations in the financial security level at a certain time horizon on the other. The term "monitoring" means the constant observation of the process in order to determine whether it corresponds to the desired result or initial preferences. The International Organization for Standardization (ISO) defines monitoring as systematic checks, oversight, surveys, and status determination conducted to identify changes in the required or expected level of functioning.

2 METHODOLOGY AND RESULTS

By analogy with the definition of the term "economic security monitoring" (Senchagov & Ivanov, 2015), financial security monitoring means the process of continuous observation of the financial system and financial markets, including data collection, monitoring the financial security indicators dynamics, identifying trends in economic development, and threat forecasting.

National financial security monitoring is aimed at preventing and reducing losses in the financial sector by observation, analyzing and forecasting the time dynamics of certain indicators, and informing interested organizations and individuals about the occurrence of negative phenomena. In a broad sense, national financial security monitoring should not only perform an information function, but also be a basis for developing measures to reduce the probability of systemic financial crises and mitigate their social and economic effects.

It is possible to distinguish some financial security monitoring functions in the system of ensuring national financial security based on this understanding:

1. The preparatory and preventive function, which includes the definition of threats and risks, data collection and processing, and threats and vulnerabilities assessment
2. The forecasting and analytical function, which includes prognosis making and assessment of risk and possible losses

3. The information function, which includes data spread to target groups, interests, and organizations

Financial security monitoring relies on observation, analysis, and forecasting of macroeconomic (principally) and aggregated microeconomic indicators. The use of the latter in monitoring is less common, because their application in empirical studies is limited by data availability. For example, foreign researchers emphasize the important role of such indicators as short-term foreign currency liabilities, as well as the significant value of the capital adequacy ratio in monitoring financial security.

The financial security monitoring tool can be divided into two groups, depending on the approach used.

The first one, named "signal," studies the behavior of indicators behavior during the noncrisis, precrisis, and crisis periods, on the basis of which conclusions are made about the indicator to the crisis or macroeconomic shock sensitivity and the applicability of the indicator for monitoring the change in the state of financial security. This is about the indicator dynamics, studying, for example, the stock index in different economic periods. Retrospective data of the indicators are used: the dynamics in a stable period (noncrisis), in the period preceding the already known crisis that occurred (pre-crisis), and accordingly the dynamics in the crisis period. If the character or the direction of the dynamics indicators change during these periods, then this makes it possible to consider any economic indicator as a signal indicator. In addition, the analysis of the behavior of indicators can be conducted:

- At a qualitative level (by visual comparison of charts), which allows identification of general patterns predicting a change in the state of financial security
- At a statistical level (by econometric analysis), which allows identification of the critical areas or intervals, which, in turn, will mean a threat to financial security.

The second approach is the estimation of econometric models, where the binary variable is used as the endogenous one. Unlike the "signal" approach, this one takes into account the behavior of all indicators simultaneously and allows assessing future financial security threats, which undoubtedly is its advantage. Yet when using the method of estimating econometric models, a number of difficulties arise, including:

- The heterogeneity of crises and macroeconomic shocks, both in one and in different countries, inhibits generalization of the experience.
- Some financial system and financial market vulnerabilities cannot be measured quantitatively.
- The determinants of crises and macroeconomic shocks can vary significantly over time.

National financial security monitoring means regular, systematic comparison of values of financial security indicators with:

- The certain threshold values—for assessing the state on a certain date
- Values for the previous periods (that is, the deviation of the indicator from a certain trend) to evaluate the processes in the financial system and to forecast the situation.

There are various approaches to the definition of threshold values (normative, analogy method, expert judgment, division of indicators into threshold, one-threshold and two-threshold indicators). The plurality of approaches makes it very difficult to assess the financial security level, despite the fact that it is the threshold values that are the decisive parameter in determining the financial security threat. In this regard, we support the position of the authors' team of S. M. Drobyshevsky, P. V. Trunin, A. A. Paliy, and A. Yu. Knobel (2006), who believe that "the same indicators dynamics may indicate both crisis probability increase, so and on the normal financial system development," so they propose considering each indicator and explaining its dynamics in the context of the overall economic situation.

A certain role in financial security monitoring is the observation of people's financial market operations.

The World Economic Forum held in 2015 in Davos singled out the main innovations on which humanity should focus in the future, including innovations that would be challenges to national financial security related to the financial behavior of people, such as:

- The world of cashless payments
- Strengthening of the role of retail investors and consumers
- Crowdfunding
- Expansion of alternative lending
- Changes in consumer priorities

Russian data analysis shows the urgency of changing consumer priorities as a challenge to Russia's financial security. In fact, the use of external sanctions in 2014–2015 gave rise to the problem of finding domestic financial resources for the development of the Russian economy. Studies have shown that households' financial resources can become such a source. However, the consequence of the macroeconomic shock in 2014 and the subsequent recession was the worsening of the financial well-being of Russian households, which was reflected in the change of the financial behavior model. The financial behavior savings model began to dominate the credit one, a tendency described in detail by Alifanova et al. (2017). As a result, at the end of 2017 it became evident that the Russian people could not be a large-scale source of financial resources for the development of the Russian economy, but also actively act as borrowers. The bank consumer lending growth in recent years leads to the fact that the banks' credit resources that could be directed to the real economy lending sector are given to the people.

In these conditions, the task of increasing the Russian people's financial literacy level becomes extremely important. The necessary factors in the strategy for increasing people's financial literacy are presented in Table 1.

The subjective factors, as well as the generally low level of financial culture and discipline, play a key role in the Russian Federation, taking into account the statistics presented in Figure 1.

At the same time, financial services are becoming more convenient and more complex; in addition, the practice of informing clients on the financial market risks of a particular financial product is not sufficiently developed.

Consistent work is being carried out to increase financial literacy in order to prevent negative trends among the Russian people. Thus, in 2011–2018, the Ministry of Finance of the Russian Federation, in cooperation with the Federal Service for Surveillance on Consumer Rights Protection and Human Wellbeing (Rospotrebnadzor), the Central Bank of the Russian Federation, Ministry of Education and Science of the Russian Federation, and Ministry of Economic Development of the Russian Federation, implemented the project "Assistance in raising the people's financial literacy level and the financial education development in the Russian Federation" (under the auspices of the World Bank). The main directions of the implementation of the project are:

- Developing an improving financial literacy strategy and monitoring and evaluation of people's financial literacy level
- Creation of personnel potential in the field of increasing people's financial literacy
- Educational programs and public awareness
- Improvement in consumer rights protection in financial services

Table 1. Factors in the formation of national strategies for increasing people's financial literacy.

Objective factors	Subjective factors
Economic crisis phenomena	Mismatch in people's knowledge level
Complications in the services offered	mismatch
Easing of access to financial services and their promotion	Low level of financial discipline
Addressing the need for reliable information on the financial services	

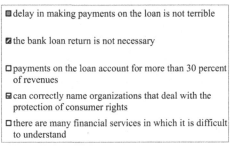

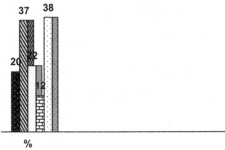

Figure 1.   Russian financial culture characteristics.

In 2018, the Central Bank of the Russian Federation, as the financial market mega-regulator, launched The Strategy of Increasing Financial Literacy in the Russian Federation for 2017–2023. The main directions of the program implementation are:

- Creation of public communication mechanisms for delivering to the people practically meaningful information
- Systematization of current activity and the transformation of individual practices into a comprehensive work

The key task of the strategy is to form a "financially literate citizen" who is able:

1. To monitor his or her own finances: income and expenditure planning, formation of long-term savings for unforeseen circumstances, debts incommensurate with income and avoidance of nonpayments on them, knowledge and compliance with tax obligations, preparation for retirement
2. To use the necessary financial information for the financial services selection taking into account all the market risks and legal financial services consumer rights, to recognize financial fraud

By 2023, as the result of the Strategy of Increasing Financial Literacy in the Russian Federation for 2017–2023, the following are planned:

- Improvement in the state policy for people's financial education
- Implementation of an interdepartmental approach to the problem of people's low level of financial literacy
- Development and implementation of financial, educational, organizational, information, and educational activities
- Availability of financial education, as well as citizens receiving information about new financial services and a popularizing guarantee of financial knowledge,
- Federation involvement in solving increasing problems of people's financial literacy

Thus, the importance of the financial literacy problem and recognition of its significance at the state level allows us to propose supplementing the group of indicators with those related

Table 2. Financial security indicators group related to people's financial behavior and their corresponding databases.

No.	Indicator	Data source
1	Growth rate of monthly individuals' deposits in rubles and foreign currency The ratio of deposits for a period up to 1 year to the total deposits volume (for individuals);	Central Bank of the Russian Federation
2	The proportion of households that kept their savings during the month preceding the survey The average savings amount deferred for the month preceding the survey (rubles) The share of households that spent savings and sold accumulated currency and jewelry The average amount of savings spent, money from the sale of jewelry and accumulated currency (rubles)	Household surveys
3	Payments for debt service and volume to income ratio of basic payments The household debt to net income ratio	Organization for Economic Co-operation and Development
4	The household debt to GDP ratio Loans to deposits ratio	the Central Bank of the Russian Federation
5	Share of households that spent money to repay loans during the past 30 days The average amount spent by the household to repay the loan (rubles) Share of households with loan debt The total amount of household loan debt average value (currency debt is exchanged into rubles)	Household surveys
6	Net volume of cash and non-cash foreign currency purchases	Russian Federation Federal State Statistics Service
7	The households spending money on the currency purchase for savings proportion, in the month preceding the survey The average amount of money expended on the currency purchase for the purpose of savings for the month preceding the survey (rubles)	Household surveys

to the people's financial behavior (Table 2). The composition of the indicators and their role in the assessment of national financial security are detailed in the work of Alifanova (2017).

3 CONCLUSIONS

In conclusion, we note that assessment and monitoring of national financial security based on quantitative indicators cannot be considered sufficient because the state of security of the financial system is also determined by factors that are not quantifiable. Qualitative indicators of financial security (including the quality of banking supervision, the adequacy of monetary policy, compliance with international standards and codes) were introduced into the analysis and monitoring practice after 1994. With this in mind, we consider it expedient to continue work on the formation and addition of qualitative indicators related to people's financial behavior that reflect the financial culture and discipline level. This will allow monitoring of financial security, taking into account threats and vulnerabilities associated with people's financial literacy level.

4 DIRECTIONS FOR FUTURE WORK

We expect further research topic development in the analysis of interrelations among risks in the financial behavior of the population and of the vulnerabilities of financial organizations in order to ensure national financial security.

REFERENCES

Alifanova, E. N., Evlakhova, Yu. S., & Nivorozhkina, L. I. 2017. Forming of the financial security indicators group reflecting households transactions in the financial market: Methodological approach. In *10th International Scientific and Practical Conference, "Science and Society,"* pp. 67–74. London: Berforts Information Press. — Available at: scieurohttp://.com/wp-content/uploads/2017/03/v1-67-74.pdf http://orcid.org/0000-0003-1610-9396 (Web of Science – ORCID).

Alifanova, E. N., Evlakhova, Yu. S., Nivorozhkina, L. I., & Tregubova, A. A. 2018. Indicators of financial security on the micro-level: Approach to empirical estimation. *European Research Studies Journal.* Special issue 1. https://www.ersj.eu/index.php?option=com_content&task=view&id=1212

Central Bank of the Russian Federation Available at: https://www.cbr.ru/eng/

"Development of a Methodology for Assessing Russia's Financial Security Based on a Study of the Impact of Macroeconomic Shocks on the Dynamics of Savings and Population Operations in the Credit and Foreign Exchange Markets," grant carried out with the financial support of the Russian Foundation for Humanities (Agreement No. 16- 02-00411).

Drobyshevsky, S. M. et al. 2006. Some approaches to the development of a system of financial stability monitoring indicators. Moscow Institute of Economy, Policy and Law.

Frankel, J. A., & Rose, A. K. 1996. Currency crashes in emerging markets: empirical indicators. NBER Working Paper No. 5437.

Gonzalez-Hermosillo, B., Pazarbasioglu, C. & Billings, R. 1997. Determinants of banking system fragility: A case study of Mexico. IMF Staff Papers, No. 44.

Order of the Russian Government No.2039-r 09. 25.2017. The Strategy of Increasing Financial Literacy in the Russian Federation for 2017–2023.

Organization for Economic Co-operation and Development. Available at: http://www.oecd.org/

Russian Federation Federal State Statistics Service. Available at: http://www.gks.ru/

Senchagov, V. K., & Ivanov, E. A. 2015. *The Modern Economic Security Structure Mechanism Monitoring in Russia.*

World Economic Forum. The future of financial services. Final report. 2015. Available at: https://www.slideshare.net/rafaeldelallamaaguir/worlrd-economic-forum-the-future-of-financial-services

Inclusive Development of Society – Lumban Gaol (eds)
© 2020 Taylor & Francis Group, London, ISBN 978-1-138-33476-2

User perception of Shari'ah compliance in PayTren

Martini Dwi Pusparini, Ayu Fatimah & Yuli Andriansyah
Department of Islamic Economics, Faculty of Islamic Studies, Islamic University of Indonesia, Yogyakarta, Indonesia

ABSTRACT: PayTren is considered the first fintech to use the Shari'ah system. It should be understood that Islamic compliance is an absolute requirement for all financial institutions with the Shari'ah system. This study aimed to find out how users perceive Shari'ah compliance in Pay Tren applications. The population sample for the study consisted of PayTren application users. Sampling was conducted by means of a snowballing technique while data collection employed questionnaires and interviews (involving 62 respondents). The study utilized three variables: users, perceptions, and compliance with Shari'ah. These variables were equipped with indicators such as opinions; selection; views; motives; needs; experiences; freedom from the elements of usury, maisir, and gharar; limiting activities; risk sharing; the desirability of materiality (real economic transaction); and justice consideration. Users responded strongly in agreeing that the PayTren application has implemented the Shari'ah compliance system properly (acceptance percentage of 95.65%). The results show that users strongly agree that PayTren works as a system of Shari'ah compliance.

Keywords: Islamic fintech, PayTren, Shari'ah compliance, user perception

1 INTRODUCTION

From a more advanced technological development comes the term fintech (Financial Technology). According to Christemsen and Bower (1995), fintech is a disruptive innovation that has succeeded in transforming and developing a system (market) through introduction of practicality, ease of access, and comfort along with economical costs (Hadad, 2017, p. 2).

The development of the financial technology business in Indonesia has been very rapid. Until June 2018, the Financial Services Authority (OJK) recorded 64 registered financial technology companies (OJK, 2018). This figure has increased compared to the previous position, where only 54 financial technology companies had registered. Furthermore, as of April 2018 the peer-to-peer lending industry (fintech lending) had distributed loans of Rp 5.42 trillion in April 2018. This value increased to 111.23% compared to December 2018. Also, the ratio of nonperforming loans (NPLs) from fintech lending reached 0.53% in April 2018. This was a decrease compared to December 2017, when it reached 0.99%. Still, the number of total borrowers in April 2018 reached 1.47 million entities, probably 468.79%.

The number of lenders reached 162,373 enti- ties, an increase of 60.86% YTD (Rossiana, 2018). Fintech itself has sprung up tremendously. For instance, there is #Cekaja #HaloMoney #Doku #Veritrans #Ruma #NgaturDuit #Jojonomic #Cermati #Aturduit #Kartuku overall, yet for fintech Shari'ah in Indone- sia there is only one: PayTren (Ardela, 2017). Many e-commerce companies have come up with various sales fields. Most of them deal with services or real products. Moreover, there is a digital mobile based on mobile banking and payment applications, a good example being PayTren. This is a product in the form of software from the company PT Veritra Sentosa International, which can be used in all types of smartphones to carry out transactions such as automatic teller machines (ATMs), mobile Banking, and payment point online Bank (PPOB), though it is valid only for closed communities that have obtained Shari'ah certificates from MUI (Andayani, 2017).

With the recognition as the first fintech to use the Shari'ah system, it is necessary to acknowledge the fact that Islamic compliance is an absolute requirement for all financial institutions that adopt Shari'ah. Compliance with Shari'ah principles is the fulfillment of its provisions in all activities carried out as a manifestation of the characteristics of the institution itself such as Islamic financial technology.

From the community's perspective, especially the users of Shari'ah technology in financial services, Shari'ah compliance is at the core of the integrity and credibility of the institution, with its existence able to meet the needs of the Islamic community to implement the teachings of Islam as a whole, including in the distribution of fintech activities.

A great deal of debate emerged regarding the clarity of Shari'ah compliance law in PayTren applications. The debates questioned whether the transactions in PayTren trade did not contain elements of gharar, maysir, and usury. In such debates, a number of Ulama opinions emerged from outside parties to PayTren. Seen from its educational background, PayTren is fairly competent (Anuga, 2017). A number of issues arise though. First, Ustadz Ammi's explanation established that PayTren contained gharar. This is attributed to the uncertainties in income users get even irrespective of the same registration fee for all users (Baits, 2017). Second, the Ustadz Erwandi explanation concluded that PayTren contained usury. The reason is said to be usury because people who buy products are not rewarded (Tarmizi, 2017).

While some people think PayTren contains ele- ments that are not in accordance with Shari'ah principles, there are also a good number that argue otherwise. K. H. Ma'ruf Amin, as Chair of the Indonesian Ulema Council (MUI), handed over the Shari'ah certificate to PayTren. Ma'ruf said the appraisal process was carried out strictly and according to procedures. Even though his relationship with Ustadz Yusuf Mansyur was very close, the requirements had to be met (Andayani, 2017). PayTren's reason for getting a halal certificate was to ensure that PayTren trades real transaction objects, the products being traded are legitimate, prices do not harm consumers, bonuses are given to members in accordance with achievements and contract and amounts, and that PayTren does not engage in any money game activities (PayTren, 2017). The aforementioned issues raised differences in people's thinking about PayTren. Public confidence and trust in the implementation of Islamic law principles were adopted in the operational rules of the Shari'ah banking law no. 21 of 2008. Where there is no compliance with Shari'ah principles, the public will lose the features they seek to influence their decision to continue using the services provided by PayTren. Noncompliance with Shari'ah principles could have a negative impact on PayTren's image and increase the chances of being abandoned by potential customers. The important objective of compliance has implications for the need to supervise its implementation. Supervision of Shari'ah compliance is an action meant to ensure that Shari'ah principles that are the basic guidelines for the operation of all the financial institutions have been implemented appropriately and comprehensively (Chrismastianto, 2017).

This differences in opinion motivated the research on PayTren's Shari'ah compliance. PayTren application users' perceptions are expected to answer the questions regarding Shari'ah compliance that are used in PayTren applications that are Shari'ah-based in financial technology. Human resource professionals who understand the principles of Shari'ah compliance are expected to solve the problems in the implementation of Islamic economic systems. Knowledge and understanding of the Islamic economic system is an absolute requirement for the actors of Islamic financial institutions ranging from leaders to employees. The purpose of the study was to determine "User Perception of Sharia Compliance in PayTren Applications in Yogyakarta." The problem statement was to establish how users perceive Shari'ah compliance in PayTren.

2 RESEARCH METHODS

This was descriptive research, with the following objective: to find general and universal descriptions that apply to a number of variations of situations and conditions (Yusi & Idris, 2009, p. 25). The research location was Yogyakarta and was conducted over 4 months from March 19, 2018 to July 19, 2018. The population for the study involved 62 users of the PayTren application.

The sampling technique used was snowballing, which means the sample size is initially small and then enlarges (Yusi & Idris, 2009, p. 68). There were two types of data: primary data sourced from the questionnaire and secondary data derived from the documents. Questionnaire measurement in this study used a rating scale, with two conflicting alternative answers: strongly disagree and strongly agree.

There were three variables, namely perception, users, and Shari'ah compliance with conceptual and operational definitions as shown in Table 1.

Table 1. Definition of operational variables.

Variable	Definition	Indicators
Perception	The process of selecting, organizing, and interpreting information inputs to create an overall Picture	a. Opinion b. Selection c. Views
User	A person's participation resulting from a product, system, or service	a. Motive b. Needs c. Experience
Shari'ah compliance	The whole system is buying and selling that does not conflict with Shari'ah principles	a. There is no element of usury b. There is no element of maisir c. There is no gharar element d. Limiting types of activities e. Emphasis on risk sharing. f. The desirability of materiality (real economic transaction) g. Justice consideration

Source: Primary data.

Data analysis techniques using descriptive analysis consisted of describing the problem by means of the existing data, and thereafter deductions were drawn. The results were explained according to the indicators of each variable in the descriptive analysis. The variable descriptive analysis explained the respondent's rejoinders to the variables. In the descriptive study, data analysis included steps to process data, analyze it, and then find results. Data results were divided into four categories: strongly disagree, disagree, agree, and strongly agree. After calculating the descriptive analysis, some detailed explanations were made and then summarized in the table one by one according to the indicators of each variable. The summary clearly stated the conclusions based on the results, including in the classes of strongly disagree, disagree, agree, or strongly agree and then explained by the results of the research and the facts that corresponded to the categories of each research indicator in the Discussion section.

3 RESULTS

PT. Veritra Sentosa Internasional (Treni) was established by Ustadz Yusuf Mansur on July 10, 2013, based on the Deed of Establishment of Limited Liability Company No. 47 by Notary/PPAT H. Wira Francisca, SH., MH as a contribution in supporting the lives of Indonesian people. From the observations of the habits and culture of Indonesian society, the idea to provide facilities aimed at facilitating and helping the community emerged.

By receiving a Shari'ah Certificate, the government had completed the acknowledgment. Previously, PayTren had been registered as an organizer of electronic systems and transactions from Kominfo, SIUPL (Direct Selling Business License). The certificate issuance was done so that people could feel safe using innovative products from PayTren or trying other opportunities offered. The PayTren Shari'ah certificate would be valid and used for monitoring to ensure it operates in the Shari'ah compliance corridor. A Shari'ah Supervisory Board had been established with members appointed by DSN-MUI including Adiwarman A Karim and Bukhori Muslim (Desastian, 2017).

Based on the results of the validity test on de- scriptive variables using SPSS 16.0 in 30 respond- ents, it can be concluded that all the questions in the

trial questionnaire are declared valid with an r table size of 0.361. As for the reliability test, it was found that the Cronbach alpha results from the total of all variables was 0.942, a value greater than 0.6, implying the questionnaires used by researchers were consistent.

Respondents' rejoinders from the questionnaires were then grouped and processed to find the total score of the class interval first. The average values of the acquisition of the respond- ent scores was grouped into four categories: strongly disagree, disagree, agree, and strongly agree with the formula:

Rating Scale = Highest Score − Lowest Score/Number of categories = $10 - 1/4 = 9/4 = 2.25$. From the results, the above calculation was categorized as shown on Table 2.

From the aforementioned categories, the distribution of the overall calculation of the user's perception of sharia compliance in the PayTren application in each dimension produces the mean values shown in Table 3.

From the results of the percentage of descriptive analysis of the foregoing data above, it can be concluded that the results of the assessment of user perceptions of Shari'ah

Table 2.   Assessment category.

Score	Category
1.00–3.25	Strongly agree
3.36–5.50	Disagree
5.51–7.75	Agree
7.76–10.00	Strongly agree

Source: Primary data.

Table 3.

No	Variable	Mean	Category assessment results
1	Perception	8.89	Strongly agree
2	Perception	9.13	Strongly agree
3	Perception	,	Strongly agree
4	User	9.13	Strongly agree
5	User	8.95	Strongly agree
6	User	8.61	Strongly agree
7	User	8.71	Strongly agree
8	User	8.97	Strongly agree
9	User	8.77	Strongly agree
10	Shari'ah compliance	8.87	Strongly agree
11	Shari'ah compliance	9.15	Strongly agree
12	Shari'ah compliance	9.18	Strongly agree
13	Shari'ah compliance	9.23	Strongly agree
14	Shari'ah compliance	9.1	Strongly agree
15	Shari'ah compliance	9.05	Strongly agree
16	Shari'ah compliance	8.68	Strongly agree
17	Shari'ah compliance	9.06	Strongly agree
18	Shari'ah compliance	7.19	Agree
19	Shari'ah compliance	8.66	Strongly agree
20	Shari'ah compliance	8.37	Strongly agree
21	Shari'ah compliance	8.79	Strongly agree
22	Shari'ah compliance	8.92	Strongly agree
23	Shari'ah compliance	8.39	Strongly agree

Source: Primary data.

compliance in the PayTren application that argues agree is 4.35% and those who strongly agree, 95.65%.

4 DISCUSSION

From the results of the research analysis. it can be concluded that PayTren has very good Shari'ah compliance standards. This conclusion was obtained from the calculation of descriptive analysis, in which 4.35% of respondents said they agree the PayTren application system was in accordance with the principles of Shari'ah compliance. Some, amounting to 95.65%. of the respondents, stated that they strongly agree that the PayTren application system is in accordance with the principles of Shari'ah compliance.

Based on perception variables with indicators such as opinion, selection, and outlook, user percep- tions strongly agreed that PayTren is in accordance with the concept and application of Shari'ah compliance and strongly agreed that PayTren is a Shari'ah-based financial technology. As with the products issued by the PayTren application, there are various PPOB payment services equipped with alms features. Some Islamic products are in the form of policy and technology most appropriate to be used in achieving the stated goals. This will affect the PayTren application and equate its perception with the financial needs of the technology. Users who are satisfied will always give good comments about the entity and tend to be loyal to it.

Based on user variables with motive, need, and experience indicators, user perceptions strongly agree that individuals use the PayTren application because it has good business prospects along with the user needs for Shari'ah-based products being high. These three indicators can be proven by the progress of various aspects including their utilization in terms of the banking business. PT Veritra Sentosa International can take advantage of financial technology progress by releasing a PayTren application. Moving from digging potential communities accustomed to using the latest technology to bridging the ease of payment of all the needs of the community by combining the use of gadgets and respecting obligations by establishing a PayTren company was an achievement. PayTren can license the use of applications or software where the marketing system is developed through partnership cooperation (direct selling/ direct selling) with the network concept. This is an application that can be downloaded via PlayStore on an Android-based smartphone or AppStore on iPhone products (Septiana, 2018).

Based on Shari'ah compliance variables with seven indicators, user perceptions strongly agree that PayTren is in accordance with the principles of Shari'ah compliance. Business in the Qur'an is inseparable from Islamic values. So the provisions of Shari'ah are the main values that become a strategic umbrella in business activities. In running a business, there are three elements that should be avoided: usury, gharar, and maisir. The results of research on these three elements found that the perception of users strongly agrees on PayTren application in accordance with the principles of Shari'ah compliance. The first one was proven by the calculation of SPSS in the questionnaire that had been distributed to the respondents. In the Shari'ah compliance, the variable with the indicator that does not contain the element of usury cannot harm one of the two parties to a transaction, showing a mean of 8.87, which falls into the strongly agree category. The minimum value given by the respondents ranged between 1 and 10, with a standard deviation of 1.684, which meant that the value of the data approached the mean. The second one was proven by the calculation of SPSS in the questionnaire that had been distributed. The variables of Shari'ah compliance with indicators did not contain maisir elements. The net PayTren from the element maisir and monopoly shows a mean of 9.15, which meant that it fell into the strongly agree category. The minimum value given by the respondents was between 5 and 10, with a standard deviation of 1.171. This meant that the value of the data approached the mean. The third one was evidenced by the calculation of SPSS in the questionnaire distributed to the respondents. In Shari'ah compliance, variables with indicators that do not contain gharar elements and in which the net PayTren from the gharar element and the sale of its non-existent products shows that a mean of 9.18 is categorized as strongly agree. The minimum value given by the respondents is 1 and up to 10, with a standard deviation value 1.563, which means that the value of the data ap-proaches the mean.

In muamalah, the rule explains that the legal origin of transactions is permissible unless there is a prohibition in the argument. According to Al-ashlu fil muamalah al ibahah, this rule implies that the scope of muamalah transactions is broad, innovation is highly appreciated, and new transactions can be accepted. In contrast to very narrow worship, there is no room for innovation. The transaction ban in muamalah can be simplified in three elements: usury, gharar, and maysir. Riba is an addition obtained without the consequences of the risk received and the sacrifice made. Riba can occur in buying and selling transactions or debts. Gharar is obscurity, including gharar in objects, gharar in transactions, or gharar in the time of surrender. Maysir or gambling or speculation can actually be in- cluded in the gharar principle because of lack of clarity. The maysir referred to is speculation that contains the principle of zero-sum-game or the profits obtained by one party that are inversely proportional to the losses suffered by other parties (Mardian, 2016). The foregoing explanation is based on previous research regarding muamalah. In this concept, the reason for the user's perception of Shari'ah compliance in PayTren applications is in the strongly agree category, because the results of data analysis revealed the same things as the research that was free from the elements of usury, gharar, and maisir.

In this study seven indicators were used in assessment of Shari'ah compliance in the PayTren application. Six indicators showed results strongly agreeing that the PayTren application fulfills the principle of Shari'ah compliance. The indicators included no element of usury, maisir, and gharar, limiting the type of activity, the desirability and materiality (real economic transaction), and consideration of justice. Nevertheless, one indicator showed a different value from others, agreeing with the PayTren application of Shari'ah compliance. The indicator was an emphasis on risk-sharing, with a mean of 7.19, which meant that it was in the agreed category. The minimum value given by the respondent was ranged from 1 to 10, with a standard deviation of 2.963. This meant the value of the data was far from the mean.

In the risk-sharing indicator, there is an indication that the user still has not received the risk-sharing facility in the management of PayTren's business, which means that the risk is borne more by the user. According to Chin (2010, 2015) and Soegiarto (2012), risk perception is defined as the uncertainty and inconvenience faced by consumers in the purchasing process, because their decisions cannot be predicted for negative consequences in product valuation that affect the decision-making process of purchasing products or services (Rahim, 2017). As a record, in accordance with the principle of risk-sharing management (Anwar & Edward, 2016), PayTren still has to fix it. From the results of the study, there are still users who give small values that produce data that only agree that PayTren uses Shari'ah compliance in managing risk sharing. It can be interpreted that PayTren still has shortcomings in the risk-sharing system even though only a few. Improving the system for managing risk sharing will increase the value of Shari'ah compliance in the user's perception. The risk itself is defined as the uncertainty and inconvenience faced by consumers in the purchasing process because their decisions cannot be predicted by their negative consequences in product valuation that affect the decision-making process of purchasing the product or service. The consumer's decision to modify, postpone, or avoid purchasing decisions is greatly influenced by the risks considered (Rahim, 2017).

Based on the Shari'ah perception variable, three other indicators not explained are the PayTren indicator which limits the type of activity, the desirability of materiality (real economic transaction), and consideration of justice. User perceptions strongly agreed that PayTren is in accordance with the principles of Shari'ah compliance. This is in accordance with Circular No. 12/13/DPS concerning the Implementation of Good Corporate Governance for Shari'ah Commercial Banks and Sharia Business Units, which requires banks to assign at least one employee to support the implementation of the duties and responsibilities of the Shari'ah Supervisory Board. This is the basis for PayTren to carrying out duties while implementing Shari'ah compliance. The supervision carried out by DPS included PayTren activities related to the fulfillment of Shari'ah, both operational and product. DSN (National Shari'ah Board) is a board formed by the MUI and tasked with the duty and authority of ensuring the compatibility between the bank's products, Services, and business activities with Shari'ah principles that granted halal certificates to PayTren on August 7, 2017. Certificates submitted directly by the Chairperson MUI KH Ma'aruf Amin prove that PayTren has indeed met operational and product standards in accordance with the provisions of Shari'ah compliance (Desastian, 2017). By receiving a Shari'ah Certificate, the government completed the acknowledgment. Previously PayTren was registered as an organizer of electronic systems and transactions from Kominfo, SIUPL (Direct Selling Business License) from BKPM and become a member of the APLI (Direct Selling Association). The certification is done so that people could feel safe in using the innovative products from PayTren or even trying the opportunities offered by PayTren, since the Shari'ah certificate will be valid in monitoring PayTren to ensure it runs in the Shari'ah compliance corridor. A Shari'ah Supervisory Board had been established with members appointed by DSN-MUI including Adiwarman A Karim and Bukhori Muslim (Desastian, 2017).

5 CONCLUSION

The test results on user perceptions of Shari'ah compliance on PayTren applications using indicators including opinions; selection; views; motives; needs; experience; freedom from the elements of usury, maisir, and gharar; limiting activities; application of risk sharing; desirability of materiality (real economic transaction); and fairness considerations provided a response. That is, the respondents strongly agree that the PayTren application has implemented the Shari'ah compliance system properly in a percentage of 95.65%. The users' perception of Shari'ah compliance in the PayTren application strongly agreed that PayTren has run a system of Shari'ah compliance in the company's performance.

REFERENCES

Andayani. D. 2017. Ketum MUI Serahkan Sertifikat Syariah Ke Paytren Yusuf Mansyur. *detik.com*.
Anuga. B. 2017. *Sertifikat halal PayTren: DSN MUI melawan fatwanya sendiri*. Available at: bahas bisnis.com: http://bahasbisnis.com/2017/08/11/sertifikat-halal- paytren-dsn-mui-melawan-fatwanya-sendiri/ (accessed August 16, 2018).
Anwar & Edward. 2016. Analisis Sharia Compliance Pembiayaan Murabahah Pada Gabungan Koperasi BMT Mitra Se-kabupaten Jepara. In *The 3rd University Research Colloqium 2*.
Ardela. F. 2017. *Mengenal 10 Perusahaan Fintech Indonesia*. Jakarta: Finansialku.com.
Aria.P. 2018. *Berita Terkini Ekonomi Dan Bisnis*. Available at: katadata.co.id: https://www.google.co.id/amp/s/amp.katadata.co.id/be rita/2018/06/04/kantongi-izin-bi-paytren-incar-dana- kelolaan-rp -30-triliun-per-bulan (accessed July 17, 2018).
Baits. U. A. 2017. *Hukum PayTren*. Available at: konsultasisyariah.com: http://konsultasisyariah.com/29323-hukum-paytren- bagian-01.html# (accessed August 16. 2018).
Chrismastianto. I. A. 2017. Analisis SWOT Implementasi Teknologi Finansial Terhadap Kualitas Layanan Perbankan Indonesia. *Jurnal Ekonomi Dan Bisnis*, 20 (1).
Desastian. 2017. *Ustad Yusuf Mansyur Lega PayTren Raih Sertifikat Syariah DSN-MUI*. Available at: Panjimascom: http://www.panjimas.com/news/2017/08/16/ustadz- yusuf-mansur-lega-paytren-raih-sertifikat-syariah-dsn- mui/ (accessed 23, 2018).
Hadad. M. 2017. Financial Tevhnology (Fintech) di Indonesia. *kuliah umum tentang Fintech – IBS*. Jakarta: Otoritas Jasa Keuangan.
Hasan. A. 2009. *Manajemen Bisnis Syari'ah*. Yogyakarta: Pustaka Pelajar.
Kadir. 2010. *Hukum Bisnis Syariah Dalam Al-quran*. Jakarta: Amzah.
Mardani. 2015. *Hukum Sistem Ekonomi Islam*. Jakarta: PT. Rajagrafindo Persada.
Mardian. S. 2016. Tingkat Kepatuahan Syariah Di Lembaga Keuangan Syariah. *Jurnal Akuntansi Dan Keuangan Islam* 3(1).

Maswadeh. S. 2014. A compliance of Islamic banks with the principles of Islamic finance (Shariah): An empirical survey of the Jordanian business firms. *International Journal of Accounting and Financial Reporting.* 171.

Masyrafina. I. 2017. *PayTren Dapat Sertifikat MUI.* Available at: Republika.co.id: https://www.republika.co.id/berita/ekonomi/syariah- ekonomi/17/08/07/oubb6c396-paytren-dapat- sertifikasi-mui (accessed July 23, 2018).

Nazwirman. 2008. JuPeranan Lembaga Keungan Mikro Berbasis Teknologi Informasi Dalam Mengembangan Usaha Mikro. *Jurnal The Winners,* 9(2)

Nurhisam. L. 2016. Kepatuhan Syariah (Sharia Compliance) dalam Industri Keuangan Syariah. *Jurnal Hukum IUS QUIA IUSTUM,* 23(1)

P3EI. 2013. *Ekonomi Islam.* Jakarta: PT. Rajagafindo Persada.

PayTren. G. 2017. *Perdebatan Halal Dan Haram Bisnis PayTren.* Available at: gadispaytren.com: http://www.gadispaytren.com//perdebatan-kehalalan- paytren/ (accessed August 16, 2018).

Rahim. H. 2017. Analisis Pengaruh Persepsi Resiko Dan Kepercayaan Terhadap Minat Pengguna Paytren Pada PT. Veritra Sentosa Internasional. *Jurnal EKOBISTEK Fakultas Ekonomi,* 6(2).

Riadi. M. 2012. *Teori Persepsi.* Available at: www.kajianpustaka.com/2012/10/teori-pengertian- proses-faktor-persepsi.html.

Riduwan. 2003. *Skala Pengukuran Variabel Variabel Penelitian.* Bandung: CV. Alfabeta.

Septiana. N. 2018. Strategi Komunikasi Persuasif Personal Selling Anggota PayTren dalam melakukan Network Marketing Di Pekanbaru. *JOM FISIP,* 5(1).

Setyowati. D. 2017. https://katadata.co.id/berita/2017/08/28/bi-prediksi- transaksi-fintech-naik-24-menjadi-rp-249-triliun-di- 2017 (accessed July 17, 2018).

Shihab. Q. 2015a. *Ali Imran Ayat 130.* Available at: tafsirq.com: https://tafsirq.com/3-ali- imran/ayat-130#tafsir-quraish-shihab (accessed August 16, 2018).

Shihab. Q. 2015b. *Al-Maidah Ayat 90.* Available at: tafsirq.com: https://tafsirq.com/5-al- maidah/ayat-90#tafsir-quraish-shihab (accessed August 16, 2018).

Shihab. Q. 2015c. *Tafsir An-Nisa Ayat 29.* Available at: tafsirq.com: https://tafsirq.com/4-an- nisa/ayat-29#tafsir-quraish-shihab (accessed August 16, 2018).

Siri. Fitriyani. & Herliana. 2017. Analisis Sikap Pengguna Paytren Technlogy Acceptance Model. *Jurnal Informatika, 4(1).*

Suhartono. 2017. Antisipasi Perubahan Teknologi Keuangan. *Ekonomi Dan Kebijakan Publik,* 9(21/1/ Puslit).

Sunyoto. D. 2011. *Metodologi Penelitian Ekonomi.* Jakarta: CAPS.

Tarmizi. U. E. 2017. *Hukum PayTren* [Motion Picture].

Treni. 2017a. *Company Profile.* Bandung: www.treni.co.id.

Treni. 2017b. *Kode Etik Mitra.* Bandung: PT. Veritra Sentosa Internasional.

Treni. 2017c. *Marketing Plan.* Bandung: www.treni.co.id.

Utomo. D. A. 2013. Motif Pengguna Jejaring Sosial Google+ Di Indonesia. *Jurnal E-komunikasi,* 1(3).

Waluyo. A. 2016. Kepatuhan bank Syariah Terhadap Dewan Syariah Nasional Pasca Transformasi Ke Dalam Hukum Positif. *Jurnal Penelitian Sosial Agama,* 10(2).

Yusi & Idris. 2009. *Metodologi Penelitian Ilmu Sosial Pendekatan Kuantitatif.* Citrabooks Indonesia.

Zamroni. W. R. 2016. Pengaruh Marketing Mix Dan Sharia Compliance Terhadap Keputusan Nasabah Memilih Bank Umum Syariah Di Kudus. *Jurnal Ekonomi Syariah,* 4(1)

Inclusive Development of Society – Lumban Gaol (eds)
© 2020 Taylor & Francis Group, London, ISBN 978-1-138-33476-2

Peers' social support and academic stress among boarding school students

Miftahul Hidayah & Fitri Ayu Kusumaningrum
Department of Psychology, Islamic University of Indonesia, Yogyakarta, Indonesia

ABSTRACT: This research aimed to determine the relationship between peer support and academic stress on boarding senior high school students. The hypothesis proposed was that there existed a negative relationship between peer support and academic stress on boarding senior high school students. The researchers used a peer support scale developed by Thompson and Mazer (2009), the Student Academic Support Scale (SASS), and an academic stress scale adapted from Sun et al. (2011), the Educational Stress Scale for Adolescent (ESSA). Both scales were given to 100 boarding high school students. The results showed a correlation coefficient of $r = -0.38$ with value of significance $p = 0.000$ for peer support and academic stress; thus the hypothesis was accepted.

Keywords: academic stress, boarding school students, peer support, senior high school

1 INTRODUCTION

A relevant education promotes academic achievement. According to Warsito (2012), improving the quality of education will have positive effects on a child's academic achievements. One form of education is boarding school. According to Fathonah et al. (2017) boarding school is a term known in the education system and refers to provision of education to students living within the institution's premises, such as Islamic boarding schools, church schools, or schools in official educational institutions.

Academic stress is a pressure related to academic activities caused by higher academic demands, poor academic results, or poor academic environment (Weidner et al., 1996). The causes of academic stress include high academic standards (Olejnik & Holschuh, 2007), academic burdens (Misra et al., 2000), fear of not getting a place in college, school exams, too much teaching material, hard to understand subjects, too much homework, and crammed school schedules (Shahmohammadi, 2011). For boarding school students, academic stress occurs due to various factors including a curriculum that is different from schools in general (Maslihah, 2011). Further, students' independence is taken away, making them unable to solve problems (Zaid et al., 2016)

Results of researchers' interviews conducted with female student identified by the initial A indicate that she had some difficulties adapting to the new school environment. She was a migrant from outside the city and had difficulties with several subjects that had never been studied back in her native country, one of which is a Javanese language. She felt less confident because of being last in the class as a result of getting low grades in Javanese language on the midterm exam. In addition, the burden of many subjects and school activities sometimes made her feel uncomfortable in school, and even led to stress, anxiety, cold sweating, need to cry, and difficulty in concentrating; sometimes she even felt too lazy to go to school. She also admitted that she was often exhausted due to crowded activities at her boarding school. After attending classes and doing course work at school, she had to take part in extracurricular activities until noon. At night she was engaged in the activities held in the dormitory, as well as having to do homework. Based on the results of the interviews, she experienced several aspects of academic stress as mentioned by

Sun et al. (2011): pressure from the academic workload, worry about grades, self-expectation, and despondency.

The influence of the environment becomes a social support and protective factor for students in dealing with academic problems and inhibits academic stress. One of them is social support from peer groups. Research conducted by Puspitasari et al. (2010) states that peer social support has a negative correlation with anxiety in students. The higher the peer social support, the lower is the level of anxiety. Anxiety is an indicator of stress (Agolla & Ongori, 2009).

According to Weidner et al. (1996), academic stress is the stress associated with educational activities due to academic demands. Olejnik and Holschuh (2007) define academic stress as a response that arises because of too many demands and tasks that must be done. Wilks (2008) stated that academic stress results from a combination of educational demands on individuals beyond their capacity. Based on the foregoing definition, causes of academic stress according to Sun et al. (2011) include (1) pressure from study including feelings of distress caused by the burden of study at school; (2) workload, including students being subjected to too much work; (3) worry about grades, especially if the students' previous performances have been below GPA; (4) inability to meet self-expectations; and (5) despondency, an indicator of academic stress that results from feeling less confident and experiencing a large number of difficulties when studying at school. Factors that influence academic stress are divided into two: internal and external. In this study, researchers focused on peer social support as factors that influence academic stress because they wanted to know whether peer social support received can influence the level of academic stress experienced by students.

Peer social support is a source of support (Sarafino, 2006). Santrock (2007) stated that peers are children or adolescents whose age and maturity level are more or less the same. Peer support according to Mead et al. (2001) is a form of granting and receiving assistance based on certain principles such as shared responsibility and mutual agreement among them about what is useful. According to Thompson and Mazer (2009) in the academic environment there are four forms of support that can be given by peers: (1) informational support to overcome academic problems including helping out with school assignments and providing advice; (2) esteem support, which is emotional and can create comfort, confidence, and increased self-esteem for students; (3) motivational support that encourages students to complete their assignments, which is an important strategy for students to cope with the more challenging school work; and (4) venting support as the main strategy used to overcome frustration with teachers and classes.

The hypothesis proposed in this study was "the presence of a negative relationship between social support and academic stress on boarding high school students." The higher the social support, the lower the academic stress experienced by boarding high school students and vice versa.

2 RESEARCH METHODS

The study aimed to determine the relationship between peer social support and academic stress among boarding high school students. The sample population for the study included 100 10th and 11th grade students in boarding high school. The research measured academic stress by means of the Educational Stress Scale for Adolescents (ESSA) adopted by Sun et al. (2011). To measure peer social support, the Student Academic Support Scale (SASS) by Thompson and Mazer (2009) was used. The researchers used statistical analysis to determine the relationship between peer social support and academic stress with the help of the Statistical Package for Social Science (SPSS) 22.0 for Windows as a data analysis method. The analysis used in this study was the Spearman Product Moment correlation test.

3 RESULTS

3.1 Descriptive statistics

Descriptive Research data from the peer social support scale and academic stress scale are shown in Table 1.

Table 1. Descriptive statistics of academic stress and peer social support.

Categories	Academic stress		Peer social support	
	Frequency	%	Frequency	%
Very low	0	0	5	5
Low	15	15	28	28
Average	58	58	30	30
High	24	24	22	22
Very high	3	3	15	15
Total	100	100	100	100

Source: Primary data.

Based on the data in Table 1, it can be seen that on the academic stress scale there were 15 students in the low category (15%), 58 in the average category (58%), 24 in the high category (24%), and 3 in the very high category (3%). On the peer social support scale it can be seen that there are 5 in the very low category (5%), 28 in the low category (28%), 30 in the medium category (30%), 22 in the high category (22%), and 15 in the very high category (15%).

3.2 Assumption test

The assumption test was done as a prerequisite that must be fulfilled before hypothesis testing. It was conducted by means of a normality and linearity test through calculation of statistics with the help of SPSS version 22 for Windows.

A normality test was done to determine the distribution of data from each research variable using the one-sample Kolmogorov–Smirnov test on SPSS 22 for Windows. Generally, the data have a normal distribution if they have a coefficient of $p > 0.05$, while the data have an abnormal distribution if they have a coefficient value of $p < 0.05$. Table 2 shows the result of the normality test of academic stress variables and peer social support.

Table 2. Results of the normality test.

Variable	KS score	p	Category
Academic stress	0.067	0.200	Normal
Peer social support	0.084	0.078	Normal

Source: Primary data.

A linearity test was done to determine whether the two variables have a linear relationship. Ideally, the relationship between the two variables in the study is linear if $p < 0.05$ whereas the relationship between the two variables is nonlinear when $p > 0.05$. The linearity assumption is stronger if the significance value obtained from F deviation from linearity is greater than 0.05 (Sig > 0.05). Table 3 is the result of academic stress variables and peer social support after the linearity test was done.

Table 3.   Results of the linearity test.

Variable		F	p	Category
Academic stress and peer social support	Linearity	17.192	0.000	Linear
	Deviation from linearity	1.097	0.369	

Source: Primary data.

3.3 *Hypothesis test*

After testing the normality and linearity, it was found that the data in this study had a normal distribution and were linear. Furthermore, a hypothesis can be tested to prove whether or not the research is accepted. This is done by the Pearson correlation test because in the assumption test, the data result is normal and linear.

Table 4.   Results of the hypothesis test.

Variable	R	p	Information
Peer social support	-0.380	0.000	Significant

Source: Primary data.

The analysis results of the correlation coefficients between academic stress and peer social support showed the values of $r = -0.380$ and $p = 0.000$ ($p < 0.05$). This shows that there was a significant relationship between peer social support and academic stress. The relationship was negative and so the higher the peer social support that is received, the lower the academic stress felt by boarding high school students. Therefore, the research hypothesis was accepted.

4   DISCUSSION

In this study, it was found that the effective contribution of peer social support variables to academic stress variables was 14.4%, meaning that 85.6% of effective contributions to academic stress variables are influenced by other factors that were not examined in this study. The results showed that out of 100 sample students, 58 (58%) were in the category of average academic stress. Furthermore, social support received by 30 of them (30%) who constituted the majority also posted an average result. The research conducted by Marhamah and Hamzah (2016) also found similar results.

The relationship between aspects of peer social support in the form of informational support has a negative correlation with academic stress variables. This was indicated by the correlation coefficient value of $r = -0.563$ with a coefficient of significance of $p = 0.000$. Informational support has a significant relationship to academic stress, with a coefficient of determination of $r^2 = 0.317$, or 31.7%. According to research conducted by Gonzalez et al. (2015), there is a significant relationship between informational support and stress.

The next aspect of peer social support, that is, esteem, also has a negative correlation to academic stress variables with a correlation coefficient of $r = 0.575$ with a coefficient of significance of $p = 0.000$. Esteem support has an impact on academic stress, with a coefficient of determination of $r^2 = 0.33$ or 33%. Research conducted by Andharini and Nurwidawati (2015) established that there is a significant negative relationship between social support (esteem support) and academic stress.

The next aspect of peer social support, motivational support, also had a negative correlation with academic stress variables, having a correlation coefficient of $r = -0.587$ with a coefficient of significance $p = 0.000$. Further, motivational support has a significant relationship to academic stress, with a coefficient of determination of $r^2 = 0.344$ or 34.4%. The relationship between aspects of peer social support in the form of informational support has a negative correlation with academic stress variables. This was indicated by the correlation coefficient value of $r = -0.563$, with a coefficient of significance of $p = 0.000$. Informational support has a significant relationship to academic stress, with a coefficient of determination of $r^2 = 0.317$, or 31.7%. According to a research conducted by Gonzalez et al. (2015), there is a significant relationship between informational support and stress.

The next aspect of peer social support, that is, esteem, also has a negative correlation to academic stress variables with a correlation coefficient of $r = 0.575$, with a coefficient of significance of $p = 0.000$. Esteem support has an impact on academic stress, with a coefficient of determination of $r^2 = 0.33$, or 33%. Research conducted by Andharini and Nurwidawati (2015) established that there is a significant negative relationship between social support (esteem support) and academic stress.

The next aspect of peer social support, motivational support, also had a negative correlation with academic stress variables, having a correlation coefficient of $r = -0.587$, with a coefficient of significance of $p = 0.000$. Further, motivational support has a significant relationship to academic stress, with a coefficient of determination of $r^2 = 0.344$, or 34.4%.

5 CONCLUSION

This study establishes that there is a negative relationship between peer social support and academic stress among boarding high school students. The higher the peer social support received, the lower is the academic stress. Conversely, the lower the peer social support, the higher is the academic stress. Therefore, it is imperative to provide boarding high school students with peer social support. It is equally important to reduce academic stress experienced by students. Nevertheless, there is a need for further research. Studies on similar topics should (1) pay attention to other variables that have an effect on academic stress, especially among boarding high school students, and (2) be conducted using qualitative methods.

REFERENCES

Agolla, J. E., & Ongori, H. 2009. An assessment of academic stress among undergraduate students: The case of University of Botswana. *Educational Research and Reviews*, 4(2):63–70.

Andharini, A. J., & Nurwidawati, D. 2015. Hubungan Antara Dukungan Sosial dengan Stres pada Siswa Akselerasi. *Jurnal Character*, 3(2).

Fathonah, D. Y., Hernawaty, T., & Fitria, N. 2017. Respon psikososial siswa asrama di Bina Siswa SMA Plus Cisarua Jawa Barat. *Jurnal Pendidikan Keperawatan Indonesia*, 3(1):69–77.

González, L., Hernández, A., & Torres, M. V. 2015. Relationships between academic stress, social support, optimism-pessimism and self-esteem in college students.

Marhamah, F., & binti Hamzah, H. 2017. The relationship between social support and academic stress among first year students at syiah kuala university. *Psikoislamedia: Jurnal Psikologi*, 1(1).

Maslihah, S. 2011. Studi tentang hubungan dukungan sosial, penyesuaian sosial di lingkungan sekolah dan prestasi akademik siswa SMPIT Assyfa Boarding School Subang Jawa Barat. *Jurnal Psikologi*, 10(2):103–114.

Mead, S., Hilton, D., & Curtis, L. 2001. Peer support: A theoretical perspective. *Psychiatric Rehabilitation Journal*, 25(2):134.

Misra, R., & McKean, M. 2000. College students' academic stress and its relation to their anxiety, time management, and leisure satisfaction. *American Journal of Health Studies*, 16(1):41.

Olejnik, S. N., & Holschuh, J. P. 2007. *College Rules! How to Study, Survive, and Succeed in College*, 4th ed. New York: Ten Speed Press.

Puspitasari, Y. P., Abidin, Z., & Sawitri, D. R. 2010. *Hubungan antara dukungan sosial teman sebaya dengan kecemasan menjelang ujian nasional (UN) pada siswa kelas XII reguler SMA Negeri 1 Surakarta*. Doctoral dissertation, UNDIP.

Santrock, J.W. 2007. *Perkembangan Anak. Jilid 1 Edisi kesebelas*. Jakarta: PT. Erlangga.

Sarafino, E. P. 2006. *Health Psychology: Biopsychosocial Interactions*, 5th ed. Hoboken, NJ: John Wiley & Sons.

Shahmohammadi, N. 2011. Students' coping with stress at high school level particularly at 11th & 12th grade. *Procedia-Social and Behavioral Sciences*, 30:395–401.

Sun, J., Dunne, M. P., Hou, X., & Xu, A. 2011. Educational stress scale for adolescents: Development, validity, and reliability with Chinese students. *Journal of Psychoeducational Assessment*, 29(6):534–546.

Thompson, B., & Mazer, J. P. 2009. College student ratings of student academic support: Frequency, importance, and modes of communication. *Communication Education*, 58(3):433–458.

Warsito, H. 2012. Hubungan antara self-efficacy dengan penyesuaian akademik dan prestasi akademik (Studi Pada Mahasiswa FIP Universitas Negeri Surabaya). *Pedagogi: Jurnal Ilmu Pendidikan*, 9(1):29–47.

Weidner, G., Kohlmann, C. W., Dotzauer, E., & Burns, L. R. 1996. The effects of academic stress on health behaviors in young adults. *Anxiety, Stress, and Coping*, 9(2):123–133.

Wilks, S. E. 2008. Resilience amid academic stress: The moderating impact of social support among social work students. *Advances in Social Work*, 9(2):106–125.

Wilks, S. E., & Spivey, C. A. 2010. Resilience in undergraduated social work students: Social support and adjustment to academic stress. *Social Work Education*, 29(3):276–288.

Zaid, S. S., Saam, Z., & Arlizon, R. 2016. Pengaruh layanan informasi terhadap manajemen stres siswa di kehidupan asrama (*boarding school*) kelas x kehutanan negeri pekanbaru tahun ajaran 2014/2015. *Jurnal online mahasiswa Fakultas Keguruan dan Ilmu Pendidikan Universitas Riau*, 3(1):1–5.

Inclusive Development of Society – Lumban Gaol (eds)
© 2020 Taylor & Francis Group, London, ISBN 978-1-138-33476-2

Accession of students' human capital through higher education

A.L. Blokhin & S.V. Kotov
Southern Federal University, Rostov-on-Don, Russia

N.S. Kotova
Russian Presidential Academy of National Economy and Public Administration, Rostov-on-Don, Russia
Southern Federal University, Rostov-on-Don, Russia

A.A. Kizim
Kuban State University, Krasnodar, Russia

ABSTRACT: This study investigates the accession of students' human capital by means of higher education, which is directly related to the research of experts from the World Economic Forum (WEF). In particular, the article examines the following aspects: the quality of higher education and methods for assessing the criteria and indicators of student's individual human capital. The study used several methods: the method that uses natural (temporary) assessment, measurement of human capital (education); methods of analysis (principles and methods of system analysis, logical-structural and comparative analysis); the calculation of the human development index, the method of integrated assessment of individual human capital of students, based on the use of computer programs. Based on the test data of relevant criteria and modules, obtained from the calculations of test program results, we could create an individual trajectory of the high school student's human capital expanding, which in the future will create a favorable development of human capital in general, namely, the development of innovation and productivity, and allow to master the necessary production or management skills.

Keywords: *accession, human capital, higher education, test program*

1 INTRODUCTION

In 2017, the World Economic Forum conducted a study of The Human Capital Index, which showed the results of human capital development across various states of the world. The fundamental idea of the research was to show that the main resource of human capital is people's knowledge and skills that allow the creation of economic values. In the global world, the disclosure and development of students' potential along with quality education are considered as strategic factors that determine the long-term prospects and competitiveness of each country.

"It is the qualitative education and work that provide people with a livelihood, enable them to realize their ambitions, contribute to the development of society. The introduction and dissemination of innovations, and labor productivity directly depend on the skill level of employees, their possession of the necessary production or management skills" [Kotova, Kotov, Blohin, Ogannisyan, Borsilov, 2017].

This attitude provides the topicality of our research to study the methodological basis of the accession of students' human capital through higher education.

The conducted pedagogical research proved that the solution of the personnel policy of a developing state and personality, the problem of accession of students' human capital, which are actual for the theory and practice of higher education, is achieved by the use of progressive views on the subject of research that have developed not only in pedagogy and psychology, but also in philosophy, sociology, economics and other sciences.

2 RESEARCH METHODS

According to the UNESCO document, there are three aspects of the quality assessment of educational activities which are applied in two models of assessing educational activities, the "French" one (the external evaluation of the university in terms of its responsibility to society and the state, through certification, accreditation, and inspection), the second model is "English", which is based on the internal self-assessment of the university academic community:

a) quality in higher education is a multidimensional concept that must cover all its functions and activities: curricular and academic programs, research and scholarships, staffing, students, assignments, physical infrastructure, equipment, work for the benefit of society and academic environment;

b) quality also requires that higher education has an international dimension: knowledge sharing, the creation of interactive networks, the mobility of teachers and students, international research projects, along with the consideration of national cultural values and conditions;

c) the following components are of particular importance for the achievement and quality assurance at the national, regional and international levels (the careful selection of teachers and personnel in higher education, the continuous improvement of their qualifications by expanding relevant programs in the field of professional development, including methodology teaching and learning, teacher's mobility between countries, between higher education institutions, as well as between the latter and the world of work, along with the mobility of students in the countries and between them, due to their impact on the acquisition of knowledge and skills new information technologies became the important means of mobility).

3 DISCUSSION

The quality of education depends on the level of educational prestige in the public consciousness and the system of state priorities, financing and physical infrastructure of educational institutions, modern technologies of their management [Petrova, Mareev, Pivnenko, Kotov, Kotova, Kharchenko, Gulyakin, 2016]. The environment created by higher education institutions affects the perception of students' qualities. Students who study at universities, where there is a high-level quality of education, tend to have a higher level of expectation) of the quality of education [Husain Salilul Akareem & Syed Shahadat Hossain, 2016].

In the international standards ISO 9000, quality is understood as the degree to which an object (product, service, process) corresponds to certain requirements (norms, standards).

Having analyzed the concepts and definitions of "qualitative education", we came to the conclusion that qualitative education is ensured by the professionalism of the teaching staff, the integration of science and the educational process, the connection of the university with enterprises, business the diversity of innovative pedagogical technologies used in educational process, various types of highly organized practical attachments.

The idea of this study is to determine whether students receive a sufficiently high-quality education through the training [Goryunova, Kotov, Akopyan, Ogannisyan, Borzilov, 2018].

In general, to meet the challenges of quality assurance of education, European and many Russian universities are creating internal systems to ensure the quality of educational services, while being guided by the standards and recommendations of the European Association for Quality Assurance in Higher Education (ENQA).

The standards and recommendations of ENQA were developed on the basis of the experience of assessing the quality of education in Western Europe, without taking into account the educational practices of the countries of Eastern Europe and Russia, as at the first stage the latter were not presented in the EAQA. Nevertheless, the standards and recommendations are worthy of detailed study and orientation on them.

One of the ENQA standards is the "Guarantee of the quality and competence of the teaching staff". It considers the teacher as the «main resource of the learning process» and determines the direct dependence of the graduates quality on the quality of teaching.

To date, there are several approaches used in the procedure for assessing the level of professionalism of university teachers:

- formally normative approach;
- the LEADER evaluation system, which involves peer review;
- evaluation of teachers by students;
- competence approach [Competence approach in MSIIR].

Therefore, the question of teachers' professional competence seems the most relevant. The professional competence of a teacher can be defined as a system of knowledge, skills and personal qualities that form the basis of his/her professional pedagogical and scientific activity.

With due regard for the multidirectionality of the subject studies of different universities, one can approach the description of the main components in the structure of the professional competence of university teachers. It can serve as a basis for assessment and self-assessment of teachers, their work, status, place in the faculty.

One of the central elements in the structure of the professional competence of higher education teachers is the cognitive component, or the intellectual-pedagogical component. It provides an opportunity to overcome the established stereotypes of pedagogical activity and master new ways of professional self-realization. This component is characterized by the following criteria: the availability of relevant integrated knowledge, the ability to continually improve it, creativity, flexibility and critical thinking; ability to analyze the professional situation and reflection. It also includes the knowledge and application of new teaching technologies at university, fully adapted to their experience and the subject specifics; focus on the connection between theory and practice in order to develop an active professional position and effective thinking of future specialists; provision of feedback in learning through various types of control and self-control.

Evaluation of teachers according to the criteria of these components can give an answer about their professional competence - the basic qualitative characteristic of a university teacher.

The study of the pedagogical essence, the formation and improvement of human capital in the system of higher professional education has not previously been the subject of independent scientific research in the context of the implementation of a new higher professional education roadmap.

Human capital, especially its intellectual resource, has a decisive influence on the rates of economic growth and the level of national welfare. In these theories, the real driving force of progress is a man, then growth is, first of all, the function of developing the opportunities laid and uncovered in man. The role of higher education at the present stage of any country's development is determined by the tasks of its transition to a democratic and lawful state, to a market economy, the need to overcome the danger of the country lagging behind the world trends of economic and social development.

Within the framework of providing effective education at the Southern Federal University (SFU), in order to improve the monitoring of students 'achievements, the formation of students' portfolio and the appointment of higher academic scholarships, an information analytical system "Personal rating of the learner" was developed.

For the implementation of the system "Personal rating of the learner" we carried out an approbation of an independent assessment of students' knowledge. The comparative analysis of the research results was assessed and conducted in the control (250 people) groups on the gender and age composition, which were identified as identical to the experimental group, which students were trained in accordance with Federal state Educational Standard of Higher Education and the traditional (deepening certain aspects of the disciplines studied) set of special courses of the variable part of the curriculum. The data obtained at the control stage were compared with the data of the control group, as well as the data of the experimental group, which had been obtained before the experiment. On the basis of the data thus obtained, it was possible to prove the effectiveness of the applied pedagogical technology.

A total of 85 students of different SFU courses. Four teachers took part in the approbation as independent experts from the Don State Technical University (DSTU).

Students passed exams to teachers of their department SFU in the presence of independent experts - teachers who did not conduct classes in these groups (DSTU). Credits took the form of interviews - answers to ticket questions.

Further, the experts carried out an independent assessment of the quality of the students' preparation, followed by the analysis of the results [Kotova, Esenskaya 2015].

The procedure for an independent assessment of the quality of students' activities was carried out by assessing the students' activities, as well as the forms and methods of testing.

Based on the results of the experiment on approbation of an independent assessment of students' knowledge, the following results were obtained: 83 students passed the test, which is 97% of the participants in the experiment, 2 students (3%) did not pass the test.

The discussion of the content and forms of conducting an independent assessment of the quality of students' activities in separate academic disciplines and the opportunities for disseminating this experience led to the following conclusions:

- the evaluation of students' activities with the involvement of external experts contributes to improving the quality of teaching, makes the evaluation procedure more objective;
- the participation of external experts has a positive impact on the processes of self-preparation of students, increases the responsibility of students, promotes more productive independent work.

We determine that the theory of human capital, which originated in economic science, can also have a pedagogical meaning.

Firstly, the economic approach as a research method can be used for scientific research in the sphere of education, including pedagogical and psychological ones.

Secondly, the conclusions obtained by G. Becker and other authors of the theory of human capital, represent value not only for theory, but also for the practice of education.

Thirdly, the concretization of the "human capital" allows us to determine the current goals, objectives and criteria for assessing the quality of education.

Therefore, in order to fully understand the connection between human capital and higher education, we want to expand the pedagogical thesaurus with the concept of "accession" of human capital into higher education, which has a great potential of modern educational practice, especially vocational education.

The term "accession" is considered by us in the pedagogical context as a general idea of accession or addition. It can refer not only to a legal term, as described in the Large Law Dictionary, but also to economics, business, diplomacy, government, sociology and even the papacy. A student, a teacher and a state enter into certain legal relations in the process of education, what happens to the student at the university is entirely the law and standards. Thus, the more knowledge, skills and abilities accumulated by a person in the process of upbringing, formal and informal education, the higher the qualitative level of human capital, the greater the probability of obtaining a higher income and a higher standard of living.

In terms of pedagogical science and practice, human capital is a combination of qualities and characteristics that allow us to determine the individual trajectory of the development of competitive human qualities: from the business qualities of an individual person to professional skills.

Measuring a student's human capital is a complex process that cannot be assessed with the help of only indicators of academic achievement though they are significant. The most important function of a higher education institution (HEI) is the development of general cultural capital, laying the universal basis for further professional activity and subsequent development. At the macro level, the highly educated workforce is the human potential of the country. [Burgess, Simon M., 2016].

To date, the following tasks are required for the accrual of human capital through higher education:

- the formation of students objective value guidelines;
- the introduction of methods of teaching and evaluation, adapted to the perception of the environment by modern students;
- the formation of a wide range of internal freedom of students;
- improving the quality of education.

To solve one of the tasks set, in the specific case, this is improving the quality of education at the Southern Federal University, a grading system has been developed for assessing students' academic performance, the goal of which is to improve the quality and productivity of training, knowledge control and competence. The Evaluation of knowledge, skills and competencies in the discipline is carried out on a 100-point scale. The final evaluation is made up of a ratio of 60/40 (electronic testing/hands-on training).

The final score, taking into account the summation of points, is translated into a 5-point one as follows: 90-100 - excellent; 76-89 - good; 60-75 - satisfactory/credited; less than 60 - unsatisfactory/not credited.

In order to improve the monitoring of students' achievements, the formation of a portfolio of students, the increased academic scholarships, the University developed an information and analysis system called "Personal Rating of the Learner", which is a tool for preserving students' achievements, quantifying them, scoring and assessing students. The personal rating of the learner is made up of data for academic, scientific, public, creative and sporting achievements. The introduction of a scoring and rating system gives full opportunities for individualization and motivation for learning.

The teacher should create the best line for applying scientific knowledge, using modern educational technologies aimed at creating conditions for successful teaching, ensuring the quality of education, and not just giving knowledge to the student.

One of the indicators of the university's activity in the field of ensuring the quality of training future specialists is the degree of students' satisfaction with various aspects of the educational process at the university.

In this regard, in 2017 we conducted a study - monitoring the satisfaction of graduates of specialists, bachelors and masters of the Southern Federal University of the Academy of Psychology and Pedagogy. The purpose of the study was to determine the quality of the educational services provided by the university [Aibatyrov K.S., Aibatyrova M.A., Annikova L.V., Bakhutoshvili T.V., Kotov S.V., Kotova N.S., Kharchenko L.N., 2017].

To conduct our study, a questionnaire was developed. A total of 120 students took part in the study. Based on the results of the questioning of students, we saw that 85.5% of respondents were satisfied with the quality of the educational services provided, 4.5% were not satisfied with the quality of the educational services provided and 10% were at a loss to answer.

Satisfaction of students with the quality of the educational services provided is due to many factors. Let us single out some of them: sufficiency of knowledge for future effective professional activity, satisfaction with the university as a whole (the prestige of the university itself), the main factor in choosing a university/direction of preparation. It is these factors that are considered to study the formation of human capital in higher professional education.

The success of the formation of the educational basis of modern human capital directly depends on the following qualities, abilities and skills of the teacher:

- natural intelligence (an individual with individual genetically determined inclinations and abilities, a certain level and characteristics of the intellect, acquired knowledge, skills and skills and having the opportunity to improve and develop human capital);
- multi-faceted competence;
- confidence (value orientation of the social characteristics of human capital, reflecting a person's motivation, which regulates person's choice in the process of implementing this motivation);
- ability to communicate;

- ability to trust relationships (it forms the criteria, the most significant from the point of view of human capital);
- readiness to support in all endeavors of students (measures to increase student participation, which strengthen their sense of responsibility and manage their own endeavors and destinies, and support students' research potential.);
- the availability of adaptive thinking (the adaptation of certain criteria of human capital to new environmental conditions, which guarantees the quality of specialists produced);
- the ability to positively interpret (positive assessment of the compliance level of the graduate's human capital with the requirements of the employer and society).

4 RESULTS

Our main research, which is outlined in the title of the article, is the development of the electronic test program "Database - Individual Student's Human Capital", which provides a comprehensive assessment of the level of individual human capital of students, measuring the values of human capital indicators. The program is designed to conduct testing to determine the individual human capital of students in the university. In this program, students are tested according to the following criteria (most significant from the point of view of human capital):

1. The capital of health. To date, the state of health of students is defined as satisfactory, but at the same time we have found out during the questionnaire that more than 50% of students who come to study at the university already have different types of physical disabilities - weak vision, scoliosis, lack of sufficient muscle mass. In the health capital we have included such modules as "healthy lifestyle", "health status" and "health value".
2. The capital of education is the inculcation of values to knowledge, the ability to accumulate them. In the capital of education we have included the modules "educational achievements", "attitude to learning activity", "learning motives".
3. Cultural and moral capital (general culture, behavior, moral values and principles) [Karpova, Mareev, Akopyan, Guterman 2017].
4. Labor capital (professional tests and additional education, professional orientation, the value of professional activity).
5. Intellectual capital (creativity, interest and ability to innovate, motivation for change).
6. Organizational and entrepreneurial capital (leadership qualities, entrepreneurship, motives of entrepreneurial activity).
7. Social capital (communicative qualities, social activity and adaptability, social motives and values).

"Database - Individual student's human capital" is designed to preserve the test results of university students in order to determine individual human capital. The Data is used for subsequent analysis and output of results using a special program in the form of a petal diagram.

The structure of the database provides for the immediate preservation of information about students undergoing testing (the year of enrollment, the course, the level of the education received), as well as the retention of answers to the test questions by 7 criteria (health capital, education capital, cultural and moral capital, labor capital, intellectual capital, organizational and entrepreneurial capital, social capital).

5 CONCLUSIONS

After conducting individual testing of the student, and keeping his information in the database, we determined by what criterion or criteria he understated human capital. And we give him the right to choose elective courses for a particular criterion chosen. To this end, in the curriculum (in each course), they already provide and include elective disciplines aimed at developing competencies for specific criteria of the electronic test program. The student, after completing the training for the selected courses, again undergoes electronic testing. After retesting, the student's human capital became higher by an average of 10-15% by all criteria, confirming our studies on the significance of the accentuation of the human capital of students by means of higher education.

Here are the main values of using the electronic test program "Database - Individual Student Human Capital" developed by us in the accession of the human capital of students by means of higher education:

1. economic effect - allows to reduce unjustified expenses of study time, promotes the purposeful development of human capital by means of education;
2. pedagogical effect - allows to realize the personality-oriented and individual approaches to the training of students;
3. psychological effect, increases the motivation of students' education, develops their subjectivity and improves the psychological and pedagogical interaction of teachers and students.

The accumulation of knowledge becomes more and more actual, which is traditionally associated with their obsolescence, but, most importantly, with the need for professional training, professional adaptation. Everyone should treat their educational capital as a businessman whose job is to move capital from less profitable to more profitable spheres of labor.

The practical results, obtained during the research, indicate that:

- the set of measures to optimize the process of accession and approbation of the technology of accession of students' human capital by means of higher education was implemented;
- the technology of accession of students' human capital allows to solve problems associated with the low quality of group human capital of students: if, at the first stage of the experiment, only 9% of the surveyed students had positive human capital, then, at the final stage, positive human capital was fixed more than 50% of students;
- the goal of the experiment, which was to determine the effectiveness of the author's technology of accession of students' human capital, was achieved.

REFERENCES

Burgess, Simon M. (2016). "Human Capital and Education: The State of the Art in the Economics of Education". *IZA Discussion Paper No. 9885. 2016. Available at: https://ssrn.com/abstract=2769193 (accessed 31.05.2018)*.

Douglas C. Bennett. (2001). Liberal Education. "Assessing Quality in Higher Education" *Association of American Colleges & Universities*. Spring 2001, Vol. 87, No. 2.

Goryunova L., Kotov S., Akopyan L., Ogannisyan, Borzilov Yu. Efficiency Study of Cultural Identity. *ASTRA SALVENSIS-RevistĈ de istorie şi cultu*r 2018r - Establishment in Schoolchildren within Municipal Indexation.

Karpova N.K., Mareev V.I., Akopyan M.A., Guterman L.A. (2017) Organizational Conditions of Fuctioning of the Adaptive System of Distance Education on the Basis of Using Open Educational Resources. Man in India. 2017, 20(97): 447–460.

Kondaurova I.A. (2016) "Knowledge and human capital in the system of production factors of the new Economy" *Public and municipal administration. Scientific notes of SKAGS, 2016, 4: 136-142*.

Kotova N.S., Esenskaya T.V. (2015) Educational Security as a Base Element of the System of Russian National Security *Public and municipal administration. Scientific notes of SKAGS, 2015, 3: 209-2013*.

Kotova N.S., Kotov S.V., Blohin A., Ogannisyan L., Borsilov Yu. (2017) Human Capital – the procpects of development in Higher Education System 4th International Multidisciplinary Scientific Conference on Social Sciences and Arts SGEM 2017, www.sgemsocial.org, SGEM2017 Conference Proceedings, ISBN 978-619-7408-21-8/ISSN 2367-5659, 24-30 August, 2017, Book 3, Vol 4, 271-276 pp, DOI: 10.5593/sgemsocial2017/34/S13.035.

Petrova N.P., Mareev V.I., Pivnenko P.P., Kotov S.V., Kotova N.S., Kharchenko L.N., Gulyakin D. V., (2016) The Higher School Teacher Matrix of Competences. *The Social Sciences (Pakistan)*. 2016, 11(18): 4539–4543.

Inclusive Development of Society – Lumban Gaol (eds)
© 2020 Taylor & Francis Group, London, ISBN 978-1-138-33476-2

Study of interpersonal interaction of educational process subjects in social and pedagogical space of the university

L.A. Ogannisyan, M.A. Akopyan, D.N. Misirov, Y.P. Borzilov & S.V. Semergey
Southern Federal University, Rostov-on-Don, Russia

ABSTRACT: This article presents an analysis of the effectiveness of interpersonal interaction of subjects of the educational process in the pedagogical educational space of the Southern Federal University (SFU). In the course of empirical research, the authors have revealed pedagogical conditions of interaction formation between the sides of the educational process. The results of the interrogation are analyzed and discussed, and methodological support of the educational process at SFU is provided on the basis of tables containing the results of the social investigation in the form of public questionnaires.

1 INTRODUCTION: THE RELEVANCE OF THIS RESEARCH

Changing the content and forms of social relations, wide information, and cultural exchange lead to an intensive search for new, subject-oriented technologies of interpersonal interaction in the educational space. At the present stage of modernization of education, special attention is paid to the problems of development of interpersonal communication of subjects of the educational process in connection with the tasks of innovative education so as to enhance such personal qualities as sociability, openness, tolerance, and the formation of social experience in the educational space of the university. Because in interpersonal interactions of everyday life interpersonal communication disappears altogether or is replaced by communication contacts that have minimal or no effect on the personality, the subjects of the educational process necessarily enter into interpersonal communication. As a result of their interaction there is an exchange of information, moral values, interests, and experience of communication in general. It accelerates development of such qualities necessary for successful communication as tolerance, politeness, goodwill, tactfulness, restraint, and respect for others that essentially define the orientation of the personality, moral and ethical orientations necessary for communication, acquisition of the corresponding communicative abilities, and mastering of the culture of communication.

Communication is the transfer of information from one individual or group to another. Communication is a necessary basis for all types of social interaction. In the context of face-to-face interaction, communication is carried out through the use of language, gestures, and body movements to understand what people say and do. With the development of writing, and then with the advent of electronic means of mass communication such as television, radio, and computer networks, communication to some extent lost the nature of direct social interaction. "Communication" is the main social process of creation, preservation, maintenance, and transformation of social realities.

The problem of interpersonal interaction in a modern university is becoming increasingly important. The organization of the process of interaction between the teacher and the student has a direct impact on the development of the subjects of the educational process. Research shows that pedagogical activity is carried out through interpersonal interaction. With ineffective interaction, there is a nonproductive transfer of knowledge and conflicts and difficulties in communication, which do not contribute to the full development of the individual. It is understood that in the interpersonal interactions of the subjects of the educational process cooperation is generated by the needs of joint activities. In the process, a person learns about and transforms the world, manifested in the establishment and development of contacts between people in the formation of interpersonal relations (Benin et al., 1999). Interpersonal interaction of subjects is considered a form of communication with peers, which plays an exceptional role in the formation of personal experience.

2 LITERATURE REVIEW

The interaction is the basic category of philosophical concepts and approaches (I. Kant, V. I. Soloviev, M. M. Bakhtin, etc.). Now it is possible to allocate several main directions of scientific research of this problem in pedagogical science. Interest in the problem of interaction increased in the past decade in pedagogy as a result of many factors (Karpova et al., 2001; Abulkhanova-Slavske, 2015).

The culture of communication is an important means of improving the effectiveness of interpersonal communication, determines the nature of the relationship of interlocutors, and is a complex system of ethical and communicative means and ethical orientations and communication skills of the individual, the implementation of which ensures the success of pedagogical interaction. If in everyday life interpersonal communication proceeds in this manner, then in purposeful pedagogical interaction it turns into a special task.

Despite the significant contribution of domestic and foreign researchers in addressing issues of education and the culture of interpersonal communication, there are certain contradictions. On the one hand, there is a growing need within educational institutions and society as a whole for effective interpersonal communication, which is realized through the prism of moral etiquette, and on the other there is a lack of preparedness of students for such communication.

Research by A. Leontyev, A. Luria, and S. Rubinstein in the field of communication activities and communication defined the following stages of formation of communicative activity of the person as criteria of communicative competence:

- Orientation to the choice of communicative intention, the basis of which is the motivation and selection of means of communication activities in accordance with interpersonal interaction
- Planning of certain statements using both verbal and nonverbal means
- Implementation of communication (verbal and nonverbal)
- Self-control over the statements and the correspondence of speech to the communicative plan
- Evaluation of the result of the communication act (satisfaction or dissatisfaction with the result)

Therefore, analyzing the foregoing material, it can be concluded that communicative competence is manifested in communicative activity, through the use of various communication tools, overcoming communication barriers, and communicative interaction of the individual with others. The source of communicative activity of the student is his or her communicative needs and motivation to communicate; through communicative activity a person seeks to create such conditions for the implementation of communicative activities that it meets his or her needs for communication and enriches the inner world, creating a sense of inner harmony.

Personal communicative activity (Bondarevskaya, 2001) manifests itself in one's own communicative style, which is understood through certain communication tools and techniques, the activation of which can help one achieve high efficiency in communication, including interpersonal. Communicative activity is influenced by the level of speech development and characterological personality traits. The activity fulfills a motivating function in communicative activities. A communicative activity manifested in externally motivated, targeted, and instrumental interactions of the individual forms a flexible dynamic system.

3 METHODOLOGICAL FRAMEWORK OF THE STUDY

The purpose of this research is theoretical substantiation and experimental verification of pedagogical conditions of formation of interpersonal communication of students of the educational process in pedagogical interaction (Benin et al., 1999).

The objectives of the study were to determine the psychological and pedagogical basis of the formation of interpersonal communication; to identify the essence of the phenomenon of "interpersonal communication of students"; and to develop and experimentally test a practice-oriented model of the formation of interpersonal communication of participants in the educational process in pedagogical interaction.

The initial stage of the experiment aimed to identify the current status of the development of interpersonal communication of the students. At this stage, the following tasks were set:

- To analyze the methodological support of the educational process in the Southern Federal University (SFU) to identify the reflection in its content
- To develop criteria and determine the levels of formation of interpersonal communication of students
- To identify the impact of the family on the formation of the communicative culture of students

The experiment was conducted with second-year students of Rostov-on-don SFU. The experiment involved 35 male and female students of the same age.

The system of cognitive principles was taken as the basis of definition stage of experimental work. This approach was predetermined by the fact that educational work should be carried out in a certain system. The system approach is the direction of methodology of scientific knowledge and social practice. Its essence lies in the complex study of large and complex objects (systems) as a whole with the coordinated functioning of all elements and parts.

The system–activity approach involves the development of models of certain activities and the allocation of certain components of it, making it possible to comprehensively explore any sphere of human activity and is manifested in

- Various activities, including general pedagogical and special components
- Content that contains psychological, pedagogical, methodological, and special knowledge, skills, and abilities
- Practical activities aimed at the implementation of the acquired knowledge and skills in the educational process, extracurricular educational work, and work with parents of students

The goal of the system–genetic approach is to reveal the conditions of origin, development, and transformation of the system.

Cognitive principles are based on general philosophical theory and are a methodological basis for many sciences. Especially effective is the study of the dynamics of science and its relationship with society and justifying the leading value of knowledge in the behavior of the individual. It should be borne in mind that the analysis of the formation of knowledge requires the study of practical and theoretical human activity in relation to its social aspect. In the center of the problem under study, there is a person as a member of society, a psychological subject, a subject of speech communication, a communicant.

Thus, the results of a survey of 35 teachers-practitioners on the concept of "interpersonal communication of students" showed that 75.6% of respondents formulated it too broadly, sometimes without age. To confirm this, we give examples of answers: "... humane interaction of individuals," "comprehensive development and formation of universal values." There were also answers (24.4%) in which teachers pointed to only one or more components of interpersonal communication, for example, "moral qualities," "moral behavior," "the ability to listen to the other," "ethical interaction," and "moral relations."

That is why 19.2% of teachers could not name the features of interpersonal communication. The rest of the respondents, again, did not take into account all the components. That is why the response to the item "characteristic of interpersonal communication of high school students in accordance with the peculiarities of speech culture" was 78.3%; "moral values," 3.8%;

"communication in accordance with generally accepted norms," 11.5%; "the ability to take into account the interests of the other," 3.8%; "the ability to listen to the other," 7.6%.

In determining the criteria for the manifestation of respondents we used interpersonal communication of older students as the observed awareness of the average level. So 1.5% of teachers mixed the concept of "criteria" with "qualities." In particular, a positive attitude to the moral norms that regulate communication was expressed by 9.6%; the ability to hear and listen, understand, and express their attitude to the speech culture of others, 20%; knowledge of the essence of behavior in the process of communication, 24.2%.

In identifying the most typical positive moral and ethical standards for students the following were determined: tact, 21%; sociability, 19%; responsiveness, 15%; politeness, 13%; goodwill, 12%; sympathy, 11%, restraint, 9%.

The negative features inherent in students that teachers listed include inattention, conflict, indifference, tactlessness, rudeness, impatience, etc.

On the question of education of any moral qualities of the personality of that modern high school students paid the most attention to, teachers were able to identify the highest percentages for goodwill, 31%; sensitivity, 22%; and courtesy, 18%.

The analysis of teachers' answers to the question about the effectiveness of the means of interpersonal communication showed that they considered projects and creative tasks as having an impact on the formation of ideas about the culture of speech and fostering a positive attitude toward the moral norms that regulate communication. This was noted by 52% of respondents, arguing: "means that form communication skills," "thanks to them, you can cultivate a culture of relationships," "with their help, you can teach folk wisdom," etc. The highest percentage of the channels used went to creative projects (86.5%), quizzes (69.2%), traditions and rituals (29.6%), but 73% of practitioners did not explain their answer at all, indicating a lack of awareness of the value and effectiveness of the impact of the selected channels on the understanding of the behavior of pupils.

During the identification of problems faced by teachers focused on the formation of interpersonal communication in the practice of formation of modern students, an interesting pattern was established. Of modern teachers surveyed, 61.5% focused on the shortcomings of families in which students are brought up; at the same time their ignorance, lack of experience and unwillingness for self-improvement is not taken into account. This is evidenced by 51.9% of the missing responses. We can only agree with the reflection of 14.5% of teachers about the loss of communication between generations in the transfer of experience of educational influences.

The answers of the interviewed teachers confirmed the correctness in the choice of pedagogical conditions focused on the effectiveness of the formation of interpersonal communication of students.

Among the methods of ethical education, as a basis for the formation of interpersonal communication used by teachers, respondents for the most part answered as follows: projects, 50%; explanation, 36.5%; story, 13.5%; exercise, 19.2%; games, 3.8%; habituation, 23.1%; order, 15.4%; promotion, 13.5%; approximately 44.2% of the requirement, 25%; competition, 3.8%; praise, 15.4; penalty, 3.8%.

The analysis of the content of the introduced forms of moral and ethical education showed that in the course of carrying out their work teachers more often convinced, instructed, criticized, and offered students ready rules of culture of interpersonal communication. However, the effective influence of forms, methods, and means of formation of interpersonal communication of students, according to theoretical assumptions, can be achieved only under the condition of systematic and purposeful use, which was defined by us as a shortcoming of individual teachers.

Only 30.8% of the students did not have relevant proposals to the question on the ideas of improving the professional competence of teachers in the formation of interpersonal communication. Among these, 94.2% were unambiguous: "to acquaint with pedagogical experience on methodical and methodological associations, seminars, courses, lectures"; 73.1% of respondents expressed their desire to engage in self-education, and 9.6% to participate in relevant trainings and master classes. To determine the degree of effectiveness of the formation of the culture of interpersonal communication of the subjects of the educational process, we had to identify the criteria, indicators, and levels of interpersonal communication.

The analysis of theoretical sources and modern pedagogical practice allowed comparison and systematization of the researched approaches and on their basis to allocate criteria of interpersonal communication of students. As modern researchers believe [Aibatyrov K.S 2017 et. al., Kotov S.V. 2018 et al.] that interpersonal communication may have an impact on the identity formation of the unity of moral and ethical consciousness, moral feelings, and moral and ethical behavior, we identified the following components of interpersonal communication, which we took as the basis for the selection of criteria levels: cognitive, emotional, regulatory, activity.

The aforementioned components of the structure of interpersonal communication became the basis for the selection of the following criteria and their indicators:

- Informative
- Emotionally sensitive
- Behavioral

On the basis of the established criteria and indicators, the following levels of interpersonal communication of students were determined: high, sufficient, medium and low.

A sufficient level of interpersonal communication characterizes those students who are characterized by adequate but not enough complete understanding of the moral and ethical standards. Students at this level are not very active in communication; they are characterized by goodwill and politeness in relationships, but they carry them out mainly at the request of an authoritative adult.

The average level of interpersonal communication is inherent in students with a certain amount of knowledge about the content of individual ideas about moral and ethical norms, with, however, insufficient awareness of them. This group understands "other people's feelings," but independently shows politeness in communication only with friends and parents; with all others, it depends on own mood, desire, or through encouragement.

The experiment involved 35 students of the Southern Federal University. In particular, we determined the presence of knowledge and ideas of students about moral and ethical qualities, ways of their manifestation, understanding of the essence of certain concepts, and the ability to use moral value judgments. The moral qualities of the participants of the survey are explained by the new social situation of personal development. The answers to the first item of the questionnaire, an awareness of the content of the norm "tact," with students giving their own interpretation, showed that almost the same percentage of students surveyed (8% of control groups, 7% of experimental groups) are at the average level of formation of interpersonal communication.

Regarding the answers to the second question, the results were as follows: 5% of the control groups and 2% of the experimental groups showed an average level of awareness of what kind of person we call benevolent. Interesting, in the context of our study, were the answers indicating the perception of students friendly with the close people who financially encourage them. In the absence of such individuals or financial incentives, students do not consider people benevolent, as evidenced by the critical response "no one is benevolent." Ten percent of the students of the experimental groups could not give any answer to this question. The first question caused some difficulties for the majority of students. Although we attributed 4% of the answers to the criterion of understanding the content of the concept, they did not contain enough specific explanations.

We conducted this study, which is an organic component of interpersonal communication, to identify the formation of the sensory and cognitive sphere of students.. To this end, the study was conducted emotionally valuable and cognitive sphere of students using the method of projects. The task was to perform design tasks of different genres and gain an idea of which were the most pleasant and the worst for them. The job was generally simple, and its purpose was to obtain specific information. It turned out that the task was well received and comprehended, because without additional questions the participants independently completed creative projects.

Table 1. Results of students' awareness of the content of moral and ethical norms for 35 students (first part).

No.	Indicators	Tact Percentage	%	Goodwill Percentage	%	Courtesy Percentage	%
1.	Awareness of the content of norms	8	16	6	12	10	20
2.	Own interpretation of the rules	3	3	5	10	7	14
3.	Lack of awareness of ethical norms	9	18	8	16	4	8
4.	Lack of response	11	22	10	20	17	34

Table 2. Results of students' awareness of the content of moral and ethical norms for 35 students (second part).

No.	Indicators	Tact (Ageev, 2014) Percentage	%	Goodwill (Ageev, 2014) Percentage	%	Courtesy (Ageev, 2014) Percentage	%
1.	Awareness of the content of norms	7	14	3	6	5	14
2.	Own interpretation of the rules	2	4	2	4	4	8
3.	Lack of awareness of ethical norms	3	6	4	8	4	8
4.	Lack of response	4	8	5	10	7	14

Table 3. Input comparative indicators of interpersonal communication.

Levels of inter-personal com-munication	Criteria of interpersonal communication of students					
	Informative		Emotional		Behavioral	
	Control group	Experimental group	Control group	Experimental group	Control group	Experimental group
High	24.5	25.1	3.3	5.1	41.7	41.3
Sufficient	9.3	8.6	5.5	6.9	23.4	35.7
Medium	26.6	26.3	34.1	33.4	12.9	8.7
Low	39.6	40	57.1	54.6	22	14.3

It is worth noting that the survey was conducted confidentially; the answers and their arguments were recorded in Tables 1–3. Thus, the answers to the first question were 70% for control and 58% for experimental groups, which would seem to indicate a high level of awareness on the part of the trainees of the ethical representation of actions.

At the same time, an indicator of the approval of the students of this action was insignificant (12% in the control and 16% in the experimental groups). However, according to the real picture of the level of formation of interpersonal communication of students, we were able to note by analyzing the arguments of the proposed answers to the second question. The indicator turned out to be low (20% in experimental groups), and this phenomenon is scientifically justified, since the majority of students (in Colberg) are at the "moral level" of development of value consciousness.

The results are presented in Tables 1–3 and focus on the censure of adults: the norms of moral and ethical behavior and their absence. These data indicate that a high level of interpersonal communication demonstrated approximately the same number of students in the first

(KG) and second (EG) groups (see Table 3). In fact, the same indicator is at the sufficient level. However, more than a third of students are at a low level of interpersonal communication. Consequently, the results of the survey demonstrate that it is characteristic that there is insufficient attention to the issues of moral and ethical education of students, in particular, to the formation of their interpersonal communication. That is why students' awareness of moral and ethical standards and culture of interpersonal communication is low and unequal. Taking into account this fact, we have conducted targeted studies to identify the impact of the family on the formation of interpersonal communication of students. To this end, a special program was developed, the content of which was aimed at studying the individual as a member of the family microcollective (Artyukhova et al., 2007).

Regarding the priorities to which parents give priority (Bondarevskaya, 2001) in the education of students, they can be distributed in the following areas: health, 59.3%; success in learning, 35.6%; interests, hobbies, inclinations of children, 19.2%; culture of behavior and communication with other people, 18.7%; politeness, respect for adults and peers, 8.4%; the status of peers, 6.8%.

In the process of studying the personal qualities that, according to parents, are first and foremost important to teach, we obtained the following data: independence, autonomy, confidence in yourself, 83.9%; determination, perseverance, the ability to act in all circumstances, 80.2%; obedience, 79.0%; respect toward parents, 69.1%; and, love of work, hard work, 65.4%; religiousness, 61.7%; thrift, 54.3%; skills for a healthy lifestyle, 50.1%; respect for traditions, awareness of moral and ethical norms of communication, 43.1%; patriotism, 42.0%; honesty, integrity, 5.7%; tolerance and respect for others, 4.6%.

4 CONCLUSIONS

During the control stage of the experiment, we received confirmation that the purposeful work of the university in the educational space of Rostov-on-don, associated with the provision of conditions of interpersonal communication of students, really develops certain personal qualities and values and promotes cultural communication of pupils.

In conclusion, this article argued that the development of personal qualities and values of students contributes to the development of interpersonal communication of students in the socio-pedagogical space of the university (Petrova et al., 2016). Based on the objectives of the study, the levels of interpersonal interaction of subjects were determined, and the general patterns of communication development at the present stage of development of society were revealed.

The effectiveness of many tasks depends not only on the individual qualities of the individual, but also on his or her ability to engage in joint activities at the level of cooperation and partnership. The communicative space is dynamically expanding (Goryunova et al., 2018), which cannot but affect the educational process: it involves various representatives of society, performing different social roles and functions. Significant changes in the content and forms of social relations entail the breaking and restructuring of previous stereotypes and cause social tension, which, in turn creates the ground for the emergence of various kinds of contradictions and conflicts. The problem of communication is given attention in the "Strategy of Modernization of the Content of Education" of the Ministry of Education of the Russian Federation, which emphasizes that "…an important place in the content of education should be taken by communication, intercultural understanding, willingness to cooperate, the development of the ability to creative activity, tolerance, tolerance for other people's opinions, the ability to discuss to seek and find meaningful compromises" (Ageev, 2014). Thus, the interaction of subjects of the educational process becomes a priority area of this research.

REFERENCES

Abulkhanova-Slavske, K. A. 2015. Personal aspect of communication. In B. F. Lomov (ed.), *Communication Psychology*, pp. 218–241. Moscow: Science.

Ageev, V. S. 2014. *Intergroup Interaction: Social and Psychological Problems.* Moscow: Moscow State University Publ.

Aibatyrov, K. S., Aibatyrov, M. A., Annikova, L. V., Bakhutashvili, T. V., Kotov, S. V., Kotova, N. S., & Kharchenko, L. N. 2017. The model of education in a modern university. *Turkish Online Journal of Design Art and Communication*, 7:1955–1961.

Artyukhova, I. S. 2007. Values and education. Pedagogy. № 4. P. 117–121.

Benin, V. L., & Fatykhova, P. M. 1999. Humanization of interpersonal relations in the context of dialogic culture. Izv. Ural, nauch. –educated center of RAO, 1:122–128.

Bondarevskaya, E. V. 2001. Meanings and strategies of personality-oriented education. *Pedagogy*, 1:18–24.

Goryunova, L. V., Kotov, S. V., Akopyan, M. A., Ogannisyan, L. A., & Borzilov, Y. P. 2018. Study of the efficiency of cultural institutions of the individual pupils within municipal index. *Astra Salvensis*, 6:735–744.

Karpova, N. K., Mareev, V. I., Goryunova, L. V., Akopyan, M. A., & Guterman, L. A. 2001. Organizational conditions of functioning of adaptive system of distance education on the basis of use of open educational resources. *Man in India*, 97(20):447–460.

Petrova, N. P., Mareev, V. I., Pivnenko, P. P., Kotov, S. V., Kotova, S. N., Kharchenko, L. N., & Gulyakin D. V. 2016. Matrix of competences of the teacher of high school social sciences (Pakistan). 11(18):4539–4543.

Inclusive Development of Society – Lumban Gaol (eds)
© 2020 Taylor & Francis Group, London, ISBN 978-1-138-33476-2

Strategies of decision-making in financial markets

I.V. Tregub
Financial University under the Government of the Russian Federation, Moscow, Russia

A.A. Dautova
LUKOIL, Moscow, Russia

ABSTRACT: This article continues the investigations of application of Granger causality tests but considering an application of this test to the analysis of the current cointegration processes between Russian and Asian financial markets. We found that over the past 5–7 years, Russia has made significant progress in establishing relations with Asian countries, in particular with China, Japan, and the Republic of Korea. The increase in trade volumes, large-scale investments in Russian projects, and large deliveries of Russian crude oil have given us reason to assume that the convergence of relations will also affect the cointegration between the Russian and Asian exchanges and, accordingly, financial instruments. It was found that the Moscow Exchange index has a cointegration relationship with the Shanghai index, with the best values obtained using the opening prices for the Moscow index and closing prices for the Chinese index that minimize the nonsynchronous trading effect. The best value, identified by the sampling method, is equal to 8 days. Moreover, Granger causality was found with the Japanese Nikkei index 225. A cointegration model based on the Granger causality test was then used for forecasting of the future quotes of the Moscow Exchange index.

1   INTRODUCTION

The frequent shift of economic leaders, lightning-fast development of technologies, and pervasive globalization are the main characteristics of the modern world. Disagreements that led to war between countries just several decades ago (Shefov, 2006) are now a closed chapter and don't cramp the creation of unions and alliances as a means of protection against new dangers and risks. We see such a situation in the Asia-Pacific region, where during past years Russia actively tried to harmonize relations with Asian neighbors (Trenin, 2017). This will have a large positive impact in terms of investments, increased commodity turnover, and sales of crude oil. But what negative sides are behind this? Is there a risk that Russia will be involved in the next financial "bubble," which can destabilize our developing financial market if it bursts?

This article investigates this question. The current situation between countries will be analyzed, and then the cointegration model of interaction of financial markets will be composed and econometrics tools such as Granger causality test will be applied, which will show us the presence or absence of cointegration between financial markets of Russia and Asian countries.

2   LITERATURE REVIEW

The concept of cointegration was suggested by Granger in 1981 (Granger, 1981). Moreover, before that, in 1969 he developed the Granger causality test (Granger, 1969), which allows identifying whether the changes of a supposed "cause" for an explained variable precede the changes of that explained variable. The Granger causality test is an important but not sufficient prerequisite for a cause-and-effect relationship, because there is a possibility of "false" causality, which happens when Granger causality is identified but the identified "cause" doesn't have a real influence on explained variable; for example, there is another variable that influences both of the previously considered variables.

Hereafter other scientists started to develop their works following this direction, for example, Engl (1987), Johansen, Philips, and others. Engl and Granger were awarded the Nobel Prize in Economics in 2003 "for the analysis method of economic time series with time-depend volatility." Johansen developed his own approach to the testing of integration, where the vector of autoregression can have the same role as the presence of unit roots in the corresponding regression (Johansen, 1991).

At a later point of time, Baumohl and Vyrost investigated the cointegration concept with respect to nonsynchronous trading effects during the Granger causality test (Baumohl, 2010).

In addition, there is some research that covers the Granger causality test for Asian markets. Some such notable works are Atmadja (2005), Kosapattarapim (2017), and Xiangyun (2018).

3   FINANCIAL MARKETS OF ASIAN COUNTRIES AND RUSSIA

Recently in addition to the improvement of political relationships between Russia and Asian countries, we see an observable improvement in the economic sector, so there is a strengthening of an interaction of countries in investments and as a consequence in financial markets.

According to the research by EY, which was presented in St. Petersburg's International Economic Forum SPIEF-2018, China is a leader by the volume of investments in Russian projects (Tregub, 2019).

Moreover, there is attention to Russian projects from the whole Asian region: the presence of Asian countries has increased by 2.5 times from 30 to 76 projects. Among top-10 leaders, we can see Japan and South Korea (invested in 17 and 12 projects accordingly).

European countries conversely started to lose their investment appetite for Russian projects. The number of projects that are of interest in Germany has decreased from 43 to 28, and French investors have decreased their portfolio from 20 to 11 projects. Participation of the United States has decreased by 50%: American partners invested in only 19 projects compared to 38 in 2016.

The main reason for such a shift of the leader can be due to the geopolitical situation. Sanctions forced many companies from the EU and the United States to stop their participation in Russian projects, while companies from Asian countries didn't have such limitations because their governments didn't participate in these sanctions.

Also, there is an effect from "turning to the East" on the part of Russia in addition to the search by Asian companies for new directions for investments in developing countries.

The strong interest of investors in finance and financial technologies is fully justified. Russia has a big, even in terms of worldwide scale, market with strong local key players. Moreover, there is a high growth in finances connected with improving the availability of loans and incremental recovery of the economy.

Despite the positive dynamics described in the foregoing, there is a slight decrease in demand for Russian agriculture. The number of projects has decreased from 41 to 38 compared to 2016. But the most probable explanation would be not the decreased attractiveness for the investors, but the stabilized growth rate in agriculture after the active growth period.

As a consequence, the interdependence of the financial markets of Russia and Asian countries should also theoretically increase, which, on the one hand, is a positive moment, because Asian financial markets are more developed. On the other hand, however, it increases the risk that if a financial or economic crisis occurs in the Asian financial markets, the "wave" will come to the Russian financial market and it is not yet known which economy will suffer more: a well-developed, albeit possibly the overheated economy of China or just developing and getting out from the period of stagnation Russia?

For the more accurate answer to the question whether the financial markets of Russia and Asian countries are already quite strongly interconnected, many economists and financiers not only want to rely on expert opinion but also try to assess the degree of market influence through various models.

4 METHODOLOGY OF INVESTIGATION

Among the most frequently used models, econometric models occupy a special place. Despite the fact that currently there are plenty of the different types of models, linear models are still the most used both in terms of the type of the explaining variable and in terms of the statistical structure (a linear form of the model parameters: there is only one parameter for each explanatory variable):

$$f\left(x;\vec{\beta}\right) = \beta_0 + \beta_1 x_1 + \beta_1 x_2, \qquad (1)$$

Granger proposed the idea of Granger causality in his 1969 article to describe the "causal relationships" between variables in econometric models. Before this, econometricians and economists understood the idea of "causal relationships" as asymmetrical relationships. Causal relations are studied because policymakers need to know the consequences of the various actions that they are considering taking. For example, given a relationship between output and the price level, we need to know whether this relationship is expected to hold if actions controlling output are implemented, when actions controlling the price level are implemented, or when both of these cases occur.

The idea of Granger causality is that a variable X Granger-causes variable Y if variable Y can be better predicted using the historical data of both X and Y than it can be predicted using the historical data of Y alone. This is shown if the expectation of Y given the history of X is different from the unconditional expectation of Y.

$$E\left(Y \mid Yt - k, Xt - k\right) \neq E\left(Y \mid Yt - k\right), \qquad (2)$$

Granger causality tests observe two time series to identify whether series X precedes series Y, Y precedes X, or if the movements are contemporaneous. The notion of Granger causality therefore does not imply "true causality," but instead identifies whether one variable precedes another. We can, therefore, use Granger causality tests to test for things we might have assumed to occur from elsewhere or that we have taken for granted.

Granger causality is normally tested in the context of linear regression models. For illustration, consider a bivariate linear autoregressive model of two variables X_1 and X_2:

$$X_{i,t} = \sum_{j=1}^{p} A_{i,j} X_{1,t-j} + \sum_{j-1}^{p} A_{i2,j} \cdot X_{2,t-j} + E_i(t), \ i = 1,2 \qquad (3)$$

where p is the maximum number of lagged observations included in the model (the model order), the matrix A contains the coefficients of the model (i.e., the contributions of each lagged observation to the predicted values of $X_1(t)$ and $X_2(t)$), and E_1 and E_2 are residuals (prediction errors) for each time series. If the variance of E_1 (or E_2) is reduced by the inclusion of the X_2 (or X_1) terms in the first (or second) equation, then it is said that X_2 (or X_1) Granger-(G)-causes X_1 (or X_2). In other words, X_2 G-causes X_1 if the coefficients in A_{12} are jointly significantly different from zero. This can be tested by performing an F-test of the null hypothesis that $A_{12} = 0$, given assumptions of covariance stationarity on X_1 and X_2. The magnitude of a G-causality interaction can be estimated by the logarithm of the corresponding F-statistic.

After determining the presence or absence of cointegration dependence, we can choose the strongest cointegration interrelationships and for models chosen on this basis determine the parameters of the model by finding the cointegration vector.

Then, based on the obtained parameters, we can make forecasts.

5   DATA COLLECTION AND ACCEPTED ASSUMPTIONS

Based on the aforementioned information, theoretically, we can estimate whether the Asian and Russian markets really have a cointegrating influence on each other.

To do this, we consider a sample of quotations of IMOX of the Moscow Exchange, SHCOMP of the Shanghai Stock Exchange, and NIKKEI_225 (Nikkei) of the Tokyo Stock Exchange indices in the period from January 1, 2015 to August 31, 2018. Initial data are intraday and daily quotes available on the official websites of the exchanges. First, closing prices will be analyzed, which is the standard course of the study. Further testing will be conducted for the data, which are the closing prices for Asian exchanges but the opening price for the Russian exchange. This action will be taken in order to minimize the nonsynchronous trading effect described in Engl (1987). However, it should be noted that in both cases quotes will be subjected to additional processing in the form of deleting quotes of those days in which trade was absent from at least one of the exchanges, for example, holiday and other nontrading days. This fact should not lead to the reappearance of the effect of nonsynchronous trading, since Granger causality test allows testing nonstationary series, i.e., excluding part of the quotations will not entail the accumulation of indicators, as it would be, for example, for a difference of the first order (profit).

Two periods will be considered separately: (1) from January 1, 2015 to December 31, 2017 and (2) from January 1, 2018, to August 31, 2018, because authors suppose that earlier periods have different characteristics if compare with the last data. The reason can be the improvement of relationships between Russia and the Asian markets that have been gaining momentum only in 2018. The year 2014 will not be considered because it was full of extra events in Russia such as the Ukrainian conflict, accession of Crimea, high volatility of exchange rates for the ruble, and so on, that in the opinion of authors can distort the result. Further, after finding the values of the parameters for the cointegration model, a forecast will be built for the period of September 2018. Then the model will be verified.

6   GRANGER CAUSALITY

In general, the cointegration model will take the following form:

$$\widehat{IMOX}_t = a_0 + a_1 \ SH_t + \ + a_2 \ IMOX_{t-1} \qquad (4)$$

$$\widehat{IMOX}_t = C_0 + C_1 \ NKY_t + C_2 \ IMOX_{t-1} \qquad (5)$$

First, we run several Granger causality tests to determine which prices—opening or closing —better demonstrate Granger causality. Also by using the trial-and-error method, we will determine the most likely time lag for use in the model.

To begin with, we consider the simplest and standard case, when we take only closing prices for all three indices. To determine what period better fits, first we will prepare the Granger causality test for the period from January 1, 2015 to December 31, 2017. The best results for SSE and Nikkei indices for the time lag = 1 are 4.81% and 2.63% accordingly. It means that the probability that the Nikkei 225 and SSE indices do not have a Granger causality to the Moscow Exchange index is quite low, but not remote.

Now we will conduct a similar test for the sample, in which the data of the MOEX index are the opening prices but the Asian prices continue to represent the closing prices. Please note that in this way we will try to correct the nonsymmetric trading effect because Asian time zones differ significantly from the Russian time zone.

This time we have the lowest result for Nikkei 225 index of 7% (lag = 1) and for SSE index of 3.42% (lag = 2). Now we will calculate the Granger causality test for the period from January 1, 2018 to August 31, 2018.

We have found that the best value for the Shanghai index is obtained at a time lag of seven (trading) days. In this case, we get the probability of an error that the Shanghai Index is causally related to the Granger for the MICEX Index of only 3.16%.

For the Nikkei 225 index, the best result is obtained at a time lag of 10 days, which is a very strange result, given the fact that the information in the markets spreads like lightning.

Perhaps the reason lies in the different "behavior" of the indices, namely in the level of volatility. To check this we compare the movement of quotations of indices using the expert method for the period under review.

The results show that the volatility of the Nikkei 225 index was significantly higher than that of the MOEX index. Therefore, Granger causality at a 10-day lag can be a "false" causality that has developed as a result of other factors or in a random way.

By translating various time lags, we have found that the best values fall on the following: (1) For the Shanghai index on a lag of 8 days, the probability of erroneous cointegration is only 0.63%, although we have to mention that there are appropriate results (nearly 1%) also for 7- and 9-day lags. (2) For the Japanese index on an 11-day lag the probability of an error is only 0.65%, which again looks strange in the sense that the volatility of this index significantly exceeded the volatility of other indices. Because tests with the opening prices of the MOEX index showed the best results, we can take them as the basis for the cointegration model.

Moreover, because the results for the Japanese index look extremely suspicious and may represent a "false causality," this index is excluded from the cointegration model. Now it is a question of what data period to use for the cointegration model. From one side, we see better results for the period from January to August 2018, but from the other side, this time period is too short and there is a high probability that previous prices can improve our model. It was decided to make calculations for both variants. First we received results for the model based on January–August 2018 data (see Table 1). We obtained quite interesting results. On the one hand, we have DW-statistics >2, which indicates the absence of autocorrelation. But on the other hand, the probability for the second parameter is 36.3%, which is an extremely low and undesirable result, and R^2 says that the model covers only 63.11% of the time series, which is not a high rate. Thus, we get a model to be tested:

$$\text{MOEX} = C_1 + C_2 \, \text{SSE}(-8) + C_3 \, \text{MOEX}(-1) \tag{6}$$

To verify the model, using the model parameters obtained in the previous step in the Eviews package, we calculate the theoretical value of the MOEX index quotes in September 2018.

Table 1. Results of model 1.

	Coefficient	Std. error	t-Statistic	Prob.
C_1	509.8635	126.3700	4.034687	0.0001
C_2	−0.008276	0.009075	−0.91197	0.3633
C_3	0.788226	0.052084	15.13383	0.0000
R^2	0.631194	Mean dependent variable		2284.561
Adj. R^2	0.625926	S.D. dependent variable		41.70631
S.E. of regression	25.50825	Akaike information criterion		9.336637
Sum squared resid	91093.90	Schwarz criterion		9.398794
Log-likelihood	−664.5695	Hannan–Quinn criteria.		9.361894
F-statistic	119.8019	Durbin–Watson stat		2.051581
Prob (F)	0.000000			

Next, we will calculate the confidence intervals according to the formula:

$$C_t^{\pm} = C_t \pm t_{crit} \text{ Stand err} \qquad (7)$$

The values of the critical t and standard error statistics are calculated using Excel functions.

When testing a sample for September 2018, we obtain the results shown in Table 2. The table shows that the model works for a rather short period of time, within two weeks. After this period, actual quotes are out of forecasted interval.

Of course, this result can be due to the small value of R^2 of the model and the high probability for the second parameter of the model, but on the other hand, this result stimulates us to try to include 2015–2017 data in cointegration model 2. Based on experience, we found that the best results correspond to the model (Table 2):

$$\text{MOEX} = C_1 + C_2\text{SSE}_{-1} + C_3\text{MOEX}_{-1} \qquad (8)$$

We see even more interesting results. Despite the fact that R^2 is much better (near 99%; see Table 2), the forecasting variable is almost fully dependent on its previous amount. It means that errors from each step of forecasting are accumulating that give us even worse results than from the first attempt with short-period data.

7 CONCLUSION

To conclude, we found out that over the past 5–7 years, Russia has made significant progress in establishing relations with Asian countries, in particular with China, Japan, and the Republic of Korea. The increase in trade volumes, large-scale investments in Russian projects, and large deliveries of Russian crude oil have given us reason to assume that the convergence of relations will also affect the cointegration between the Russian and Asian exchanges and, accordingly, financial instruments.

However, after conducting several Granger causality tests, we found that this was not the case. The Moscow Exchange index has a cointegration relationship only with the Shanghai index, with the best values obtained using the opening prices for the Moscow index and closing prices for the Chinese index that minimize the nonsynchronous trading effect. The best value by the sampling method was identified for the data period from January to August 2018 and is equal to 8 days.

Moreover, Granger causality was found with the Japanese Nikkei index 225, but this causality was considered potentially false, because expert analysis of the charts of quotes revealed a strong volatility of the Nikkei index, while the movements of the other two indices looked identical.

Table 2. Results of model 2.

	Coefficient	Std. error	t-Statistic	Prob.
C_1	35.677	12.775	2.79	0.005
C_2	-0.004	0.0021	-2.13	0.032
C_3	0.9901	0.0040	246.3	0.000
R^2	0.9902	Mean dependent variable		1963
Adj. R^2	0.990202	S.D. dependent variable		226.4
S.E. of reg	22.42006	Akaike information criterion		9.06
Sum sq.	377,999.7	Schwarz criterion		9.08
Log-likelihood	−3,417.81	Hannan–Quinn criteria.		9.06
F-statistic	38,100.19	Durbin–Watson stat		2.05
Prob (F)	0.000000			

In addition, the Granger causality test was prepared for data from January 1, 2015 to December 31, 2017. The results have demonstrated that there was less possible causality than in the year 2018.

Further, the parameters of the cointegration model 1 were calculated, and also the forecast of the index movements for September 2018 was made.

However, a comparison with actual values showed that the model provides results only for the next 2 weeks after learning sample.

We linked this fact with low values of R^2, as well as with a high probability for the second parameter of the model. After this, we tried to improve our model by including data for 2015–2017, but the results were even worse than prove our assumption that causality between Asian and Russian market appeared only recently. All this leads us to the conclusion that despite the seemingly obvious interaction between the Asian and Russian markets, the empirical data show that so far only with the Shanghai market is there an observable cointegration relationship. Moreover, the current data allow us to build a model that would predict data only for a short period with an acceptable level of error. This interesting result let us to suppose that there are other factors and processes that influence the movement of quotes, not just Granger causality with the Shanghai index. This result gives us a good start for further investigations of the interaction of Russian and Asian markets that will help us to predict future prices and maybe financial crises.

REFERENCES

Atmadja, A. S. 2005. The Granger causality tests for the five ASEAN countries stock markets and macroeconomic variables during and post the 1997 Asian financial crisis. *Jurnal Manajemen dan Kewiraussahaan*, 7(1).

Baumohl, E. 2010. Stock market integration: Granger causality testing with respect to nonsynchronous trading effects *Finance a úvěr-Czech Journal of Economics and Finance*.

Engl, F. R., & Granger, C. 1987. Co-integration and error correction: Representation, estimation, and Testing. *Econometrica*, 55(2):251–276.doi:10.2307/1913236.

Granger, C. 1969. Investigating causal relations by econometric models and cross-spectral methods. *Econometrica*, 37(3):424–438.

Granger, C. 1981. Some properties of time series data and their use in econometric model specification. *Journal of Econometrics*, 16(1):121–130. doi:10.1016/0304-4076(81)90079-8.

Johansen, S. 1991. Estimation and hypothesis testing of cointegration vectors in Gaussian vector autoregressive models *Econometrica*, 59:1551–1580. doi: 10.2307/2938278.

Kosapattarapim, C. 2017. Granger causality between stock prices and currency exchange rates in Thailand. *AIP Conference Proceedings* 1905, 050025. doi:10.1063/1.5012244.

Shefov, N. A. 2006. Russian battles. Moscow: ACT.

Tregub, I. V. 2016. On the applicability of the random walk model with stable steps for forecasting the dynamics of prices of financial tools in the Russian market. *Journal of Mathematical Sciences*, 5: 716–721.

Tregub, I. V. 2018. Econometric analysis of influence of monetary policy on macroeconomic aggregates in Indian economy. *Journal of Physics: Conference Series*.

Tregub, I. V. 2019. Digital economy: Model for optimizing the industry profit of the cross-platform mobile applications market. *Advances in Intelligent Systems and Computing*.

Trenin, D. 2017. Russia's evolving grand Eurasia strategy: Will it work? Carnegie Moscow Centre, July 20.

Turkina, S. N., & Tregub, I. V. 2014. Menges model applied to the economy of the USA. *Marketing I menedzhment innovacij*, 4:128–135.

Xiangyun, G. 2018. Modelling cointegration and Granger causality network to detect long-term equilibrium and diffusion paths in the financial system. *Royal Society of Open Science*, 5(3):172092 doi: 10.1098/rsos.172092.

Inclusive Development of Society – Lumban Gaol (eds)
© 2020 Taylor & Francis Group, London, ISBN 978-1-138-33476-2

Goal-setting theory and gamification in mobile fitness app engagement

P.G.M.A. Pg Arshad, N. Zaidin, R. Baharun, M.S.M. Ariff, N.Z. Salleh & F.S. Ahmad
Azman Hashim International Business School, Universiti Teknologi Malaysia, Johor, Malaysia,

ABSTRACT: Previous scholars have demonstrated the efficiencies of mobile app usage in fostering engagement among game players. Nonetheless, their focus was merely on the gaming and education contexts. Using goal-setting theory (GST), this research enhances these investigations by examining the effect of the introduction of extrinsic forces (gamification) for the purposes of engagement in a non-game context. Thus, this paper discusses the application of GST and gamification as extrinsic forces on mobile fitness app engagement. According to GST, two antecedents explain goal setting—goal core and goal mechanism—while engagement is the actual behaviour. This model was pre-tested by industrial experts as well as by scholars using interviews. The pre-test results showed that the antecedents, consisting of goal core and goal mechanism, were identified as intrinsic forces, and should include individual objectives and direction planning attributes. Gamification is the extrinsic force, and should consist of game design and game mechanics. Both these forces, extrinsic and intrinsic, were hypothesised for the purpose of engagement with mobile fitness apps.

Keywords: Goal-Setting Theory, Gamification, Engagement, Mobile Fitness Apps, Extrinsic Intrinsic Forces

1   INTRODUCTION

In the modern age, the advancement of mobile technology and applications in mobile devices such as smartphones and PC tablets has had a great impact on society, changing the lives of millions of people around the globe (Jusoh, 2017). Both academics and practitioners have found that mobile technology such as mobile applications (mobile apps) can attract and motivate users to make health-conscious choices (Lim & Noh, 2017). Several mobile fitness apps that are fitted with a number of persuasive features have been introduced to the market to encourage users to adopt them, which leads people to feel the benefits of experiencing physical exercise (Higgins, 2016).

In 2015, mobile health (m-health), such as mobile fitness, is one of the fastest growing app categories, and more than half of all mobile device (such as smartphone) users downloaded mobile fitness for the purposes of improving their health and life styles (Byun, Chiu, & Bae, 2018). Krebs and Duncan (2015) concluded that 58% of smartphone users have downloaded at least one of mobile fitness onto their mobile devices. Because of the increasing demand of mobile fitness apps, several sports brand providers, such as Nike and Adidas, have developed their own mobile fitness (Gibbs & McLaren, 2017). Most mobile fitness apps on the market focus on consumers, and the advancement of mobile apps has encouraged practitioners to develop a mobile fitness app.

These mostly emphasise consumers' requirements, which has contributed to the growth of the mobile fitness market today (Kai-Kao & Leibovitz, 2017). Like other new technologies, mobile fitness app developer faced the fundamental challenge of how to encourage users, especially newcomers, to try out the apps.

Gamification concepts were among the approaches they used to foster engagement (Lim & Young-Noh, 2017). Nonetheless, gamification must truly be a motivational factor, otherwise it leads instead to disengagement. An example of this would be the gamified Nike mobile fitness apps, which encountered severe user dissatisfaction and disengagement because Nike removed certain features of game mechanics, such as badges (Welch, 2016). Reward is one of the important features in gaming. In addition, the developer should include the user's opinion based on their experience (Wolf, Weiger, & Hammerschmidt, 2018).

In academic research, the citations for gamification in mobile fitness apps are few. Although gamification has helped developers to generate $2.8 billion (2016), there is little evidence in academic research and literature to show that gamification has improved user engagement in mobile fitness apps (Peterson, 2016). Reviews of past academic literature revealed some limitations of gamification effectiveness in improving mobile fitness app user engagement behaviour. Therefore, the information discussed above sheds light on the research problems and provides an opportunity for conducting the research which discussed in the next section.

2 LITERATURE REVIEW

2.1 *Engagement*

The subject of engagement has attracted many practitioners as well as academics in recent years. Technological innovation in mobile devices has increasingly required practitioners to engage their customers at all possible touch points (Islam & Rahman, 2016). Browden (2009) proposed that engagement views are a psychological process that involves cognitive and emotional aspects. Van Doorn et al. (2010) argued that engagement is a primarily behaviours to the specific customer activity based on types or patterns. Based on these two arguments, it is clear that engagement conceptualisation still generates mixed opinions, which requires an extension of the engagement concept. A number of scholars have considered the consequences of engagement, including the concepts of perceived value (Kim, Kim, & Wachter, 2013) and brand image (Greve et al., 2014). However, few scholars have extended the study of gamification constructs to explore the consequences of user engagement, although they have provided details on how gamification could conceptualise user engagement in mobile applications such as mobile fitness apps (Hofacker et al., 2016). Thus, engagement studies in the field of information technology have faced challenges in convincing stakeholders that users would engage and accept.

2.2 *Gamification*

The development and innovation of multiple technology applications featuring games has brought about a new trend known as gamification. Gamification has become a fast-emerging business practice in industry worldwide. The term 'gamification' was initially introduced by Nick Pelling in 2002, and it started to gain popularity in information systems academics around 2010 (Liu, Santhanam, & Webster, 2017). The co-author of the book *Game-Based Marketing*, Gabe Zichermann, defined gamification as 'the art of turning your customer's daily interaction into gaming experience that serve the business purposes' (Zichermann & Liner, 2010).

Gamification has obtained great attention in the area of fitness and healthcare (Johnson et al., 2016). The evidence suggests that gamification is one of the extrinsic forces that stimulates enjoyment, engagement and compliance with health and fitness activities. This positively affects health behaviour outcomes and reinforces the development of advanced fitness digital platforms by developing mobile fitness apps that include gamification. Thus, there is great potential to add positive experiences to users' primary fitness goals (Johnson et al., 2016; Pereira et al., 2014; Sardi et al., 2017).

Based on insights from the group of scholars and an extensive literature review on gamification, the group of scholars extended gamification justification in different contexts. Early scholars like Deterding, Dixon, Khaled and Nacke (2011) studied gamification as the practice of applying game elements such as gamefulness, gameful interactions and gameful design with a specific intention in mind. Werbach and Hunter (2011) studied gamification as the application of game elements and design principles into non-game contexts to create engagement relationships among users.

Hofacker, De Ruyter, Lurie, Manchanda and Donaldson (2016) suggested that gamification should be explored by looking at human behaviour psychology perspectives and at the moderator that enhances extrinsic forces originating outside human motivation, such as rewards and points, that encourage an individual's engagement. They defined this as the practice of game elements and extrinsic forces in human psychology behaviour and reactions through enhancement of the consumer value and encouraging value creation of behavioural psychology outcomes such as engagement, greater loyalty or product advocacy.

Although gamification positively enhances engagement and acceptance behaviour, some studies discovered issues regarding the justification of gamification, such as the lack of a research context (Sanmugam et al., 2014) and small sample size (Hamzah, Ali, Saman, Yusoff, & Yacob, 2015). Scholars face major challenges in justifying gamification as an extrinsic force that could encourage individual engagement. For instance, recent research has shown that incorporating fitness activity engagement with gamification for fitness goals has a modest effect on individual engagement (Washburn et al., 2014).

Groh (2012) studied gamification usability effects on connected user goals and found that gamification did not positively affect usability, which is known as the extent to which a user can effectively achieve specific goals.

Lister et al. (2014) concluded that gamification effects on mobile fitness app engagement are not clear with regard to justifying an individual goal, and that there is a need for further research. Only a few scholars have extended the study of gamification to different contexts, such as the mobile fitness app, because it is also related to engagement behaviour, as suggested (Seman & Ramayah, 2017). Hence, this paper aims to fill this gap by studying gamification with engagement behaviour in the mobile fitness app context.

2.3 *Goal-setting theory*

Goal-setting theory (GST) was originally introduced by Locke (1968) who proposed that an individual's desired behaviour is achieved when the specific goal is clear. It has been applied for decades by scholars because it helps to explain individual motivation to perform better in related tasks and in many contexts through the setting and monitoring of goals (Locke & Latham, 2002). This theory was developed based on findings from various empirical studies that posited that user behaviour is directly related to the individual's goals for pursuing with the relevant content (the object of an action) and intensity (the amount of effort to achieve the goal; Locke & Latham, 2002).GST theorised that goal core is the immediate regulator of individual behaviour and the standard used to evaluate task performance (Kylo & Landers, 1995). Goal core is identified as being specific, measurable and committed. These are the major inputs that provide an external referent, such as time, space or increment, to track the progress of goal achievement. This reduces ambiguity and allows focus on precise actions and behaviour related to goal mechanisms that empower an individual to achieve his or her goals (Locke & Latham, 1990). Locke and Latham (2006) in their research notes on GST application recommended that GST should be extended by scholars by looking at individual behaviour such as engagement from different contexts, instead of organisation.

Parker, Jimmieson and Amiot (2009) found that autonomy as a goal mechanism improved self-efficacy and goal core, which in turn improved task performance towards reaching goals. Thus, the concept of GST helps to motivate individuals and teams to perform engagement behaviour. Although GST has explained individual motivation behaviour, this was limited to exploring intrinsic force in organisational settings such as manufacturing and human resources (Mento et al., 1987). However, a few scholars have extended the study of GST by adding gamification as an extrinsic force when studying human behaviour.

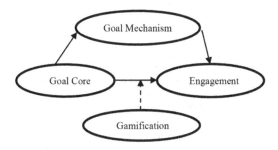

Figure 1. A proposed research framework.

For instance, Jacobs (2013) extended GST by incorporating gamification antecedents in employee engagement and found that individual goal core was positively related to engagement, but gamification did not have an effect in GST study. Further, Morschheuser et al. (2017) and Mo-ra et al. (2017) concluded that gamification antecedents should be extended when justifying individual engagement in mobile fitness apps because only a few scholars have linked gamification to GST in the mobile fitness app context. From the analyses discussed above, it was found that GST is by nature an intrinsic force. The investigation into intrinsic factors to foster engagement is well advanced.

Therefore, this research proposes a focus on extrinsic forces in the relationship, which gamification marked as the line with dots (extrinsic factor) is identified as the variable that will moderate the relationship, as shown in Figure 1.

3 PRE-TEST RESULTS

This research viewed experts to be individuals who engaged in academic research and individuals who were experienced and related to the industry. Two experts from the Faculty of Computing and Azman Hashim International Business School (Universiti Teknologi Malaysia), three experts from industry who were familiar with fitness and mobile applications, and one expert gamer were approached to provide an expert view on the items shown in Table 1.

The results shown in Table 1 suggest that, to help respondents have a better understanding, abbreviated terms should not be used. Some advice was given by the experts, such as, provide a brief explanation about mobile fitness apps and their relation to gamification. The experts also suggested that the researcher should rephrase some of the words and be more specific, so the respondents can understand the questions.

Table 1. Summary of pre-test results.

Pre-Test Results		
Items	Academician	Practitioner
Goal Core	The 'play' phrase needs to be reviewed because it creates a diversion.	The phrase of "Play" and "Goal Core" need to rephrase
Goal Mechanism	Some of the phrases need to be corrected, such as the differences between 'goal' and 'goal core'.	Included 'the nutrition features' as the measurement of user mobile fitness app engagement.
Gamification	The item requires some references and how it can the best mobile fitness app and tested as the moderator	Internal motivation should be added.
Engagement	All items accepted but need some refinement.	Social media aspects such as 'post results' should be included.

4 CONCLUSION

The application of GST and gamification studies has been extended by scholars looking at engagement behaviour in various research contexts. However, most applications of GST are being conducted in organisational settings, and gamification in education engagement. Surprisingly, the studies combining GST and gamification with engagement behaviour are limited in academic research. Hence, it is advisable for future research to extend the study of GST and gamification to examine engagement in mobile applications such as mobile fitness apps. Instead of using a quantitative method, an experimental method is encouraged because it can provide accurate results and support the justification for the research model. Applying an experimental design can help future researchers to draw causal conclusions and obtain better comparison findings (Katule et al., 2016). It would be well worth extending this research because the results could contribute to the body of knowledge in engagement research, especially regarding individual engagement justification in mobile applications such as the mobile fitness app.

REFERENCES

Bowden, J. L. H. (2009). The process of customer engagement: A conceptual framework. *Journal of Marketing Theory and Practice*, 17(1), 63–74.

Byun, H., Chiu, W., & Bae, J. S. (2018). Exploring the adoption of sports brand apps: An application of the modified technology acceptance model. *International Journal of Asian Business and Information Management (IJABIM)*. 9(1), 52–65.

Deterding S., Dixon D., Khaled R., & Nacke L. (2011). From game design elements to gamefulness: Defining gamification. *Proceedings of the 15th International Academic MindTrek Conference: Envisioning Future Media Environments*. ACM.

DeWalt, D. A., Davis, T. C., Wallace, A. S., Seligman, H. K., Bryant-Shilliday, B., Arnold, C. L., & Schillinger, D. (2009). Goal setting in diabetes self-management: Taking the baby steps to success. *Patient education and counseling*, 77(2), 218–223.

Gibbs, S. (2016). Runkeeper bought by Asics in latest sports brand app acquisition. *The Guardian*. Retrieved from

Goodwin, E., & Ramjaun, T. (2017). Exploring consumer engagement in gamified health and fitness mobile apps. *Journal of Promotional Communications*, 5(2).

Greve, G. (2014). The moderating effect of customer engagement on the brand image–brand loyalty relationship. *Procedia-Social and Behavioral Sciences*. 148, 203–210.

Hamzah, W. M. A. F. W., Ali, N. H., Saman, M. Y., M., Yusoff, M. H., & Yacob, A. (2015). Influence of gamification on students' motivation in using e-learning applications based on the motivational design model. *International Journal of Emerging Technologies in Learning*, 10(2), 30–34.

Hofacker, C. F., De Ruyter, K., Lurie, N. H.,Manchanda, P., & Donaldson, J. (2016). Gamification and mobile marketing effectiveness. *Journal of Interactive Marketing*, 34, 25–36.

Jacobs, H. (2013). Gamification: A framework for the workplace (Doctoral dissertation).

Jusoh, S. (2017). A survey on trend, opportunities and challenges of mHealth apps. *International Journal of Interactive Mobile Technologies*. 11(6), 73–85.

Kao, C. K., & Liebovitz, D. M. (2017). Consumer mobile health apps: Current state, barriers, and future directions. *PM&R*, 9(5), S106–S115.

Krebs, P., & Duncan, D. T. (2015). Health app use among US mobile phone owners: a national survey. *JMIR mHealth and uHealth* 3(4).

Kylo, L. B., & Landers, D. M. (1995). Goal setting in sport and exercise: A research synthesis to resolve the controversy. *Journal of Sport and Exercise Psychology*, 17(2), 117–137

Lim, J. S., & Noh, G. Y. (2017). Effects of gain versus loss-framed performance feedback on the use of fitness apps: Mediating role of exercise self-efficacy and outcome expectations of exercise. *Computers in Human Behavior*, 77, 249–257.

Liu, D., Santhanam, R., & Webster, J. (2017). Toward meaningful engagement: A framework for design and research of gamified information systems. *MIS Quarterly*, 41(4).

Locke, E. A., Cartledge, N., & Koeppel, J. (1968). Motivational effects of knowledge of results: A goal-setting phenomenon? *Psychological Bulletin*, 70(6p1), 474.

Locke, E. A., & Latham, G. P. (2002). Building a practically useful theory of goal setting and task motivation: A 35-year odyssey. *American Psychologist*, 57(9), 705.

Locke, E. A., & Latham, G. P. (2006). New directions in goal-setting theory. *Current Directions in Psychological Science*, 15 (5), 265–268.

Islam, J. U., & Rahman, Z. (2016). Linking customer engagement to trust and word-of-mouth on Facebook brand communities: An empirical study. *Journal of Internet Commerce*, 15(1), 40–58.

Peterson S. (2014). Gamification market to reach $2.8 billion in 2016. Retrieved from.

Sanmugam, M., Mohamed, H., Zaid, N. M.,Abdullah, Z., Aris, B., & Suhadi, S. M. (2016). Gamification's role as a learning and assessment tool in education. *International Journal of Knowledge-Based Organizations*, 6(4), 28–38.

Seman, S. A. A., & Ramayah, T. (2017). Are we ready to app?: A study on mHealth apps, its future, and trends in Malaysia Context. In *Mobile platforms, design, and apps for social commerce* (pp. 69–83). IGI Global.

Van Doorn, J., Lemon, K. N., Mittal, V., Nass, S., Pick, D., Pirner, P., & Verhoef, P. C. (2010). Customer engagement behavior: Theoretical foundations and research directions. *Journal of Service Research*. 13(3), 253–26.

Wolf, T., Weiger, W. H., & Hammerschmidt, M.(2018)Gamified digital services: How gameful experiences drive continued service usage.

Zichermann, G., & Linder, J. (2010). Game based Marketing: Inspire Customer Loyalty through Rewards. *Game-Based Marketing*, 240.

Inclusive Development of Society – Lumban Gaol (eds)
© 2020 Taylor & Francis Group, London, ISBN 978-1-138-33476-2

Factors affecting Malaysian Muslim Millennials' choice of Indonesia as their halal tourism destination: A conceptual discussion

A. Hanafiah
Faculty of Economy and Business, Universitas Mercu Buana, Jakarta, Indonesia

F.S. Ahmad, N. Zaidin, M.S.M. Ariff, R. Baharun & R.M. Nor
Azman Hashim International Business School, Universiti Teknologi Malaysia

ABSTRACT: This study focuses on a conceptual discussion and framework to examine factors affecting Malaysian Muslim Millennials in choosing Indonesia's halal tourism destinations (HTDs). Based on the theory of planned behaviour (TPB), which relies on attitude, subjective norms and perceived behavioural control in influencing intention, the study expands the concept to include antecedents of internal stimulation of religiosity and external stimulation of marketing promotion. The model is theoretically supported based on diverse earlier studies that incorporate new variables in TPB to anticipate consumer behaviours. It aims to provide answers to the industrial concerns of the Halal Tourism Acceleration and Development Team of the Indonesian Tourism Ministry regarding their marketing stimulations and promotional efforts in attracting Malaysian Muslim Millennials. The findings include details on the type of factors influencing visit intentions, theoretical understandings and market development aspects for a reliable empirical investigation. This paper concludes with recommendations for continuous development of Indonesia halal tourism and development of marketing knowledge in conceptualising a robust and reliable research framework to capture meaningful data and information from the market.

Keywords: Halal Tourism Destinations, Muslim Millennials, Visit intentions and Consumer Behaviours

1   INTRODUCTION

1.1   *Background and problem statement*

Like other emerging countries, Indonesia is striving to make tourism its leading sector to create an economic multiplier effect for the country. It is predicted that by 2030 the majority of tourists will have shifted to the emerging country and be positively contributing to the economy. The United Nations World Tourism Organization (UNWTO; 2016) reported that tourism has become the biggest and quickest developing global financial area, ranking third after the food and automotive industries.

A growing tourism subset is halal tourism, and this discussion focuses on halal tourism destinations (HTDs) of Indonesia as a preferred destination among Malaysian Muslim Millennials (MMMs). Halal tourism is defined by Sureerat (2015) as 'offering tour packages and destinations that are particularly designed to cater for Muslim considerations and address Muslim needs'. According to the Global Islamic Economy Indicator (2017),

Indonesia ranks fourth for halal tourism, with the favourite top four HTDs being Aceh, Lombok, West Sumatra and Jakarta.

Indonesia's Ministry of Tourism in 2015, under the newly formed Halal Tourism Acceleration and Development Team (HTADT), has projected an increase in Muslim travellers, particularly from Indonesia's neighbouring countries. With a projection of 108 million Muslim travellers around the world in 2014 being converted into 150 million visitors in 2020 (Mastercard-Crescent Rating, 2015), the ministry targeted 20.4% average growth to reach 5 million international Muslim visitors in 2019. Hence, it began implementing strategic initiatives emphasising the marketing and promotion pillar (branding, advertising, selling) to increase awareness among potential Muslim visitors (Rahmiasri, 2016). Nonetheless, by 2017, numbers of Muslim inbound tourists to Indonesia were below the yearly target of 3.1 million Muslim visitors (Kaul, 2018); this number is very small compared with the projection of 18 million Muslim tourists in ASEAN and 156 million globally (Global Muslim Travel Index [GMTI], 2018).

Consequently, HTADT sensed the urgency to focus on the biggest demographic segment of the market, the millennial generation, also referred to as Gen Y, that is, anyone born between 1980 and 1999. This generation is regarded as the fastest growing consumer segment globally. The MMM segment, with a population of approximately 6 to 7 million, seems to be an attractive target for Indonesia to achieve 5 million international Muslim visitors by 2019. HTADT believed that the world's largest archipelago, with more than 300 distinct native ethnicities, 746 languages, and natural and cultural resources to offer, makes Indonesia unique. Curiosity grew regarding the effectiveness of the marketing and promotional effort as the external stimuli to intervene with the MMM's intention to visit Indonesia identified HTDs. However, at the same time, it is important to understand whether their religiosity plays role in their attitudes, social norms and behavioural control.

1.2 *The importance of Malaysia Muslim Millennials in the research*

Malaysia is the second major market share to Indonesia's inbound tourists from the Asia region. From OIC members' category, Malaysia contributes the highest number of visitors to Indonesia and it has become one of the strongest contributors to the halal tourism market. The GMTI (2017) reports the younger Muslim population as one of the key drivers of growth in the Muslim travel market. It is predicted that Muslim millennial travellers will begin to develop their travel preferences and enter their peak earning, spending and travelling life stages within the next 5–10 years (Mastercard-HalalTrip, 2017).

Although some industry reports have attempted to study the millennial potential market, no single earlier industry report has focused on Muslim millennials' travel. Only recently, the Muslim Millennial Travel Report (MMTR; 2017) by Mastercard-HalalTrip pointed to the huge opportunities that lie in this subset of millennial travellers. The report reveals country-wise contribution to the overall Muslim millennial travellers in general, Malaysia being one of the top countries (Table 1).

Table 1. Muslim Millennial Travellers (Mastercard--HalalTrip, 2017).

Top Countries for Muslim millennial travellers	
OIC countries	non-OIC countries
1. Saudi Arabia	1. Germany
2. Malaysia	2. Russian Federation
3. Turkey	3. India
4. Kazakhstan	4. United Kingdom
5. Egypt	5. China
6. Indonesia	
7. Oman	
8. Iran	
9. UAE	
10. Qatar	

Recent development of Indonesia's halal tourism has highlighted the need to find a more reasonable approach to achieving the target. Researchers deliberated on Malaysian demographic data of the MMM population born in the 1980s and 1990s. Studies show they mostly have significantly fresh global outlooks and a high degree of fluency in the digital economy. In terms of attitude, they like to be surrounded by different people, cultures, ideas and lifestyles, with a high preference to explore the world around them. They are estimated to be about 24% of the Malaysia population size of 31 Million. (Hakuhodo Institute of Life and Living ASEAN, 2017; The Edge Market, 2017).

As reported by Kaul (2018), in HTADT's recent interview with the media, it was acknowledged that there were only 2.7 million Muslim international visitors in 2017, which is well below the expected target of 3.1 million. Therefore, the recent development of Indonesia's halal tourism has highlighted the need to find a more reasonable approach to achieving the target. MMMs form the basis for Indonesia HTDs that aim to capture this millennial segment, with a population of approximately 6 to 7 million people. They represent the low hanging fruits for Indonesia to fill the halal tourism industry gap in terms of yearly Muslim visitors to Indonesia.

2 LITERATURE REVIEW

2.1 Theory of planned behaviour: The underpinning theory of the study

The theory of planned behaviour (TPB) was used to underpin this research to provide a general understanding of consumer behaviour, and specifically to investigate the MMMs' visit intentions to Indonesia's HTDs. Three TPB predictors are attitudes, subjective norms and perceived behavioural control. Ajzen and Madden (1986) argued that, to perform the behaviour, a person must have both the opportunity and resource, which results to some degree in performing the behaviour. Thus, when an individual is confronted with a decision-making situation, he or she takes into consideration the likely consequences of the existing options (known as behavioural beliefs), evaluates his or her referenced or group expectations (known as normative beliefs), and takes into account the requisite sources and possible hardships or difficulties (known as control beliefs). The consideration of beliefs results in the creation of attitude, subjective norms and perceived behavioural control. Figure 1 illustrates the TPB with factors predicting behavioural performance.

Earlier studies reported the application and adoption of new variables of TPB in predicting intention and behaviour in varied social science research, including social media usage (Cameroon et al., 2011), hotel guest satisfaction (Berezan et al., 2013), the green hotel industry (Kang et al., 2012) and (Chen and Tung, 2014), the e-WOM impact on destination choice (Jalilvand & Samiei, 2012) and many others. In a similar manner, the application of TPB as the model theory was also used in past halal-related marketing studies, such as studies on

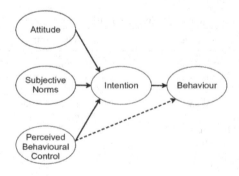

Figure 1.   Theory of Planned Behaviour (TPB) (Ajzen, 1985).

halal product concerns (Golnaz et al., 2009), the acceptance of Malaysia as a halal hub (Rahman et al., 2012), halal and marketing components (Azis & Chok, 2012), the preference for JAKIM-certified products (Mohayidin & Kamarulzaman, 2014), information process and food labelling (Latiff et al., 2015; Razzaque & Chaudry, 2013), and the hospitality and tourism industry (Quintal et al., 2010). These studies discovered a positive relationship between the predictors of TPB and that the behaviour of interest in the respective domains strengthened the reliability of TPB. Nevertheless, there are concerns over the variables that are not included in TPB, based on the peculiarity of the market segments and other possible industry strategic marketing initiatives. In the context of this study, the Indonesia Tourism Ministry is concerned to understand the internal and external drives towards HTDs among MMMs (targeted segment) based on the internal factor of religiosity and the external market stimulation.

2.2 *Attitudes, subjective norms and perceived behavioural control*

Attitude towards a behaviour denotes the degree of a person's assessment, favourable or otherwise, of performing the behaviour in question (Ajzen, 1991). Therefore, attitude acts as a function of beliefs (behavioural beliefs) or a psychological tendency that dictates the behaviour in question (Haque, Sarwar, Yasmin, Tarofder, & Hossain, 2015; Bonne et al., 2007; Ajzen, 1991; Lada et al., 2009). Attitude in this study will be operationally defined as the Muslim millennials' level of favourability regarding visit intention to his or her chosen halal tourism destination, which is in accordance with Islamic principles (known as leisure in halal tourism). Finally, based on that phenomenon, the study examines how attitude mediates the relationship between religiosity and intention to visit HTDs, as well as the relationship between marketing and promotion and intention to visit HTDs.

The second variable of TPB is subjective norms, which generally consists of two normative beliefs that might influence a person's intention: injunctive and descriptive norms (Ajzen, 1991; Hagger & Chatzisarantis, 2005). Accordingly, such norms would result in peer or social pressure towards a person. Hence, to operationalise subjective norms in this study, the MMM decision about his or her intention to choose an HTD must comply with the desires of a spouse, family and/or other important and close persons under consideration. This study posits that subjective norms mediate the relationship between religiosity and intention to visit an HTD, as well as the relationship between marketing and promotion and intention to visit HTD.

Perceived behavioural control, as the third variable, is described as the awareness of the degree to which the behaviour is being perceived to be controllable and easy (Ajzen, 1991; Ajzen, 2005; Ajzen & Madden, 1986). It relates to the form of beliefs concerning whether a person has access to the necessary resources and opportunities to perform a behaviour successfully, weighted by the perceived power of each factor to facilitate or impede the behaviour, which is referred to as control beliefs. This underlying belief includes external and internal control. The external control factors are opportunities, dependence on others and barriers, while the internal control factors are information, personal deficiencies, skills, abilities and confidence (Conner & Armitage, 1998).

2.3 *New variables adoption in TPB: Religiosity and marketing promotions*

Since researchers refer to halal as any practice or activity that is 'permissible' according to Islamic teaching, it is important to understand the Muslim consumer's intention in selecting the halal option to fulfil their daily needs. For example, a study was conducted to explore Malaysia consumer intentions in choosing Islamic home financing products (Ibrahim, Fisol, & Haji-Othman, 2017) and adopting Islamic home financing in Malaysia (Bassir, Zakaria, Hasan, & Alfan, 2014). The authors found that religiosity is the strongest determinant of consumer intention. Further, according to Alam et.al. (2012), the significance of religiosity is due to its capability to affect an individual's cognition and behaviour.

Similarly, Ajzen (1985, 1991), the proprietor of TPB, considered religion as one possible factor that affects both individual intention and behaviour. Additionally, along with Fishbein (Ajzen & Fishbein, 2005), he pointed out that the behavioural, normative and control beliefs people hold about the performance of a given behaviour are influenced by a wide variety of other factors, which become possible antecedents, such as individual personality, mood, emotion, intelligence, education, age, gender, income, religion, race, ethnicity, culture information, knowledge, media and other interventions (Ajzen & Fishbein, 2005 p. 194).

According to Temporal (2011), Islamic branding in business should link to at least three main sources: Islamic countries (with Muslim-majority populations), Islamic companies, and Islamic products or services. Hence, to exploit the opportunity in the halal industry, Temporal (2011) suggested the need for a brand decision-maker, to understand the typology or architecture of Islamic branding according to its sources, as depicted in Figure 2.

Basically, advertisement is needed to educate Muslim consumers and create awareness about the brand's halal products and services, but also to position the brand. In other words, ideally, the first step of awareness creation will lead to the connection of internal emotional factors of consumers, known as positioning of brand image, and which becomes a true differentiator (Temporal, 2011).

Consequently, the proposed conceptual framework of the study was devised and is presented in Figure 3 below, which covers all the variables related to the proposed study.

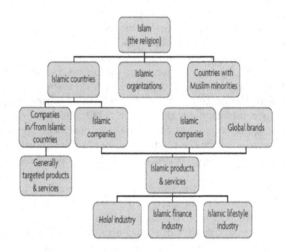

Figure 2.   A typology of Islamic brands (Temporal, 2011).

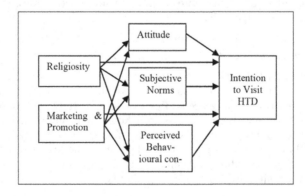

Figure 3.   Intention to visit HTD—conceptual framework.

302

3 METHODOLOGY

3.1 *Data collection for future research*

The purpose of this study is to empirically investigate the relationship between external stimuli of religiosity and marketing and promotion and the intention to visit HTDs. With the TPB approach, both variables are thought to be possible antecedents in determining individual intention. Data will be obtained from internet-based questionnaire distribution, because this is easier to distribute and ensures the researchers' absence when the respondent is completing the questionnaire (Sue and Ritter, 2007). Moreover, the online-based questionnaire is suitable for millennials who are technology savvy. However, if necessary, hard copy questionnaires will be distributed and collected, if the required sample is difficult to achieve.

Before the questionnaires are distributed, pre-test procedures will be conducted. Pre-test is defined as the activity of developing questionnaires and their measurements to be used in the actual survey or research (Anderson & Gerbing, 1991 After completion of the data collection stage, statistical procedures will follow to preserve an objective perspective in analysing the data.

The population of this study is composed of MMMs. Details in the profiles of the samples include those of Malaysian Muslims born in the 1980s and 1990s. The current age in 2018 is between 19 and 38 years old, and was selected based on a non-probability convenient sampling technique. Since the data will be analysed using structural equation modelling (SEM), the research will adopt Hair et al.'s (2014) suggestion in using the G Power software version 3.1 to address this issue. Thus, according to the G power software calculation, and with parameters of effect size (f^2) = 0.15, α error probability = 5%, statistical power = 95% and number of predictors = 5, the required sample size is 138. However, to anticipate reduction after the data screening process, additional sampling will be used, and the approximate initial sample size for this study is 400

3.2 *Proposed technique for data analysis and model validation*

SEM is a type of statistical model that seeks the relationship between variables. Moreover, this statistical technique combines factor analysis and multiple regression aspects to simultaneously evaluate the relationship of a series of interrelated dependences. Gefen et al. (2000) noted that an SEM approach can find answers to some interrelated research questions by concurrently modelling the relationships among multiple independent and dependent constructs. Therefore, according to Diamantopoulus (2011), independent and dependent variables will no longer be differentiated in the SEM analysis, but rather between exogenous variables, which are equivalent to the independent and endogenous variables explained by the relationship in the model.

Moreover, as noted by Urbach and Ahlemann (2010), these techniques have two interrelated models, which are assessed concurrently. First, the outer model is known for the measurement model, which refers to the relationship between the latent variables and empirically observable indicators. Second, the inner model is known for the structural model, which contains the variable relationships.

To use SEM, there are two general approaches. The first approach can be utilised with the help of software such as AMOS and LISREL, which suits covariance-based structural equations (CB-SEM). The other approach for SEM is variance-based and is known as PLS-SEM (for instance, PLS Graph and PLS Smart). According to Hair et al. (2011), comparison of CB-SEM and PLS-SEM is aimed at maximising the explained variance of the dependent latent constructs, that is, a causal modelling approach. Urbach and Alhemann (2010) argued that PLS-SEM, practically, can be used for theory confirmation throughout the theory development process.

This study used the PLS-SEM approach for a few reasons related to expert explanation and research. First, referring to Hair et al. (2011), PLS-SEM is a powerful analysis method, despite its minimal demands on sample size and multivariate normality of data, with better prediction

Table 2. 2012–2014 tourism foreign exchange earnings of asian countries (UNWTO, 2015).

Tourism as Foreign Exchange Earning (US$ million)				
No	Asia Countries	2012	2013	2014
1	China	50,028	51,664	56,913
2	Macao (China)	43,860	51,796	50,815
3	**Thailand**	**33,855**	**41,780**	**38,437**
4	Hong Kong (China)	33,074	38,934	38,376
5	**Malaysia**	**20,250**	**21.496**	**21,820**
6	Singapore	18,939	19,301	19,203
7	Japan	14,576	15,131	18,853
8	South Korea	13,429	14,629	18,147
9	Taiwan(pr. Of China)	11,770	12,323	14,618
10	**Indonesia**	**8,324**	**9,119**	**9,848**

capability. It maximises the explained variance of endogenous latent variables through partial model relationships estimated in an iterative sequence of ordinary least squares regression (Hair et al., 2011).

4 FINDINGS FROM THE LITERATURE

A literature review or secondary research was useful in establishing the conceptual foundation, as well as the research structural model to be validated by the future data gathered.

First, the literature review allowed the researchers to understand the magnitude of the issues in the development of HTD in Indonesia. First, Indonesia recorded tourism foreign exchange earnings below its neighbouring countries (see Table 2).

The foreign exchange earnings described above in Table 2 are basically the proceeds from the export of goods and services of a country, and the returns from its foreign investments, denominated in convertible currencies. The data in 2015 from UNWTO reported that Indonesia earned less than half the earnings of Malaysia, a quarter of Thailand's earnings and less than one-fifth of China's. Taking an optimistic outlook, Indonesia's tourism minister sees room for improvements and is keen to ensure tourism is the biggest contributor of Indonesia's foreign exchange earnings by overlapping other main economic sectors, such as oil, coal and rubber (Sofyan, 2016). This confidence is reinforced by Indonesia's natural and cultural resources, for example by being the world's largest archipelago with natural scenery of mountains, oceans, coastlines and colourful coral reefs. Therefore, the target of 20 million international visitors by 2019 (16.4% average growth), which includes 5 million Muslim visitors, is challenging yet at the same time reasonable.

The intensive review also allowed the researchers to construct an appropriate questionnaire for data collection, which will be validated and tested for reliability before the future data collection. The theoretical exploration, which includes various arguments and theory critiques, as well meeting industrial practical needs, helped the researchers to identify the right variables, namely, religiosity and marketing and promotions, to be integrated into the established TPB for a more robust and holistic approach in understanding the target market—the MMMs' visit intentions to HTD.

5 CONCLUSION

Managing the nation's tourism development is a challenging process for any government. In the context of the halal tourism industry, the rapid establishment of HTDs spells the need to assess the market capacity and factors influencing consumers' decision-making for organisations' competitiveness. The current study proposes to fill the gap regarding the obvious possibility of influences such as religiosity and marketing efficiency. The examination of these factors is appropriate to assist policy makers as well as to enhance current marketing knowledge of consumer behaviour.

In terms of practical contributions, this study will benefit tourist-related organisations to understand how to implement appropriate strategies to respond to possible market information to be gathered. In terms of future studies, researchers should adopt the same framework in different contexts or cultures because this may produce different results. Future researchers could also integrate other relevant antecedent, mediating or moderating variables to enhance the current research framework

This conceptual framework for Indonesian HTD's visit intention and its model validation process will shed more light on the current status and directions of the tourism and hospitality industry of Indonesia and equip marketers with more relevant strategies to cater to the dynamic markets of Muslim millennials, particularly those from Malaysia.

REFERENCES

Ajzen, I. (1985). From intentions to actions: A theory of planned behavior. In J. Kuhl & J. Beckman (Eds.), *Action-control: From cognition to behavior* (pp. 11–39). Heidelberg, Germany: Springer.

Ajzen, I. (1991). The theory of planned behavior. *Organizational Behavior and Human Decision Processes, 50*(2), 179–211. doi: https://doi.org/10.1016/0749-5978(91)90020-T

Ajzen, I., & Madden, T. (1986). Prediction of goal-directed behavior: Attitudes, intentions, and perceived behavioral control. *Journal of Social Experimental Psychology, 22*, 453–474.

Bassir, N. F., Zakaria, Z., Hasan, H. A., & Alfan, E. (2014). Factors influencing the adoption of Islamic home financing in Malaysia. *Transformation in Business & Economics, 13*(1).

Hair, J. F., Ringle, C. M., & Sarstedt, M. (2011). PLS-SEM: Indeed a silver bullet. *Journal of Marketing Theory & Practice, 19*(2), 139–151.

Haque, A., Sarwar, A., Yasmin, F., Tarofder, K., & Hossain, M. (2015). Non-Muslim consumers' perception toward purchasing halal food. *Journal of Islamic Marketing, 6*(1), 133–147.

Ibrahim, M. A., Fisol, W. N. M., & Haji-Othman, Y. (2017). Customer intention on Islamic home financing products. *Mediterranean Journal of Social Sciences, 8*(2), 77.

Kaul, R. (2018, April 26). *Indonesia to introduce halal travel index to bolster Muslim arrivals.* Retrieved from https://www.ttgasia.com/2018/04/26/indonesia-to-introduce-halal-travel-index-to-bolster-muslim-arrivals/

Mastercard-Crescent Rating. (2015). Retrieved from https://www.crescentrating.com/halal-muslim-travel-market-reports.html

Mastercard-Halaltrip. (2017). *Muslim millennial travel report 2017.* Retrieved from https://www.halaltrip.com/attraction/downloadlink/?file=ht-muslim-millennial-travel-report-2017.pdf

Temporal, P. (2011). *Islamic branding and marketing: Creating a global Islamic business.* John Wiley & Sons.

United Nations World Tourism Organization. (2011). *Tourism towards 2030: Global overview.* Retrieved from http://media.unwto.org/sites/all/files/pdf/unwto_2030_ga_2011_korea.pdf

Inclusive Development of Society – Lumban Gaol (eds)
© 2020 Taylor & Francis Group, London, ISBN 978-1-138-33476-2

A cross-case analysis approach to heritage tourists' behavioral intention to revisit: Toraja context

Hendrikus Kadang
Faculty of Economics and Business, Universitas Atma Jaya Makassar, Indonesia

Rohaizat Baharun & Umar Haiyat Abdul Kohar
Azman Hashim International Business School, Universiti Teknologi Malaysia, Johor, Malaysia

ABSTRACT: The factors that influence tourist behavioral intentions to revisit heritage destinations such as Toraja in South Sulawesi, Indonesia remain a contemporary subject for
extensive studies that could bring new findings in terms of discovering innovative factors that
influence tourists' behavior. Noticeable declines of visitors' arrivals from 1997 to 2008,
coupled with extreme fluctuations during the last seven years, are obviously a phenomenon in
the Toraja's tourism industry. This article explores the factors that are significant in influencing tourist behavioural intentions to revisit heritage destinations in Toraja. The study
adopted an exploratory research design in surfacing potential influencing factors that could be
embedded in the broad tourism policy of the government and eventually boost tourist figures
to revisit heritage destination sites. Purposive sampling of 15 participants in 3 categories was
conducted through semistructured interviews and the data obtained from the interviews were
analysed using cross-case analysis. Thus, the study generated nine factors that could influence
heritage tourists' behavioral intentions to revisit the heritage destinations in Toraja.

1   INTRODUCTION

1.1   *Tourism*

Generally, the tourism industry influences various aspects in society as well as individual
members of the community (Wells et al., 2016). In Indonesia, with its numerous heritage tourism destinations spread all over the archipelago, the tourism sector has beneficially contributed to the improvement of communities in and around the tourism destination areas
(Walpole & Goodwin, 2000). For instance, the communities around the Borobudur Temple
continuously gain economic benefits from the incessant tourist arrivals by selling various
locally made handicrafts as well as locally produced tourism souvenir items (Canny, 2013).
However, given the serious concern of declining heritage tourist arrivals as well as its economic impact on the communities, the government of Indonesia viewed it as a serious concern. Thus, realizing this critical issue, the national government has afforded some policy
shifts such as simplifying the policies on tourist visa application in order to attract more visitors to the various heritage tourism destinations (Sugiyarto et al., 2003; Cultural and Tourism
Minister of Indonesia, 2006; Tourism Law of Indonesian Government No. 10, 2009).

1.2   *Toraja heritage site*

One of the most famous heritage tourism sites in Indonesia can be found in Toraja Regency in
the province of South Sulawesi (Adams, 2007). The Toraja heritage tourism site is considered
the second most famous heritage tourism destination in Indonesia after Bali because of its
unique cultural and various traditional attractions, such that it has become the "prima
donna" of South Sulawesi (Adams, 2003). The numerous areas of tourism interest in Toraja

include the distinctive and original ritual in holding funeral ceremonies, the age-old tradition of burying the dead, traditional folk dances, ancient dwellings, unique handicraft, and the old practice of carving statues or building traditional houses and other traditional cultures; it also includes a natural hill rock formation around the residential communities (Adams, 2012; Lestari et al., 2012; Indratno et al., 2013).

Toraja was nominated as one of the World Heritage Sites in Southeast Asia owing to its various attractive destinations such as the natural scenery of the highland formation and the uniqueness of cultural, burial, or ceremonial grounds (Adams, 2003; King, 2013). However, despite comparative advantage in tourism over other areas in Indonesia, the attractiveness and the distinctiveness of Toraja have not augured well in translating those good features into a positive influence on heritage tourist arrivals (Toraja Tourism Departement, 2014). Mapaliey (2009) emphasized that there was a gradual decline of visitor arrivals during the period 1997–2007 and the Toraja Tourism Department (2016) concluded that there were extreme fluctuations of tourists' arrival during the past seven years. It is therefore interesting to explore what those influencing factors are that impinge on visitor behavior to visit Toraja heritage site.

Numerous studies were conducted on tourists' behavioral intention to revisit tourism sites, while studies concerning heritage sites were too few (Rodríguez et al., 2012; Canny, 2013). Other studies were focused on investigating the factors that influence visitors' future behavioral intentions (Canny & Hidayat, 2012; Som et al., 2012). In relation to Toraja's heritage sites, there is a need to conduct an exploratory study to bring out potential factors that could influence tourists' behavioral intention to revisit its heritage sites. Accordingly, the result of this study could generate some factors that would influence the heritage tourists' behavioral intention to revisit the heritage sites in Toraja to address the problem of declining tourist arrivals and possibly come up some innovative influencing factors aside from the established ones.

2 LITERATURE REVIEW

2.1 Heritage tourism

A heritage tourism site refers to the legacy of a place or a site, including its built-up structures having significance to the country's history, that can be considered as a unique and attractive site that can be considered a legacy destination (Chen & Chen, 2010). Accordingly, these characteristics are reflected in various heritage tourism sites (Poria et al., 2004), while Hodur (2010) emphasized that a heritage tourism site must have a distinct characteristic that would serve as its trademark and can draw better appreciation from the visitors.

These distinct heritage tourism characteristics can be experienced in Indonesia. Some of the country's heritage tourism sites such as Borobudur Temple in Java Island and Komodo National Park in Nusa Tenggara Island Province provide a sense of a historical timeline. The former provides a glimpse into the early Indonesian civilization, while the latter is a habitat of rare animals. These World Heritage Sites have become famous heritage tourism destinations in Indonesia (UNESCO, 2013). In terms of the number of visitors, the Borobudur Temple attracted approximately 2.9 million tourists in 2008, while Komodo National Park received 36,429 tourists during the first six months of 2014 (Manuella, 2014). Other heritage destinations in Indonesia are the Toba Lake in North Sumatera Province, Bunaken Sea Park in North Sulawesi Province, and the Kalimutu Mountain in East Nusa Tenggara Province, among others (Wardiyatmo, 2012).

Heritage tourism sites that have become famous may have significant implications for many other sectors. A number of scholars (Schubert et al., 2011; Antonakakis et al., 2014) found that the tourism sector has delivered multiple beneficial impacts to communities and to the broader society. This sector is considered to be an engine of economic growth and a poverty-alleviating mechanism because the contribution of this sector is very important to increase incomes of people in or around the destination (Richardson, 2010). Similarly, these positive impacts to the society could also be experienced within the heritage tourism sector. The impact of heritage tourism can be seen through increased tourism-related establishments such as restaurants, accommodation facilities, and employment (Vong & Ung, 2012). Schubert et al. (2011) stated that tourism in general increases trade, while Antonakakis *et al.* (2014) cited its impact on various aspects in the community. Hence, all types of tourism activities, including heritage tourism, provide similar positive impacts to people and establishments having the opportunity to be involved in this sector.

2.2 *Heritage tourist*

Generally, the term "heritage" can be attributed to a legacy, which means something that is transferred from one generation to the next (Nuryanti, 1996). Zhao and Timothy (2015) indicated that a particular heritage characterizes essential features of a specific people and their associated values, norms, and beliefs. On the other hand, Rhone and Neil (2001) contended that heritage tourism is analogous to exploration of destinations with historical significance. These historical connotations include interesting narrative of places, their people and their cultures, as well as existing cultural icons such as buildings and edifices that tell stories of the rich historical past. This comprehensive description professes a clear differentiation between heritage tourism and other types of tourism activities. In other words, heritage tourists are those who come and visit historical sites specifically to get pleasure from the attractiveness and uniqueness of the sites and to discover the historical past of the destinations.

Furthermore, although there are some shortcomings of heritage tourism, the profound advantage of heritage tourism includes generating employment for the people and communities located near the site (Wells et al., 2015), the potential infrastructure improvements associated with the destination will have an impact on higher living standards for the people (Yankholmes et al., 2009), the accompanying public awareness campaign will encourage the appreciation of the heritage (Endresen, 1999; Rahmawati et al., 2014), and heritage tourism development would essentially provide a positive economic impact to other areas (Weaver, 2011). Relatedly, Rhone and Neil (2001) suggested that efforts are being made to focus on heritage tourism necessities to consider the natural friendliness and courtesy of the local community destinations, maintenance of security in the destinations, establishment of a suitable atmosphere, and the awareness and consideration to upgrade the facilities around the destination for visitors to highly consider revisiting the heritage site.

2.3 *Behavioral intention to revisit*

Previous studies have been conducted relating to the behavioral intention of tourists to revisit, such as those by Rodríguez Molina et al. (2012) and Canny (2013). Ashworth (2015) indicated that heritage tourist behavior was closely related to the heritage place. In other words, a heritage tourist would visit the site once that heritage site provided attractive and interesting features; thus, visitors would aspire to returnbecause they had gained satisfaction during their first visit. This means that the satisfaction of tourists during their initial visit to heritage sites is one of the most important considerations in heritage tourism (Boukas, 2013). Furthermore, the heritage tourism sites should be prepared to consider the needs of heritage tourists in order to gain mutual benefit for both the heritage tourists and the service provider. Satisfaction as a result of the availability of quality tourism attractions and services can be a challenge to the tourism service providers to improve the quality of their services.

Other studies found that some significant factors that influence a heritage tourist's behavioral intentions include the level of service quality, destination image, promotion (Rhone & Neil, 2001), quality experience, visitor satisfaction, and perceived value (Chen & Chen, 2010). All these factors that influence a heritage tourist's behavioral intention are envisaged to emerge in this exploratory study.

3 RESEARCH METHOD

To achieve the objectives of this study certainly needs an appropriate research method so that research results can be significantly accounted for. This research is an exploratory study using the cross-case analysis approach, which requires semistructured interviews in data collection (Laforest, 2009).

The interview process was performed by first determining the categories of the participants and subsequently deducing from the established category the number of respondents, finally coming up with 15 participants (Creswell, 2014). This study used purposive sampling to select the participants. Thus, the composition of the participants included six tourists, three representatives from government agencies, three representatives from travel agencies, and three representatives from nongovernment organizations. This study used a thematic analysis with cross-ase analysis approach to manage the data (Blondiau, 2015; Seijger et al., 2015).

4 FINDINGS

This study adopted cross-case analysis in support to the thematic analysis. It consists of three processes that include open coding, axial coding, and selective coding. The open coding process generated around 114 codes through cross-case analysis based on the data obtained from

Table 1. Axial coding results for the factors influencing Heritage Tourists' behavioral intention to revisit.

No.	Axial code results
1	Tangibles
2	Responsiveness
3	Communication
4	Consumable
5	Empathy
6	Experiences to enjoy the site
7	Experiences on Toraja daily life
8	Involve the culture festival
9	Beautiful and attractiveness destination
10	Unique destination
11	Welcoming people
12	Toraja had a good image
13	Nature of scenery
14	Involve in traditional culture activities
15	Learn and find new things
16	Money spent for new things at Toraja heritage site
17	Need effective promotional material
18	Need a map
19	*Tongkonan* and funeral ceremony is unique
20	Specific culture perceived of Toraja quality
21	Preserved culture
22	Cultural uniqueness
23	Visit many destinations
24	Seeking new experience
25	Seeking new knowledge
26.	Need a proof of Toraja culture
27	Relaxing
28	Feeling satisfied
29	Feeling amazed and surprised
30	Indicator of heritage tourist behavioral intention to revisit
31	Heritage tourist's behavioral intention

Table 2. Nine main themes related to heritage tourists' behavioral
intention to revisit.

No.	Themes	Citation percentage
1	Motivation	53
2	Cultural uniqueness	87
3	Monuments	60
4	Service quality	75
5	Experience quality	60
6	Perceived value	53
7	Destination image	67
8	Heritage tourist's satisfaction	93
9	Promotion	87

semi-structured interviews. The results of the open coding were next impressed into a similar theme in the axial coding. The axial coding results were generated from the list of open coding through the combination of the same characteristic coding into a new code.

Table 1 demonstrates the axial coding results that emanated from the open coding results. Through this process, there a total of 31 axial codes were generated. Based on this result, the following step was to perform the selective coding process that aimed to find out the factors influencing heritage tourists' behavioral intention to revisit Toraja's heritage destinations.

Table 2 is a summary of the factors influencing heritage tourists' behavioral intention to revisit.

Table 2 illustrates the result of cross-case analysis process where nine significant factors emerged that could influence the behavioral intention of heritage tourists to revisit a heritage destination site. According to the order of the themes, the factors are motivation, cultural uniqueness, monuments, service quality, experience quality, perceived value, destination image, heritage tourists' satisfaction, and promotion. In terms of citation percentage, the theme Heritage Tourist's Satisfaction obtained the highest citation percentage, with 93%, followed by Cultural Uniqueness and the Promotion themes, with 87%each; Service Quality, with 75%, while both the Motivation and the Perceived Value themes had the lowest, 53% each, in terms of citation percentage. In sum, these nine themes are considered the dominant factors that influence heritage tourists' behavioral intention to revisit the heritage destinations at Toraja in Indonesia.

5 CONCLUSION

This study sought to specifically identify the factors that influence heritage tourists' behavioral intention to revisit the heritage and cultural destinations in Toraja. This study was also considered timely because the trajectory of tourist arrivals has been declining and continues to have extreme fluctuations in terms of real numbers. For Toraja Regency, including the private sector stakeholders, to sustain high tourist arrivals, the needed improvement in the tourism sector should be addressed. In the main, this study has scientifically and successfully generated nine important factors that are considered to impact significantly the behavioral intentions of heritage tourists to revisit Toraja. Although the process of simplifying visa requirements is a step in the right direction, these nine identified key factors are truly relevant concerns that if addressed properly would generate economic benefits to the various stakeholders and the communities in broader scope. Thus, with the identification of nine themes/factors such as motivation, cultural uniqueness, monuments, service quality, destination image, perceived value, experience quality, satisfaction, and promotion, a tourism renaissance in Toraja Regency is potentially feasible.

6 RECOMMENDATIONS FOR FUTURE RESEARCH

This study identified nine factors influencing heritage tourists' behavioral intention to revisit. This research result could be further conceptualized in several research framework formulations and investigations as confirmatory studies by other researchers.

REFERENCES

Adams, K. M. 2003. The Politics of Heritage in Tana Toraja, Indonesia: Interplaying the Local and the Global. 31(89).

Adams, K. M. 2007. Domestic tourism and nation-building in South Sulawesi. *Indonesia*, (November), 37–41.

Adams, K. M. 2012. Tourism and the renegotiation of tradition in Tana Toraja (Sulawesi, Indonesia). 36(4):309–320.

Antonakakis, N., Dragouni, M., & Filis, G. 2014. How strong is the linkage between tourism and economic growth in Europe? *Economic Modelling*, 44:142–155.

Ashworth, G. J. 2015. Consuming heritage places: Revisiting established assumptions. *Tourism Recreation Research*, 35(3):281–290.

Blondiau, A. 2015. Challenges for inter-departmental cooperation in hospitals: Results from cross-case analysis. *Health Policy and Technology*, 4(1):4–13.

Boukas, N. 2013. Youth visitors' satisfaction in Greek cultural heritage destinations: The case of Delphi. *Tourism Planning & Development*, 10(3):285–306.

Canny, I., & Hidayat, N. 2012. The influence of service quality and tourist satisfaction on future behavioral intentions: The case study of Borobudur Temple as a UNESCO World Culture Heritage destination. 89–97.

Canny, I. U. 2013. An empirical investigation of service quality, tourist satisfaction and future behavioral intentions among domestic local tourist at Borobudur Temple. *International Journal of Trade, Economics and Finance*, 4(2):86–91.

Chen, C. F., & Chen, F. S. 2010. Experience quality, perceived value, satisfaction and behavioral intentions for heritage tourists. *Tourism Management*, 31(1):29–35.

Creswell, J. W. 2014. *Research Design: Qualitative, Quantitative, and Mixed Method Approaches.* Thousand Oaks, CA: SAGE.

Cultural and Tourism Minister of Indonesia. 2006. A regulation of cultural and tourism, 2004–2006.

Endresen, K. 1999. Sustainable tourism and cultural heritage. 67.

Hodur, N. M. 2010. Characteristics and the economic impact of visitors to heritage and cultural tourism attractions in North Dakota. *Disertation, North Dakota State University.*

Indratno, I., Sudaryono, S., & Khudori, A. M. 2013. Refleksi Ruang Tongkonan, (2), 1–17.

King, V. T. 2013. UNESCO in Southeast Asia : World Heritage Sites in comparative perspective. (4), 1–116.

Laforest, J. 2009. *Guide to Organizing Semi-Structured Interviews with Key Informant*, Vol. 11.

Lestari, W., Soleha, M., Ibrahim, I., Ruwaedah, & Roosihermiatie, B. 2012. *Etnik Toraja Sa'dan.*

Manuella, A. 2014. Destination management organization in Flores.

Mapaliey, Y. S. 2009. *Kajian penurunan jumlah wisatawan di tana toraja tesis.*

Nuryanti, W. 1996. Heritage and postmodern tourism, (2), 249–260.

Poria, Y., Butler, R., & Airey, D. 2004. Links between tourists, heritage, and reasons for visiting heritage sites, 1–42.

Rahmawati, D., Supriharjo, R., Setiawan, R. P., & Pradinie, K. 2014. Community participation in heritage tourism for Gresik resilience. *Procedia: Social and Behavioral Sciences*, 135:142–146.

Rhone, R., & Neil, K. 2001. Heritage tourism in Black River, Jamaica: A case study. *Heritage.*

Richardson, R. B. 2010. The contribution of tourism to economic growth and food Security. *United Stated Agency International Development, Mali*, (June), 1–8.

Rodríguez Molina, M. Á., Frías-Jamilena, D.-M., & Castañeda-García, J. A. 2012. The moderating role of past experience in the formation of a tourist destination's image and in tourists' behavioural intentions. *Current Issues in Tourism*, (April), 1–21.

Schubert, S. F., Brida, J. G., & Risso, W. A. 2011. The impacts of international tourism demand on economic growth of small economies dependent on tourism. *Tourism Management*, 32(2),377–385.

Seijger, C., Dewulf, G., Van Tatenhove, J., & Otter, H. S. 2015. Towards practitioner-initiated interactive knowledge development for sustainable development: A cross-case analysis of three coastal projects. *Global Environmental Change*, 34:227–236.

Som, A. P. Mat, Marzuki, A., Yousefi, M.,& Abu Khalifeh, A. N. 2012. Factors influencing visitors' revisit behavioral intentions: A Case Study of Sabah, Malaysia. *International Journal of Marketing Studies*, 4(4).

Sugiyarto, G., Blake, A., & Sinclair, M. T. 2003. Tourism and globalization: Economic Impact in Indonesia. *Annals of Tourism Research*, 30(3),683–701.

Toraja Tourism Department. 2014. Pemerintah Kabupaten Toraja, 0423.

Tourism Law of Indonesian Government No. 10. 2009. Undang-Undang Republik Indonesia No. 10 Tahun 2009.

UNESCO. 2013. World Heritage sites participate in Earth Hour 2013.

Vong, L. T.-N., & Ung, A. 2012. Exploring critical factors of Macau's heritage tourism: What heritage tourists are looking for when visiting the city's iconic heritage sites. *Asia Pacific Journal of Tourism Research*, 17(3): 231–245.

Walpole, M. J., & Goodwin, H. J. 2000. Local economic impacts of dragon tourism in Indonesia. *Annals of Tourism Research*, 27(3):559–576.

Wardiyatmo. 2012. Sektor Pariwisata Turut Donkrak Perekonomian Nasional, April.

Weaver, D. B. 2011. Contemporary tourism heritage as heritage tourism Evidence from Las Vegas and Gold Coast. *Annals of Tourism Research*, 38(1):249–267.

Wells, V. K., Manika, D., Gregory-Smith, D., Taheri, B., & Mccowlen, C. 2015. Heritage tourism, CSR and the role of employee environmental behaviour. *Tourism Management*, 48:399–413.

Wells, V. K., Smith, G. D., Taheri, B., Manika, D., & McCowlen, C. 2016. An exploration of CSR development in heritage tourism. *Annals of Tourism Research*, 58:1–17.

Yankholmes, A. K. B., Akyeampong, O. A., & Dei, L. A. 2009. Residents' perceptions of transatlantic slave trade attractions for heritage tourism in Danish-Osu, Ghana. *Journal of Heritage Tourism*, 4(4):315–329.

Zhao, S. (Nancy), & Timothy, D. J. 2015. Governance of red tourism in China: Perspectives on power and guanxi. *Tourism Management*, 46:489–500.

Inclusive Development of Society – Lumban Gaol (eds)
© 2020 Taylor & Francis Group, London, ISBN 978-1-138-33476-2

Strategic partnership as a form of management structure improvement in the hospitality industry

I.V. Mishurova, S.N. Komarova & M.E. Volovik
Rostov State University of Economics (RINH), Rostov-on-Don, Russian Federation

D.V. Nikolaev
Southern Russian Institute of Management Branch of Russian Presidential Academy of National Economy and Public Administration, Rostov-on-Don, Russian Federation

A.V. Temirkanova
Southern Federal University, Rostov-on-Don, Russian Federation

ABSTRACT: Modern trends in the development of the hospitality industry such as globalization, the increasing influence of non-monetary factors in a competitive environment, standardization of the service processes are predetermined the wider use of management tools of the strategic partnerships. This article analyzes the peculiar workings of strategic partnerships in Russia, the competitive characteristics of hotel chains operating in Russia. Conclusions have been drawn about competitive disadvantages in comparison with foreign companies. Therefore, the benefits of strategic partnerships in the hospitality industry, as well as the influence of political, economic, social and technological factors were identified through expert methods and strategic group charts. The potential competitive advantages created within the frame of concept "strategic integration" were identified. This made it possible to distinguish some particular strategies of forming competitive advantages in the hospitality industry and to create an algorithm for the formation of strategic partnerships for small business as well.

1 INTRODUCTION

International and domestic business competitiveness in the hospitality industry has become increasingly sensitive to non-monetary factors such as service quality and product, outstanding futures, presale and post-sale services, etc. According to several studies, it can be stated that large (transnational) organizations, whose scale of production and experience gained in the hospitality industry are setting service standards and carrying out such standards around the world and remain as key leaders on mentioned positions (Burgers, W. P., Hill C. W. L., Kim W. C. 1993, Duysters G. And Hagedoorn J., 1996, John C. Crotts, Dimitrios Buhalis, Roger March 2008).

Taking into account the above-mentioned circumstances, the objective of the study is to develop methodological tools of strategic partnership management as one of the most powerful instruments in hospitality industry competitiveness.

Consequently, the main tasks of the research concern the following aspects:

- Analyze the signification of the concept "strategic partnership";
- Identify the advantages of strategic partnership development in hospitality industry;
- Determine the most significant external factors which impact the hospitality industry progress;
- Apply the chart of strategic group (competitors) method in order to analyze the competitiveness situation among the most prominent hotel chains operating on the Russian market (as a "strategic partnership representatives");

- Reveal the main competitive advantages of using "strategic partnership" concept based on the data defined and specific features of strategic partnership functioning;
- Develop a decision-making algorithm which serves for improving and structuring the process of strategic partnership establishment.

Most of the contemporary reading materials offer the concept of "strategic partnership" and "network of firms" as mutually supportive and, indeed, consider the network as one of the forms of strategic partnership (Masyuk N. N… Kulik D.G., 2014 Kuznetsov, I. D., Chernyshev, 2011). However, distinguishing these definitions in this research, it is very important to emphasize some differences.

Thus, a strategic partnership consisting of two key terms: "strategy" and "partnership", in order to achieve adopted goals implies teaming for a certain period of time on the basis of mutually beneficial conditions for all parties. Whereas, in this study the network of firms is understood as an enterprise that expands its range of activity through incorporation or creation of new facilities. In most cases, this expansion takes place within the framework of horizontal integration, without the involvement of adjacent businesses.

The main difference between a strategic partnership and a network of firms is that a strategic partnership forms by teaming of two or more companies for a certain period to achieve adopted goals; whereas a network of firms – a permanent state of the organization, which covers market in more comprehensive way.

That forms of partnership have emerged and developed due to a number of reasons, such as (Drago W. A., 1997, John C., 2008, Mishurova I.V., 2016, Volovik M. E., 2016):

- the dependence of the hospitality industry on the external environment requires the introduction of more responsive organizational structure;
- the need to combine the competitive advantages of various companies in order to achieve even greater efficiency;
- insufficient resources of firms to implement development strategies;
- growth of globalization effects and global tourism progress determines the necessity of every company (operating in hospitality industry) to respond the international trends, therefore, the development of any firm, its international and regional market expansion is relevant and potentially effective.

Therefore, the expansion of the hospitality industry is a relevant process, which is based on the consequences of globalization. The success of a large company in the current circumstances is propelled not only by economic indicators, but also by the ability to adapt to changing conditions of the external and internal environment. In this regard, such forms of development as: strategic alliances, insourcing, outsourcing, public-private partnership, progress of the franchise network are being increasingly adopted in international business and in the hospitality industry particularly (Volovik M. E, 2018).

2 RESEARCH METHODS AND RESULTS

In order to systematize the impact assessment of external environmental factors on performance and competitiveness of enterprises in the hospitality industry, in the present study, PEST analysis was chosen as a diagnostic technique of the influence of external factors and to analyze the macroenvironment for its proximity to the object of study. The analysis structure includes the most significant factors of the macro environment: P-political, E-economic, S-sociocultural, T-technological factors according to the specifics of the hospitality industry.

To establish the expert estimation of the external environmental factors and subfactors impact on the competitiveness of the enterprise sector in the field of hospitality, a questionnaire has been drafted and surveys have been carried out. Expert speakers were the managers of large hospitality enterprises in the Rostov region. Questions asked concerned influence of political, economic, sociocultural and technological factors. To analyze the competitiveness of hotel chains, has been used the method of strategic groups charts and web analytics data of large hotel chains operating in Russia.

According to experts, among these factors, with the use of the points system (from 1 to 5) the following provisions were identified:

1. factors that are profoundly affected the development of the enterprise sector in the hospitality industry;
2. risk of established factors changes.

Basing on the obtained points, the probability weighted estimation of the external factors impact on the hotel business competitiveness was calculated.

Based on the study, we can conclude that the competitiveness in the hospitality industry is influenced by the following conditions:

1. Political environment:
   * Hosting the 2018 Football World Cup
2. Economic environment:
   * exchange rates of major currencies.
3. Social and cultural environment:
   • quality requirements for products and service;
   • the level of migration and immigration sentiments.
4. Technological environment:
   * development of the internet and Internet connectivity, development of the mobile platform.

On the basis of the analysis, it was possible to build an assessment profile (Figure 1).

Research suggests that under the current circumstances political factors have a greater degree of influence, in particular the Football World Cup 2018 in Russia, this event gave a significant industry development incentive, the opening of new accommodation facilities, entry of multinational companies to the domestic market. It is, therefore, such changes have contributed to the management improvements and to the raising of the level of competitiveness of the industry in general, as well as to the effective operations of enterprises in the market.

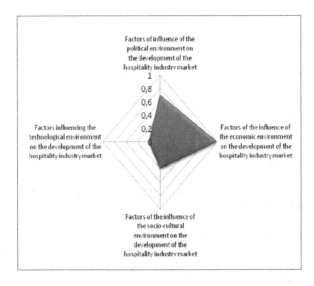

Figure 1.   Profile PEST-analysis assessment of the external factors impact on competitiveness of the hospitality industry.

PEST-analysis helped to study external factors of competitiveness and allowed to analyze the external environment identifying certain trends and features affecting the hospitality industry as well as the structural changes in the management system.

In this study, however, the specification of the circumstances of some individual competitors in the Russian hospitality sector is made, as it is considered, as one of the pillars of the hospitality industry. The following major multinational and Russian hotel chains represented on the Russian market have been chosen as the target of the study: The Rezidor Hotel Group (which includes the hotels Radisson and Park Inn), Accor Group (Mercure hotels are represented on the Russian market), Hilton Worldwide, Amaks Hotels & Resorts, AZIMUT Hotel. Hotel industry is one of the significant part of the hospitality industry. Above mentioned hotel chains are the main participants on the Russian market of hotel industry and have the highest indicators in many characteristics.

First of all, it is necessary to identify the competitive characteristics of each network (Table 1). As prime competitive characteristics of hotel chains we should stress the following: the level of price, quality, geographical reach (availability of hotels outside Russia), the number of rooms (in Russia), the number of hotels (in Russia), Russian cities where the network is represented; brands represented in Russia; additional services offered to guests.

There are pairs of factors that need to be located along the axes of the graph, that have been identified on the basis of characteristics. Competitors are marked on the graph according to performance in each of the variables. Circle has to be drawn around each competitor, and diameter should be in proportion to the share of the competitor's total sales.

Using the described algorithm, a chart of strategic groups (competitors) has been compiled.

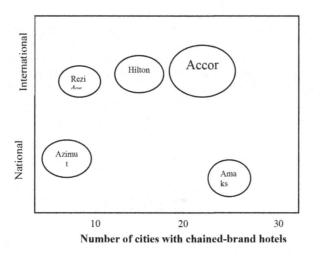

Figure 2. chart of strategic groups (competitors). Variables: geographical coverage - number of cities with chained-brand hotels.

This chart allows to determine the position of each competitor in the industry. Moreover, with the help of the chart it is possible to find out the competitive position of each market player and to analyze whether competition has increased or there is no competition – in another word, to see the advantages and disadvantages of competitors. Chart shows industry's completeness as a whole as well.

According to data received, Accor Group occupies the leading position among the chained-brand hotels: high level of service, quality and value; hotel chain is widely presented to the domestic market in different hotel categories providing guests with different types of services from business conferences to beach holidays. The chart shows that other international networks in a competition taking the leading positions for most of the characteristics: the Rezidor Hotel Group and Hilton Worldwide. High level of service, quality and value is the same as for the

Table 1. Competitive characteristics of hotel chains operating in Russia.

Characteristics	Chained-brand hotels				
	The Rezidor	Accor	Hilton	Amaks	Azimut
Quality/Price level	middle, high	middle, high	middle, high	middle, high	middle, high
Geographical reach	International	International	International	International	International
Room fund (in Russia)	>4 000	>7 500	>4 400	>6 200	>6 000
The number of hotels (in Russia)	13	41	25	26	18
The number of cities with represented hotel chains (in Russia)	6	21	19	24	5
The number of brands (in Russia)	2	6	4	1	1
Extra service	All inclusive	All inclusive	Limited	Limited	Limited

Accor Group, it was achieved by an experienced professional team of all hotel chains. However, in comparison with Accor and Hilton (which in the Russian market have both mid-range hotels and suites), the Rezidor chain has only two brands which limits the range of additional services and the number of rooms, therefore, weakens the competitive position in general.

Russian hotel chains Amaks and Azimut have less competitive advantages over international hotel chains. It can be viewed within the chart of strategic groups (competitors). These hotel chains are widely provided on the market by mid-range hotels; they lack set of differentiated brands that imply availability of a different range of services. On the other hand, benefits from the hotel room capacity speak for the scope of activities that add some bonus points over other rivals.

The analysis of external and internal factors of competitiveness of the hospitality industry allows to identify certain features of the strategy of forming competitive benefits in the industry.

1. Marketability of the hospitality industry cannot depend entirely on itself. The hospitality industry involves different types of business, various sociocultural, municipal components, as well as security aspects. That is, the formation of the competitive environment on the hospitality market is a set of policy decisions and business and the state cooperative activities. The state provides the destination with non-profitable amenities, and with a driver for growth and development directly influencing the attractiveness of the region and its competitiveness in the tourism sector as a whole.
2. The implementation of the price strategy constraining the improvement of the firm's competitive position in the hospitality industry. The goods and services provided must be commensurate with a certain level of differentiation. Organization which is able to provide a variety of products (services) of discrete level and value will be more competitive.
3. The branding strategy in the hospitality industry significantly improves the competitiveness of the organization but requires a lot of experience and bigger capital outlays to form a "name". The analysis results of competitive ability of the environment showed that the leading positions are owned by companies with a well-known brand and they have standards that comply with the requirements of specific segments of consumers. Thus, during the formation of competitive advantages in the hospitality industry should be observing the following aspects: to study the market attractiveness, i.e. the destinations; management of the differentiation strategy quality and value, branding strategy implementation. These factors directly affect the formation of competitive advantages in the hospitality industry. Such factors can have different degree of impact on the competitiveness of the environment of the studied industry. However, their interrelationship should be considered in the development and management of strategic decisions.

The development of strategic partnership is particularly prevalent in tour operator activities in the domestic hospitality industry. Thus, in the formation of a tourist product, 3 main aspects should be included – flight, ground service, insurance – tour operators cooperate and unite with partners in order to provide these services. Moreover, cooperation takes place not only through contractual agreement, when operators purchase a certain number of seats in transport or certain beds in accommodation. A company comes into a strategic cooperation agreement with air carriers, host companies, hotels – which are becoming unique for this company and the market as a whole.

For example, the tour operator ANEX Tour in 2014 started cooperation with a strategic partner airline company AZUR AIR; Hotel networks such as PGS and Swandor are operating by Pegas Touristik in many countries worldwide; Biblio Globus is the only mediator that based on long-term agreement between the companies performing selling activities of rooms of OK "Sochi-Park hotel" in the Imeritian Bay.

In addition, almost every tour operator has unique partner – the inviting company, that offers ground service for tourists to destination: transfer, excursions, accommodation, etc.

Thus, based on the data, it appears that the effectiveness of management through the generation of strategic partnerships is reflected through cooperation with partners who are promising and are able to ensure the strategic development of the whole company.

Apart from the defined aspects, the strategic partnership has other policies (Harrigan,1988, Chernyshev, 2012):

1. Common interest of all parties: cooperation should be mutually beneficial and respect the interests of all parties – only by respecting this condition that it will be possible to build effective and promising team-work. The key challenge for implementation of this principle is to identify common strategic goals.
2. Effective mechanisms for the implementation of long - term contractual relations are the most difficult and complex aspect. When adopting a strategic partnership, it implies the creation of an organizational structure within which the relative independence of the companies involved in the cooperation would be maintained, but at the same time ensuring the compatibility of reaching decisions based on common strategic objectives.
3. Willingness to sacrifice own interests for the sake of common strategic goals is a necessary condition for entering into a strategic partnership, as the general strategic plan becomes a primary objective since the partnership established, and individual objectives of the partner company may not coincide, so the contribution to the development of the entire organization as a whole, therefore, all management objects in particular should is top-priority.
4. Bilateral agreement and other documents conclusion to obtain legal support of strategic partnership. The possibility of an effective and legitimate strategic partnership should receive the support of the legislation of the country in the territory of which the unity takes place. Moreover, the specified conditions of joint activity should be established in contracts in order to avoid disputes, as well as to maintain business relations as robust as possible between all partners.

Thus, the development of forms of strategic partnership in the hospitality industry – the most effective way to develop the organization aimed at improving the quality and differentiation of services or goods. The allegation is connected with the fact that the hospitality industry is a sphere of activity that includes different types of business: transport services, hotel facilities, catering and leisure, excursions, insurance, which interact and not mutually exclusive. Therefore, long-term partnerships between these areas are necessary and directly strategically important: common goals and objectives are formed, which facilitates the process of development, formation and progress of a single complex of the hospitality industry at a certain destination.

According to the abovementioned competitive characteristics (Table 1) and determined aspects of strategic partnership benefits it is possible to identify the most actual trends of strategic partnership in hospitality industry, presented in the form of table.

Table 2 shows that the creation of a strategic partnership is a multifunctional task, which affects the interests of the principal parties of the business: the entrepreneur, the personnel

Table 2. Competitive advantages are created within the frame of concept "strategic integration".

Advantage	Entrepreneur	Personnel	Client
Well-known brand	Additional business value	Brand affiliation and quality of work	Confidence in service quality
Proposal differentiation (reach new markets, capture new segment of consumers)	Expending of marketing possibilities	Extra bonuses for the promotion of other participants	Choice possibility (price and level)
Increasing production and marketing of goods and services	- Development of client base, pricing policies - Capacity increasing	Stuff rotation possibility, carrier path build up	Service specter build up

and the client. Therefore, as part of the formation of competitiveness tools in the hospitality industry through the development of forms of strategic partnership and networks of firms, it is important to improve decision-making algorithm of the organizational structure in order to establish an integration partnership.

This algorithm can be implemented from two aspects: from the aspect of a large enterprise, whose task is to expand its activities farther; from the aspect of a medium or small business, whose task is to establish its own production and sales by joining a network or union with a partner.

Consider the steps of creating an integration partnership from the perspective of small business.

Stage 1. A comprehensive analysis of the internal and external environment of the company is performed and the main competitive advantages and disadvantages are determined – on the basis of which the strategic goals that the company intends to achieve in the future as well as the main opportunities and potential are highlighted.

Stage 2. On the basis of the information received, the organization can determine what is more relevant and acceptable to achieve its strategic goals: to join the existing network of companies, or to look for a strategic partner for a specified period. In accordance with the chosen vector of development, the company works out a certain proposal for the partner (in order to motivate), or "draws" either changes the conditions of its activities to meet network entry requirements (number of employees, area of premises, legal form of organization, etc.).

Stage 3. The conclusion of the contract is the last stage towards the establishment of one of the forms of integration partnership after which the joint activity begins.

3 INSIGHTS

The decision-making algorithm for creating integration partnerships demonstrates the most rational process of decision-making on the way to the formation of one of the forms of unity. The central plank is a gradual examination of the company's environment, as well as its strategic goals and interests. These steps allow to generate the direction and the relationship starting with the purpose of the company to the necessity of forming partnerships. This statement means that company's strategic targets are connected with the decision, if the strategic partnership is effective or not, based on the analyze of external factors and competitiveness.

The external factors and competitiveness analyze methods showed the actual conditions for business in hospitality industry which cause appropriate management tools' use. One of them is a concept of strategic partnership.

As a result of the research it should be noted that the development of forms of strategic partnership and networks of firms is a powerful tool to enhance competitiveness in the hospitality industry, which combines the resources, capabilities and competitive advantages of numerous companies while complementing each other. This aspect is particularly relevant to the versatility of the hospitality industry.

In future work data and results defined will help to develop the more effective strategic management instruments in hospitality industry in order to make the strategic planning more adaptive, especially within transforming conditions of Russian tourism market.

REFERENCES

Volovik M. E. Principles of formation of competitive advantages in the hospitality industry– Azimuth of Scientific Research: Economics and Administration. 2018. Vol. 7. № 2 (23), p. 93–96 (in Russian).

Kuznetsov, I. A. Current development trends of strategic alliances and partnerships in Russia / I. A. Kuznetsov // Transport case of Russia. -2014. - № 4 (113). - P. 181–187 (in Russian).

Masyuk N. N., Kulik D. G. a Strategic partnership of stakeholders: business networks//Fundamental research. - 2014. - № 12-10. – Pp. 2179–2184; URL: http://fundamental-research.ru/ru/article/view?id=36548 (access date: 20.07.2018) (in Russian).

Mishurova I. V. Volovik, M. E. assessing the impact of the macro on the development of the hospitality industry in the region // journal of Russian entrepreneurship. - 2016. - Volume 17. — № 22. — doi: 10.18334/rp.17.22.3706 https://bgscience.ru/journals/rp/current/ (in Russian).

Chernyshev D. A. Management of strategic alliances in the tourism business. / Scientific and practical journal "Russian entrepreneurship", № 4 (issue 2). - M.: Publishing house "Creative economy", 2011, p. 147–153 (in Russian).

Official website The Rezidor Hotel & Group [Electronic source]. – Available at: http://www.rezidor.com (access date 15.03.2018).

Official website Accor Group. [Electronic source]. – Available at: https://www.accorhotels.com/ru/russia/index.shtml (access date 15.03.2018).

Official website Hilton Worldwide. [Electronic source]. – Available at: https://www.hilton.com/en/corporate/ (access date 18.03.2018).

Official website Amaks Hotels & Resorts. [Electronic source]. – Available at: http://www.amaks-hotels.ru (access date 18.03.2018).

Official website AZIMUT Hotels. [Electronic source]. – Available at: https://azimuthotels.com (access date 25.03.2018).

Official website [Electronic source]. – Available at: http://www.anextour.com/page/491/vazhnye-daty-v-istorii-kompanii (access date).

Official website [Electronic source]. – Available at: https://pegast.ru/agency/about-company (access date 24.09.2018).

Burgers W.P., Hill C.W.L., Kim W.C. *A theory of global strategic alliances: the case of the global auto industry*. Strategic management journal. 1993. Vol. 14, No. 6. P. 419.

Drago W.A. *When strategic alliances make sense*. Industrial management & data systems. 1997. Vol. 97, Issue 2. P. 8–12.

Duysters G. And Hagedoorn J. *Internationalisation of corporate technology throught strategic partnering: an empirical investigation*. Research Policy. 1996. V. 25. P. 1–12.

Harrigan K.R. *Strategic Alliances and Partner Assymmetries*. Lexington Books, 1988, p. 42.

John C. Crotts, Dimitrios Buhalis, Roger March. *Global Alliances in Tourism and Hospitality Management*, 2008, P.49

The adaptation of the concept of fuzzy logic in the management of socioeconomic systems

L.K. Koretskaya & E.E. Lomov
Financial University under the Government of the Russian Federation, Russian Federation

A.M. Gubernatorov
Vladimir State University, Russian Federation

D.V. Kuznetsov & N.V. Yudina
Financial University under the Government of the Russian Federation, Russian Federation

ABSTRACT: The article substantiates the effectiveness of the developed architecture of the industry complex. The concept is based on harmonic analysis with elements of fuzzy logic.

1 INTRODUCTION

The world around us has changed, and these changes (as well as scientific and technological progress, which is the driver of these changes) are irreversible but they will bring society and the economy to new horizons, thus ensuring the fusion of physical and virtual reality, transforming the manufacturing and service sectors. The observed progress can be seen as a benefit to mankind, as new scientific, technical, and technological solutions reduce the damage to the environment caused by the development of civilization. This, on the one hand, is true because modern civilization, despite the fact that it incorporates different socioeconomic systems in terms of development, has a common, most important metaresource for all in the form of a global natural and climatic environment, without which the modern world-system (or super system) cannot exist. On the other hand, however, technologization, digitalization, and the building of electronic systems for creative human activity carry certain threats. The problem is that mankind has no historical experience that would allow comprehension of all consequences taking place as a result of scientific, technical, and technological progress (Tsvetkov, 2018).

The coming fourth industrial revolution could cause deeper and more radical social upheaval than the previous one, because of its speed and scope. It is natural that on the background of these remarks appear neo-Luddites, who see the new technology as a threat to society and employment. The continuation of automation of labor, on the one hand, is destructive (due to the replacement of labor with capital, an increasing amount of labor is released). On the other hand, the increase in labor productivity leads to an increase in employee incomes and producer profits, which reduces prices and increases consumer demand, thereby stimulating economic growth (Belova et al., 2018).

Mathematical fuzzy set theory (fuzzy sets) and fuzzy logic (fuzzy logic) are generalizations of classical set theory and classical formal logic.

Fuzzy logic is a branch of mathematics that is a generalization of classical logic and set theory, based on the concept of a fuzzy set, first introduced by Lutfi Zadeh in 1965 as an object with an element membership function that takes any values in the interval [0, 1] {\displaystyle [0,1]} [0, 1], not just 0 {\displaystyle 0} {\displaystyle 0} or

1 {\displaystyle 1} 1. Based on this concept, various logical operations on fuzzy sets and the concept of a linguistic variable were formulated, with the values of the variable being stupid fuzzy sets.

The subject of fuzzy logic is the study of reasoning in terms of fuzziness, similar to the reasoning in the usual sense, and its application in computer systems.

These concepts were first proposed by the American scientist Lotfi Zadeh in 1965 (Zadeh, 1965, 1976). The main reason for the emergence of a new theory was the presence of fuzzy and approximate reasoning in the description of human processes, systems, objects.

It took more than a decade since the inception of the theory of fuzzy sets before the fuzzy approach to the modeling of complex systems was recognized worldwide. It is accepted to allocate three periods in the development of fuzzy systems. Work on the measurement of the degree of fuzziness began in 1972 (De Luca & Termini, 1972). The first period (late 1960s–early 1970s) is characterized by the development of the theoretical apparatus of fuzzy sets (L. Zadeh, E. Mamdani, Bellman). In the second period (1970–1980s) there are the first practical results in the field of fuzzy control of complex technical systems (steam generator with fuzzy control). At the same time, attention was paid to the construction of expert systems based on fuzzy logic, the development of fuzzy controllers. Fuzzy expert systems for decision support are widely used in medicine and economics. Finally, in the third period, which began at the end of the 1980s and continues at the present time, there are software packages for the construction of fuzzy expert systems, and the application of fuzzy logic has significantly expanded. It is used in the automotive, aerospace, and transportation industries; in the field of household appliances; in the field of finance, analysis, and management decision-making; and many others (Kruglov & Dli, 2002).

The triumphant march of fuzzy logic around the world began after the proof of the famous Fuzzy Approximation Theorem (FAT) by Bartolomei Kosko in the late 1980s (Kosko, 1994). In business and finance, fuzzy logic gained recognition after in 1988 the expert system based on fuzzy rules for predicting financial indicators was the only one that predicted the stock market crash (Masalov). And the number of successful phase applications is now in the thousands.

In the conditions of rapid digitalization of the economy, the use of the apparatus of fuzzy logic and fuzzy logical statements in the management of socioeconomic systems of different levels is significantly expanding in comparison with the traditional methods of management. Digital transformation based on the use of fuzzy logic methods is the basis for decision-making in the issues of sustainable development of economic systems.

Fuzzy set theory uses the analysis of linguistic variables (words, terms), which, reflecting the quality of certain objects and situations, cannot be reliably estimated in accurate (objective) quantitative estimates.

2 THE TECHNIQUE OF IDENTIFICATION OF RESEARCH FRONTS

To interpret the data on all the analyzed parameters in the theory of fuzzy sets, the method of constructing the membership function is used by constructing a triangle (triangular fuzzy set), written in the form of the formula: P = (a, b, c), where the parameters a, b, c denote respectively the smallest possible, the most probable, and the largest possible values of the considered value. These three states are sometimes described (when a large value corresponds to the positive value of the considered judgment) as "pessimistic," "most probable," and "optimistic" scenarios of the factor change.

The formation of fuzzy rules is performed by the methods of expert one-dimensional scaling. The use of expert assessment of predoped lilo as the membership functions of Gaussian type (1):

$$\mu(x) = \exp\left[-((x - c)/\sigma)^2\right] \tag{1}$$

where $\mu(x)$ is the degree of membership to the fuzzy set; c is the mathematical expectation (center of the fuzzy set); and σ is the standard deviation.

Fuzzy-multiple models allow you to build a functional correspondence between fuzzy linguistic concepts (e.g., the level of control of objects of a distributed trading enterprise can be estimated as "very low," "low," "high," etc.) and special functions that express the degree of belonging of the values of the measured parameters to fuzzy descriptions.

3   TESTING OF THE TECHNIQUE AND ANALYSIS OF RESULTS

The level of sustainable development of the corporation is affected by financial, technical, technological, organizational, production, marketing, investment, social, environmental, and risk factors. The information base for the analysis and assessment of the level of sustainable development of the corporation is the data of corporate social reporting, accounting (financial) reporting, as well as production accounting data.

Denote all incoming variables as follows:

URD = Fd (EU, SU, ECU, RU)
- integral level of sustainable development of the corporation;
m = fm (x1, ..., x5)
- financial;
n = fn (x8, x7, x8)
- technical and technological;
v = fv (x9, x10, x11, x12)
- organizational;
w = fw (x13, x14, x15)
- production;
k = fk (x16, ... x19)
- marketing;
l = fl (x20, x21, x22, x23)
- investment;
CY = f (x24, ..., x27)
- integral indicator of social sustainability;
ECU = f (x28, ... x31)
- integral indicator of environmental sustainability;
RU = f (x32, x33)
- integral index of risk stability.

In the most general form, the integral indicator of environmental sustainability is formed as a linear combination of individual indicators taken with specific weights that characterize the quality of the environment.

Graphically, this situation can be depicted in the form of a hierarchical tree of logical inference, shown in Figure 1.

Figure 1 shows the structure graph of the sustainable development model, which shows the relationship between the incoming and outgoing indicators of the model. To simplify the process of constructing the model, we reduce the additional generalized linguistic initial parameters.

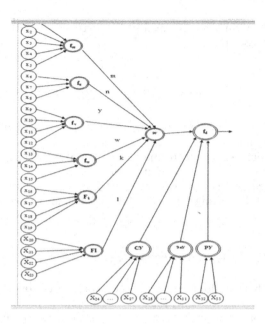

Figure 1. The tree of the logical conclusion of the assessment of the level of sustainable development of the corporation (Teslenko, 2018).

The model for assessing the level of sustainable development of a corporation, based on the theory of fuzzy logic, uses, along with quantitative variables, linguistic ones, which are fuzzy. The analysis of factors influencing the level of sustainable development of the corporation allows prediction of the dynamics of fluctuations in the ratio of various indicators.

The proposed model, based on the harmonic analysis, expands the representation of design changes necessary to improve the sustainable susceptibility of the socioeconomic system and allows optimization of the innovative components of the strategic development of the socio-economic system in terms of smoothing the spatial imbalance and improving the quality (level) of life.

The structure obtained of the corporate sustainable development assessment model shows that this model actually contains four models: (1) a model of the dependence of the level of sustainable development of a corporation on the level of economic factors; (2) a model of the dependence of the level of sustainable development of the corporation on social indicators; (3) a model of the dependence of the level of sustainable development on environmental factors; and (4) a model of dependence on risk factors.

To determine the theoretical basis for the development of logical inference algorithms in corporate control systems, consider the most commonly used methods of fuzzy inference.

The Mamdani method uses a knowledge base with Mamdani rules and provides the following sequence of actions:

1. Introduction of fuzziness. There are degrees of truth for the prerequisites of each rule.
2. Logical inference. There are levels of "clipping" for the prerequisites of each of the rules (using a minimum operation).
3. Definition of truncated membership functions.
4. Composition. The combined truncated membership functions are combined using the maximum operation, which leads to a final fuzzy subset for the output variable with the membership function.
5. Bring to the clarity of the function.

In Tsukamoto's method, the initial assumption, as in the previous method, is that the output parameter functions are monotonic.

1. Introduction of fuzziness. There are degrees of truth for the prerequisites of each rule.
2. Fuzzy inference. There are levels of "clipping" for the prerequisites of each of the rules: and then for each outgoing rule, clear values are determined by solving the equations.
3. A clear value of the variable is determined based on the method of centering the function.

Tsukamoto's algorithm is less precise than the Mamdani algorithm; the average difference is about 1%.

The Sugeno method uses a set of rules in the form: if, then for all, where there are some weighting factors.

1. Introduction of fuzziness. There are degrees of truth for the prerequisites of each rule.
2. Fuzzy inference. There are levels of "clipping" for the prerequisites of each of the rules, as well as individual rule exits.
3. A clear value of the output variable for the fuzzy set is determined based on the centering method.

The Larsen method is used in the same cases as the Mamdani algorithm. In a number of cases it turns out to be more precise than the Mamdani algorithm (with nonmonotonic input fuzzy sets), but it requires more multiplication operations.

Consider an assessment of the level of sustainable development of a corporation based on fuzzy logic using only the integral indicator of environmental sustainability (ECU).

The oil industry was chosen as an example. Thus, the level of energy intensity, that is, the volume of final output, is from 230 to 260 million tons, and the emissions in the oil industry range from 120 to 150 million tons.

It follows that there are two input parameters in the valuation model: emissions in the oil industry $(x1)$ and energy intensity $(x2)$, as well as one output parameter: the environmental pollution factor (y).

The creation of a fuzzy logic inference system simulating the dependence $y = f (x1, x2) = x1/x2$ in the region $230 \leq x1 \leq 260, 120 \leq x2 \leq 150$ will be considered, revealing the main design stages using the algorithm Mamdani (E. Mamdani) greater distribution in problems of fuzzy modeling, since it does not provide for a large amount of computational procedures.

We will design the system of fuzzy inference based on the graphical representation of the indicated dependence.

To construct a three-dimensional image of the function $y = f (x1, x2) = x1/x2$, in the region $230 \leq x1 \leq 260, 120 \leq x2 \leq 150$, we compile the program (Zadeh, 1983).

As a tool for solving problems with fuzzy numbers, it is advisable to use special software tools: for example, the Fuzzy Logic Toolbox component in MatLab (MathWorks), FuziCalc, Fuzzy for Excel, CubiCalc, etc.

As a result, the program received a graphic image, shown in Figure 2.

If we designate X1 as the level of sustainable development, and X2 is the coefficient of K, then we can see that this surface is very flat, almost flat, without strong drops and volumes.

The resulting surface has a more heterogeneous structure in the course of setting the rules than the result obtained previously in the program, which indicates the setting of certain parameters in the system, which in turn most accurately describe the situation in the oil industry. The higher the score, x1 and x2, the higher y.

It was found that the prediction model has a sufficiently large degree of reliability—the level of error is slightly less than 3%.

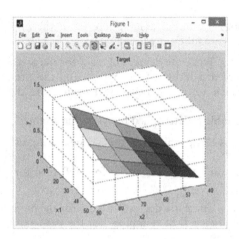

Figure 2.   Graphical representation of the reference surface (Leonenkov, 2003).

4   CONCLUSION

Using a fuzzy logic tool in building a model for analyzing, assessing, and predicting the level of sustainable development of a corporation should become an integral part of the overall methodology for managing sustainable development.

It is concluded that the main advantages of fuzzy models, in the context of the problem, are the following. Unlike traditional methods, fuzzy methods have a more flexible principle of setting the "point," carried out by introducing the concept of membership function rather than the absolute task of "point" on the principle of exclusion of the third, used by classical mathematics, including statistical methods. Unfortunately, the extension Fuzzy Logic Toolbox, which is responsible for fuzzy modeling,cannot simultaneously carry out operations of gazifikatsii and defazifikatsii. Fuzzy inference is performed for intermediate variables, followed by the transmission of clear values of these variables to fuzzy systems at the next level of the hierarchy. Therefore, on each of the intermediate values are performed operations gazifikatsii and defazifikatsii. Fuzzy methods provide the opportunity to more efficiently, i.e., verbally, describe the problem to be solved by introducing the concept of a linguistic variable, the meaning of which is characteristic of human thinking, and which is determined not by numbers, but by fuzzy concepts expressed by linguistic variables. The use of the mathematical apparatus of the theory of fuzzy sets and in particular the rules of fuzzy conditional inference allows formalization of the main elements and their interrelations of the content, i.e., economic formulation of the problem, as well as design of decision support models in companies taking into account these features, which is their undeniable advantage compared to most of the currently used techniques.

Thus, the use of fuzzy sets and fuzzy logic allows formalization of more flexible connections between the parameters, which is more consistent with the nature of the studied real phenomena and the description of interactions in natural language.

REFERENCES

Belova, L. G., Vikhoreva, O. M., & Karlovskaya, S. B. Industry 4.0: opportunities and challenges for the world economy. Available at: https://www.econ.msu.ru/science/economics/archive/2018/3/

De Luca, A., & Termini, S. 1972. A definition of a non-probabilistic entropy in the setting of fuzzy set theory. *Information and Control*, 20:301–312.

Kosko, B. 1994. Fuzzy systems as universal approximators. IEEE Transactions on Computers, 43(11):1329–1333.

Masalov. Fuzzy logic in business and finance. Available at: www.tora-centre.ru/library/fuzzy/fuzzy-.htm

Kruglov, V. V., Dli, M. I. 2002. *Intelligent Information Systems: Computer Support of Fuzzy Logic and Fuzzy Inference Systems*. Moscow: Fizmatlit.

Leonenkov, A. V. Fuzzy Modeling in the Environment of Matlab and FuzzyTech, pp. 75–80. St. Petersburg: BHV.

Teslenko, I. B. 2018. *Digital Economy*. Moscow: RUSAINS.

Tsvetkov, J., Dudin, M. N. N., & Lyasnikov In. Digital economy and digital technologies as a vector of strategic development of the national agro-industrial sector. Available at: https://www.econ.msu.ru/science/economics/archive/2018/1/

Zadeh, L. 1976. The concept of linguistic variable and its application to approximate decision-making. Moscow: Mir.

Zadeh, L. A. 1965. Fuzzy sets. *Information and Control*, 8:338–353.

Zadeh, L. A. 1983. A computational approach to fuzzy quantifiers in natural languages. *Computer and Mathematics*, 9:149–184.

Inclusive Development of Society – Lumban Gaol (eds)
© 2020 Taylor & Francis Group, London, ISBN 978-1-138-33476-2

A sequential explanatory study on factors that impede the use of MyHEALTH portal among Malaysians

S.N.M. Tobi, S.F.M. Yatin, M.R.A. Kadir & M.M. Yusof
Universiti Teknologi MARA, Puncak Alam, Malaysia

ABSTRACT: This study explored the personal experiences of nine respondents involved in the earlier study of MyHEALTH Portal (MHP) intention to use. A sequential explanatory design was employed whereby an earlier MHP online survey had been conducted and this was followed through with a set of structured email interviews on the study. The aim of the interviews was to consolidate, support, and explain the results achieved in the earlier survey data relating mainly to the nonsignificant paths. The study applied nested sampling design, and the potential respondents for the structured interviews were selected using the maximum variation of demographic background. Further interviews revealed lack of health concern as the most contributing factor in the absence of a complexity-attitude relationship, while age group preference and insufficient number of MHP promotions were the causal factors in the subjective norms-intention to use and external cues-intention to use relationship. The findings would help the Ministry of Health of Malaysia understand the critical hindrance factors, which finally could help them to boost MHP usage in the population. Some suggestions for future research within the area studied are also highlighted.

1   INTRODUCTION

The main focus of this article is on the web-based Health Information Service (WBHIS), which was created to educate health consumers regarding health issues. WBHIS is an example of online health education that operates under the pillars of ehealth. As a health information portal, the WBHIS is a subset of ehealth systems. The World Health Organization (WHO) highlighted ehealth as providing a new method for using health resources by transferring health resources and healthcare such as information, money, and medicines by electronic means (WHO, 2015). Clearly, the WBHIS has also been predicted to be a tool to support the emergence of the informed and empowered health consumer, and a factor in a shift in the balance of power between patient and professional (Hardey, 1999).

2   BACKGROUND

Health facts in Malaysia show that noncommunicable disease has been the main cause of death and the biggest contributor in terms of disability life-years (Noor et al., 2014). Noncommunicable diseases such as cardiovascular disease, diabetes, and cancer are the major causes of admissions and deaths in government hospitals, and the statistics are increasing (Health Informatics Center Planning and Development Division, 2012). The number of patients diagnosed with noncommunicable diseases will continue to rise. This has alerted the government to initiatives to change the mindset of its citizens toward adopting healthier lifestyles.

The absolute transitions of how people today find information and react toward the information received has changed how people should be perceiving information related to them. This understanding is crucial, as it greatly influence individuals' state of health in today's digital world. To cope with this, the Malaysian government has planned and executed several initiatives to improve the state of healthcare in Malaysia simply by empowering health consumers to become health literate citizens. Despite the availability of many healthcare portals and websites, governmental health portals and websites often provide higher quality and more reliable and trustworthy health information than other online healthcare sources. Likewise, MyHEALTH Portal (MHP) is a national WBHIS, which is an initiative by the Malaysian Government providing health information and a source of local health information to its citizens (Tobi et al., 2017).

MHP was introduced in Malaysia in 2005 but the outcome is known to be limited among health consumers and the Ministry as well. As of today, there was only one elementary study conducted on MHP and it is limited to the health consumers' awareness concerning the existence of MHP (Tobi et al., 2017). Furthermore, it is considered imperative to investigate the actual health consumers' intention to use MHP within the national context, as the government has spent a considerable amount of its budget for the health promotion works and there is no point in spending time and money on health education if it does not work. The statement is further supported with statistics highlighted in the elementary study of MHP that revealed only 24% of respondents knew about MHP but only 9% had used it (Tobi et al., 2017). A statement also was made by the Malaysia National Audit Department that highlighted that MOH had spent unnecessarily immense amount of budget allocation through its miscellaneous health campaigns, yet the outcomes are unachievable and remain unclear (National Audit Department, 2012). Owing to the aforementioned circumstances, the study is deemed fundamental to investigate the actual MHP behavioral intention among Malaysian health consumers while answering the alarming issues highlighted in the foregoing.

In its early years such as in 2006, the number of hits for the portal was reported as 198,271 before decreasing slightly in 2007 with around 137,226 hits. However, in 2013, the number started to jump to 1 million hits and has been steadily rising every year. Table 1 shows the statistics of MHP usage from the year 2006 until 2017, in which the data had been retrieved from an internal report of MHP statistics with the officer in charge from the Malaysia Health Education Division, a division under the Ministry of Health that is responsible for operating the portal (MHED, 2018).

There have been few studies conducted on perceived factors of MHP usage intention (IHBR, 2014; Tobi et al., 2018). For instance, findings from Tobi et al. (2018) on multiple factors influencing usage intention of MHP revealed the significant influence of MHP-diffusion specifically with respect to trialability and relative advantage constructs while complexity showed no significant influence. The authors of the study suggested that the portal's complexity level does not contribute to health consumers' attitude toward MHP usage. This result is not consistent with previous studies that had examined this variable toward users' attitude in using WBHIS. Previous studies found that complexity and ease of use had significantly influenced health consumers' attitudes toward the use of healthcare portals (Yun & Park, 2010; Lazard & Mackert, 2014). In terms of the sociocognitive aspect relating to MHP health consumers (Tobi et al., 2018), attitude has been identified to have positive influence on health consumer's intention to use MHP while health consumer's subjective norms do not contribute toward their intention to use MHP. Both results were found consistent with

Table 1.   MHP statistics for 2006–2017.

Year	Hits	Year	Hits
2006	198,271	2012	728,376
2007	137,226	2013	1,263,710
2008	257,702	2014	1,943,644
2009	176,154	2015	1,868,917
2010	552,992	2016	1,567,961
2011	583,806	2017	9,629,766

previous studies that had examined attitude and subjective norms toward users' intention to use WBHIS. Finally, the findings also showed that within the HBM-psychological frame, MHP-outcome expectations and internal cues positively influenced health consumers' intention except for external cues that have received no adequate support for MHP-intention to use (Tobi et al., 2018).

Despite the contributory factors for MHP-usage, it is also almost as important to seek in-depth clarification for factors that have been shown to lack adequate support. The reason is that, although the aforementioned findings are beneficial for the betterment of the portal's future strategic planning, it also necessary to justify the unexpected phenomenon resulting from the quantitative findings and thereby allow meaningful conclusions to be made for the study. Thus the contribution of this article is to highlight respondents' feedback relating to factors that did not contribute toward their intention to use MHP as reported in Tobi et al. (2018). Follow-up interviews then were conducted to get feedback from the respondents. The main aim of the interview was to explore respondents' personal views and their experience relating to the factors found to be nonsignificant in the quantitative phase. In addition, it also aims to provide justifications to support previous empirical findings within the area of study. Earlier in the quantitative study, it was revealed that no significant relationship was found between (1) complexity and attitude, (2) subjective norms and intention to use, and (3) external cues and intention to use (Tobi et al., 2018).

3 METHODOLOGY

The sample of users who participated in the research interview were selected from those who had participated in the online survey and had given consent earlier to be contacted for further study action. This study applied a nested sampling design, and potential respondents for the structured interviews were selected using maximum variation of demographic background and highest mean score of constructs.

A systematic two-stage case selection procedure was applied. During the first stage, potential respondents were chosen based on the average score of the summed means for each construct followed by the second stage. The "best informants" were then selected using a maximal variation strategy based on demographic information (Creswell, 2005) such as age, gender, race, family status, occupation, and education level. Data were gathered through a series of email interviews with nine participants in the MHP survey.

Three interview questions were developed to cater to the needs of the interviews. Prior to asking any questions, researchers briefly provided an overview of the study, including a definition of complexity, subjective norms, external cues, and intention to use. After that, the questions were listed and follow-ups were made in anticipation of feedbacks from the participants from time to time. All of them were required need to view and explore MHP for approximately 15 minutes so as to prompt their experience with the portal before response to the questions. This study followed the qualitative data analysis procedures proposed by Creswell (2005) and the analysis was performed at two levels: within each case and across the cases (Stake, 1995; Yin, 2003). The steps in the qualitative analysis included (1) recording, organizing, and preparing the data collected; (2) preliminary exploration of the data by reading through the email transcripts; (3) coding the data by segmenting and labeling the text; (4) using codes to develop themes by aggregating similar codes together; (5) connecting and interrelating themes; (6) cross-case thematic analysis; and (7) interpretation of the data.

4 RESULTS

4.1 Descriptive statistics

Table 2 highlights the demographic details of the participants interviewed.

Table 2. Respondents' characteristics.

Characteristics	Classification	N
Gender	Male	3
	Female	6
Age	20–29 years	5
	30–39 years	4
Race	Malay	9
State	Selangor	4
	Johor	1
	Melaka	2
	Kuala Lumpur	2
Marital Status	Single	4
	Married	5
Education	Master's degree	1
	Bachelor's degree	6
	Diploma	2
Occupational Level	Professional	3
	Support staff	2
	Student	4

4.2 *Complexity and attitude*

Results of earlier studies showed that complexity did not significantly influence the relationship with regard to the predictor of health consumers' attitude in using MHP (Tobi et al., 2018). Follow-up interviews identified six factors that emerged from the respondents' answers: lack of concern with health-related-matters, lifestyle, health practitioners' dependency, deficiency of external cues, search engine preferences, and disparity in internet access.

4.2.1 *Lack of health concern*

Of the nine respondents, five had raised the issue of lack of awareness on health concerns. This has been found as *the highest concern within the studied area. Health awareness is one of the important factors to consider, as according to WHO (2014) this is the main factor that drives individuals to have good health and quality of life by getting the right care, screenings, and treatments and taking the steps that help their chances of living a longer and healthier life. One of the respondents interviewed, respondent two (R2), highlighted this as:* "the lack of awareness regarding the health information available in the portal also influences my use towards the portal." This statement was further supported by *respondent three* (R3), who claimed that the public are not focusing on their health so much because they are less likely to have knowledge regarding their health condition. R3's statement was: "*The public are less preferred to know regarding their health condition or putting less focus on their internal health.*"

Respondent five (R5) had expressed her views on the issue by saying that people careless about reading the health information in the MHP although it is easy to use, and they have a positive attitude toward it because of a lack of health concerns and awareness on their part. This is proven by the statement given by R5: "*….lack of awareness of health, therefore they do not want to read the information given in the portal.*" Meanwhile, lack of health concern also was mainly highlighted by *respondent six* (R6) as the issue that possibly influences the low level of complexity and positive attitude of MHP. She claimed: "*This is probably due to lack of awareness in finding health information on the internet.*" This statement was also supported by *respondent seven* (R7), who said: "*most of them (users) are not so concern about their health and they also take health issues for granted.*"

4.2.2 Lifestyle

Four participants raised the issue of personal lifestyle as influencing the relationship between low complexity level and positive attitude toward MHP usage. According to WHO (2014), individual lifestyle has been emphasized as a significant determinant of health. Within this study, this factor has been found as the second highest concern within the studied area. Two respondents claimed that they had no time to search for health information in the portal because of a busy working lifestyle and occupation with other matters as well. *Respondent one* (R1) highlighted this issue and said: "*I don't have time to scroll down the portal and search for any health information like in my situation busy with working and other things*". Meanwhile, another respondent (R7) gave feedback regarding the matter and claimed: "*we prefer to focus more on other aspect of life such as their work-life.*"

Lifestyle is also expressed in both work and leisure behavior patterns in activities, attitudes, interests, and allocation of time and income (Makvana and Rohit, 2016). This was expressed by R3 through the statement: "*The public tend to spend more time on other activities such as work-life and focus more on education and socialism.*" The busy working life and socializing aspect also were mutually agreed by another respondent, R6, who mentioned that besides being busy with work, social media applications were also one of the factors that shaped this respondent's busy lifestyle.

4.2.3 Dependence on general practitioners

People are more likely to turn to their general practitioners to discuss or seek advice relating to medical matters compared to accessing a health portal. Three participants admitted that they prefer to consult general practitioners for health advices. R1 claimed: "*I prefer to see the doctor. If I see the doctor directly, I can get instant MC and medicine that I need. I think that is why I prefer to see the doctor.*" R2 highlighted: "*I think it perhaps because there are other sources which is more important for us to refer for example like the Doctor itself means I prefer more to face to face consultation.*" Meanwhile, R3 strongly believed the interpersonal aspects established from verbal consultation with general practitioners have resulted in patients' satisfaction, confidence, and trust in the doctor.

4.2.4 Deficiency of external cues

Within the psychological health discipline, external cues refer to motivations from outside regarding a specific health behavior (Gray et al., 2005). These include events or information from close others (Janz & Becker, 1984), the media (Carpenter, 2010), or healthcare providers promoting engagement in health-related behaviors (Janz & Becker, 1984). The goal is to make people conscious and aware of their health behavior by providing them with information about their health problems. Three participants at least claimed they knew less about the existence of MHP; for instance, R1 responded: "*The factors that may influence people from using this portal is because lack of information about the existing of this portal, therefore they do not want to read the information given in the portal.*"

4.2.5 Search engines

Online searching through web search engines has become a common method for obtaining information. A web search engine is a software system that is designed to search for information on the World-Wide Web (Malik, 2015). Many general search engines are available on the internet, such as Google, Yahoo, Bing, AltaVista, and MSN Search. Health consumers have been interested in using general search engines to find health information because of their super quick and easy features. R2 highlighted, "*I also prefer to use Google to find any information and this includes health information as well. For me it is more simple and direct search, easy too.*" According to Jung and Loria (2010), citizens need to be better informed of the health resources that are available to them in order for the resources to be efficiently utilized. It is suggested that MHP developers take advantage of the capability and popularity of search engines to make MHP visible by positioning it at the top list of the search screen.

4.2.6 *Disparity of internet access*

According to Benigeri and Pluye (2003), internet access disparity refers to dissimilarities in individuals' ability to connect to the internet, or in another words, gaps in terms of online connectivity. This would include the speed, geographical coverage, cost to access, and other related factors. Based on the interview, R6 simply highlighted this issue as: *"Depending on the speed of internet coverage in certain location."*

4.3 *Subjective norms and intention to use*

Interviews were conducted to elicit participants' experiences to further explore the nonsignificant relationships between health consumers' subjective norms and MHP intention to use (Tobi *et. al.*, 2018). Seven factors emerged from interviews: age group preferences, influence of subjective norms, habits and lifestyle, disparity of internet access, lack of interpersonal communication with general practitioners, search engine preferences, and incomplete health information.

4.3.1 *Age group preference*

Subjective norms refer to the beliefs of individuals that they should engage in particular behavior based on the perceived social pressure received, specifically from those people who are important to them. Though the role of subjective norms was found higher in the study of Tobi et al. (2018) with total mean = 5.437, within the study, one respondent highlighted the capacity for making decisions still rests on them rather than on people in their surrounds. This capacity relating to respondents' self-decision in using the portal is highlighted by R1:*"I agree the influence coming from the people around me but at last the decision is depends on ourselves whether we want to use it or not."* It was found further that among the issues that determined the results between subjective norms and intention to use is were age group preferences.

The age group is an important factor that influences the accessibility of online health-related information. Evidence shows that younger adults between 18 and 29 years old were more likely to search online health information than older adults aged 30 years and above (Fox, 2006). Moreover, internet-based programs also have been suggested to be increasingly feasible and desirable for younger participants (Weinstein, 2006; Monaghan and Wood, 2010; Kothe et al., 2011). This could probably be due to what is labeled the "gray gap," which refers to the tendency for older demographic groups to lag behind the younger cohorts in information and communications technology literacy (Pew Research Center, 2000). Also related to the age group factor, older individuals preferences to turn to their general practitioners to discuss or seek advice relating to medical matters, which also influences their accessibility (Mokhtar *et al.*, 2009; Zuria et al., 2013;). Three participants raised their views on age group preferences, one of whom, R1, who is 35 years old, said: *"I think this portal is suitable or meant for young generation only because older generation prefer to listen to Doctor's advices compared to IT technology."* Supporting this, R6, who is 32 years old, highlighted: *"There is a handful of people or individuals which is more preferably and satisfactorily meeting directly with the doctor compared visiting any portals."* R9, who is 29 years old, pointed out: *"...although important people in my life would recommend me to use this portal to find for health information but I prefer to see the Doctor face to face when I need advises regarding my health problem."*

4.3.2 *Influence of subjective norms*

Though the role of subjective norms was found to be higher, it did not establish any significant relationship with intention to use MHP (Tobi et al., 2018). There is a unique reasoning that revealed the important influence within subjective norms that hinder people from the use of MHP is the family factor itself. In this situation the family is not using the portal, although they think it is good and have positive attitude toward it and would encourage the respondents to use it to satisfy their health information needs. The influence of families that do not use MHP on respondents' MHP usage are portrayed in the following statements by R2: *"This is due to those family members etc. also were not using and not knowing regarding myhealth portal,"* and R8: *"Sometimes, what family are not practice we also not practice and the awareness among family is connected to this factor that can influence ourselves. Maybe if family use it I will use it too. And then each family member can share information in the portal."*

4.3.3 *Habit and lifestyle*
Mobile technologies, internet, and social media have become a social phenomenon (Lenhart et al., 2010) and had become a life habit for most people today in the developed and developing countries. Within the Malaysian population, a recent study showed Facebook and other social networking (Instagram) emerged as the most common internet motivation followed with downloading and listening to online music such as in YouTube etc. (Khan and Magdalene, 2016).

The close engagement toward these online social activities has shaped the urban lifestyle of individuals within society, which has resulted in a lack of time to engage with health-directed activity such as engaging with MHP to search their health information needs. R1 highlighted: *"It is really good actually this portal, but because it is the habits of our people which love more to read entertainment than health information. Because I think many people today tend to see posts/updates and like to read Facebook/Instagram rather than health portal."*

4.3.4 *Disparity in internet access*
As highlighted previously, disparity in internet access refers to dissimilarity in individuals' ability to connect to the internet or, in other words, gaps in terms of online connectivity (Benigeri & Pluye, 2003). This would include the speed, geographical coverage, cost to access, and other related factors. This factor is supported as influencing the primary result, with R2 responding in the interview as: *"In workplace or at home, the internet access also limits the access to the portal where at home I don't have speed broadband to browse internet smoothly."*

4.3.5 *Lack of interpersonal communication with general practitioners*
A personal communication that occurs between a patient/health consumer and a healthcare provider within a healthcare setting was found to have a significant impact on the effectiveness of the health communication process. Based on the interview conducted, MHP was seen as a healthcare medium that does not provides a personal communication or message exchange between the two parties involved, as highlighted by R3, who stated: *"If it is possible for this portal to allow a 'messenger' or 'comments' in private between users and health practitioners as this can brings confidence on users to use this portal."* MHP is seen as a healthcare medium that does not provide a personal communication or message exchange between the two parties involved. Although family members, friends, and relatives demonstrate a positive impression toward one's intention to use MHP, limited opportunities provided to interact with the health practitioners in MHP had impacted health consumers engagement with MHP. In several studies relating online communication between doctors and patients, respondents were particularly enthusiastic if there is a possibility of communicating online with doctors when seeking health advice or treatment (Kleiner et al., 2002). In another study, older patients responded that they prefer if they can use email to communicate with their physicians (Singh et al., 2009). In another study, the majority health consumers stated that the availability of online communication services would influence their choice of healthcare providers to some extent, either by meeting them physically or just relying on WBHIS (Cummings, 2006).

4.3.6 *Search engine preference*
Although their families and peers think it would be best for respondents to use MHP to search for health information, the capacity for making decision still rests on themselves rather than on the people in their surrounds. Since much evidence has shown that health consumers around the globe prefer to search for health information using general search engines, and connected to this, respondents also felt it is convenient to rely on general search engines to find their health information rather than MHP. This statement was simply mentioned by R5:

"Because I think people could find many information from other sources in the internet rather than open the portal."

4.3.7 *Incomplete health information*

Although the main aim of healthcare websites and portals is to improve health consumers' knowledge by providing information about health problems, self-care, and prevention, however, technology also possesses content shortcomings. This is due to uneven quality of medical information caused by incomplete health information contained in the portal (Benigeri & Pluye, 2003).

Several authors, in fact, considered the quality of medical information on the internet as poor (Doupi & Van der Lei, 1999; Kunst et al., 2002). This factor potentially leads to the low intention to use online health information among health consumers, which also in the long term would significantly jeopardize the goals of WBHIS. Based on the interviews conducted with regards to this factor, one respondent, (R7), said: *".......the use of this portal does not affect me because the information contained on this portal is incomplete and insufficient for the use and health problems I am currently experiencing as I need more information on the health problems that I faced."*

4.4 *External cues and intention to use*

Feedbacks and opinions from respondents were solicited during the interviews and generally revealed one main issue, that is, lack of MHP promotion due to deficiency in external cues. Suggestions and recommendations given by respondents during the interviews were also represented concerning the means of advertisements and promotions of the portal.

4.4.1 *Deficiency of external cues*

Participants' feedback had found seven out of nine respondents mutually agreed that the main factor that led to the result is the lack of promotions and advertisements regarding the existence of MHP among health consumers in Malaysia. The external cues deficiency relating to this situation is a situation in which there is an acceptable understanding that there is an insufficient number of promotions, advertisements or reminders, regarding the existence or the use of MHP. For example, R5: *"There is a possibility that many people would be more aware and become more interested in using the health portal if efforts to promote the portal are being escalated in the future,"* and R8: *"Never ever saw any promotions on it on media, TV, radio or Facebook. If it promoted very well then I will know about it and use it in the future."*

4.4.2 *Mediums of advertisement and promotion*

During the interviews, researchers discovered some motivating findings besides the issue of deficiency. It became more interesting as respondents shared their thoughts on how to effectively advertise the portal to make it known to the public. There are five important mediums thought by the respondents as very effective for advertising MHP including radio and television, social media, clinics and hospital settings, health campaigns, and offering healthcare consultation.

TV and radio are still by far considered by the public as one of the most preferred advertising mediums. This is stated by R1 and R6, who also give some suggestions regarding TV and radio promotions of the portal. R1 said: *"Ministry of health should pop up more on TV advertisements then people might more interested to see myhealth portal. Maybe like during peak hour time is good. Radio and TV play very important role."* The second medium the respondents thought best to advertise MHP is social media. Two out of nine respondents, R5 and R6, claimed social media as one effective method of promoting MHP. The third is promoting MHP by highlighting its benefits and practical usage in local healthcare settings, for instance clinics and hospitals. This can be done by placing educational pamphlets and brochures about MHP, hanging banners to show the existence and usefulness of the portals, and by circulating MHP flyers to the public who attend healthcare settings. Two out of nine respondents, R2 and R3, proposed MHP promotion in healthcare settings. For instance, R2 highlighted:

"...placed information on the portal in local clinics and hospitals so patients can know about it like hang a banner or pamphlets and even the staff or doctors at the hospitals or clinics can tell patients about its benefits and advantages to the patients."

Fourth, MHP can also be well promoted through its traditional way of promoting health activity, that is, by organizing health campaigns. A health campaign is a type of media campaign that attempts to promote public health by making information on a new health intervention or alternatives available. The aim is to inform the public regarding healthy decision making and behavior. One respondent, R7, highlighted this medium, saying, "..it can be promoted like creating campaigns on the MyHEALTH portal in school or participating in any program that is being held to enlighten the importance of health in our lives." Like in this case, MHP promotions can be greatly organized in public areas targeting different groups such as schools, universities, public seminars, and the most classic example—campaigns held in malls or shopping centers. R3 proposed that MHP can be used as an online solution that extends its skills and expertise in offering healthcare consultations to available health-related websites. Offering such health skills and consultations either to local websites or across the nation would greatly benefit MHP, as it would be significantly acknowledged and receive a reputable and eminent amount of attention. According to R3, "...the portal should...be used as an expert consultant in the websites on health issues and health care that most public do not know."

4.5 Summary of the interview outcomes

The outcomes of the email interviews have shed light on the factors that could explain the unexpected phenomena from previous results. Interestingly, with regard to the relationship between a low complexity level and positive attitude toward MHP usage, it was found that a lack of health concern and awareness was the factor contributing most toward the usage issue. This factor was posited by five of nine respondents who participated in the interviews. The next factors found influencing MHP complexity-attitude were the health consumers' lifestyles, which were further explained by several sub-factors: being occupied with work-life, education, and also socializing through social media networking. Following this factor, depending too much on health practitioners and deficiency of external cues were the third and fourth factors that most contributed to this area of study, with three respondents respectively for each factor.

The next area of investigation was in finding an explanation with regard to the relationship between subjective norms and intention to use MHP. Seven factors were identified within this area, with the majority of them highlighted age group preference followed by subjective norms influences, which are family members and peers. These two factors were rated by three and two respondents respectively while the rest of the factors identified were proposed by at least one respondent for each of the factors derived. Finally, feedback from respondents relating to the third area of investigation, that is, between external cues and intention to use, demonstrated that the majority of respondents agreed that deficiency of external cues significantly influenced the earlier finding. An insufficient number of promotions and advertisements regarding the existence or the use of MHP had been highlighted by seven out of nine respondents during the interviews.

5 RECOMMENDATIONS

The interviews revealed a lack of health concern relating to self-health as the main underlying factor for users not wanting to use MHP. There is an urgent need for exploration into this so-called knowledge-behavioral gap, specifically highlighting the community's reluctance to take ownership of their health issues. As community empowerment thus becomes instrumental, understanding this gap on a deeper level would provide practical solutions to creating a health-empowered community adopting healthy lifestyles by accessing the health portal. Thus future studies are anticipated to uncover the role of health empowerment within society to address the issue of the behavioral gap.

6 CONCLUSION

The study revealed there are several spots of attraction that help responsible parties reach target users for MHP exposure such as in hospitals and clinic settings, schools, universities, and public seminars. This effort would help minimize the gaps and barriers in utilization of MHP among the public due to their lack of awareness of the existence of the portal. The study also confirmed that the public were in mutual agreement that they experienced lack of promotions and advertisements regarding the existence of MHP. The findings also identified several channels of advertising proposed by the public that they considered effective, including the radio and television, social media, clinic and hospital settings, health campaigns, and offers of healthcare consultation. This knowledge also has contributed toward a compelling comprehension that revealed specific age group preferences for advertising mediums. Thus, the information is beneficial for responsible bodies in designing a marketing strategy that can be specifically matched with a particular group of the target users. For instance, the preferred mediums for promotions are different for younger groups than for mature adults. A future study is expected to uncover the role of health empowerment within society to address the behavioral gap issue.

ACKNOWLEDGMENTS

Profound appreciation goes to the Institute of Research Management and Innovation (IRMI) of Universiti Teknologi MARA, and Ministry of Higher Education of Malaysia for funding the research work through Fundamental Research Grant Scheme (FRGS). Grant No: 600-IRMI/FRGS5/3(44/2016). A special acknowledgment also goes to the Health Education Division, Ministry of Health Malaysia for contributing inputs and supporting the research. The authors declare that there is no conflict of interest with the brand MyHEALTH Portal, which is operated under the Ministry of Health Malaysia.

REFERENCES

Benigeri, M., & Pluye, P. 2003. Shortcomings of health information on the Internet. *Health Promotion International*, 18(4):381–386.

Carpenter, C. J. 2010. A meta-Analysis of the effectiveness of health belief model variables in predicting behavior. *Health Communication*, 25(8):661–669.

Creswell, J. W. 2005. *Educational Research: Planning, Conducting and Evaluating Quantitative and Qualitative research*, 2nd ed. Upper Saddle River, NJ: Pearson.

Cummings, J. 2006. Few patients use or have access to online services for communicating with their doctors, but most would like to. *The Wall Street Journal Online*, 5 (16),n.p.

Fox, S. 2006. The online health care revolution: How the web helps Americans take better care of themselves. *Pew Internet and American Life Project*. Available at:http://www.pewinternet.org/reports/pdfs/PIP_Health_Report.pdf (accessed September 19, 2014).

Gray, N. J., Klein, J. D., Noyce, P. R., Sesselberg, T. S., & Cantrill, J. A. 2005. Health information-seeking behavior in adolescence: The place of the internet. *Social Science and Medicine*, 60:1467–1478.

Hardey, M. 1999. Doctor in the house: The Internet as a source of lay health knowledge and the challenge to expertise. *Sociology of Health and Illness*, 21(6):820–835.

Health Informatics Center Planning and Development Division. 2012. *Health Facts 2012*. Putrajaya: Ministry of Health Malaysia.

IHBR. 2014. *MyHEALTH Portal Research Survey*. Institute for Health Behavioral Research Malaysia. Unpublished Study: IHBR.

Janz, N. K., & Becker, M. H. 1984.The health belief model: A decade later. *Health Education Behavior*, 11(1):1–47.

Jung, M. L., & Loria, K. 2010. Acceptance of Swedish e-health Services. *Journal of Multidisciplinary Healthcare*, 3:33–63.

Khan, V. T., & Magdalene, C. H. A. 2016. Internet use and addiction among students in Malaysian public universities in East Malaysia: Some empirical evidence. *Journal of Management Research*, 8(2):31–47.

Kleiner, K. D., Akers, R., Burke, B. L., & Werner, E. J. 2002. Parent and physician attitudes regarding electronic communication in pediatric practices. *Pediatrics*, 109 (5):740–744.

Kothe, E. J., Mullan, B. A., & Amaratunga, R. 2011. Randomised controlled trial of a brief theory-Based intervention promoting breakfast consumption. *Appetite*, 56:148–155.

Kunst, H., Groot, D., Latthe, P. M., Latthe, M., & Khan, K. S. 2002. Accuracy of information on apparently credible websites: Survey of five common health topics. *British Medical Journal*, 324(7337):581–582.

Lazard, A., & Mackert, M. 2014. User evaluations of design complexity: The impact of visual perceptions for effective online health communication. *International Journal of Medical Informatics*, 83:726–735.

Lenhart, A., Purcell, K., Smith, A., & Zickuhr, K. 2010. Social media and mobile internet use among teens and young adults. Pew Internet and American Life Project. Available at: http://www.pewinter net.org/Reports/2010/Social-Media-and-Young-Adults.aspx (accessed January 15, 2017).

Makvana, S. M., & Rohit, V. K. 2016. Life style among willing to take divorce male and female. *International Journal of Social Impact*, 1(2), n.p.

Malik, O. F. 2015. Effects of terrorism fears on job attitudes and turnover intentions: The moderating role of job involvement. In *2nd International Symposium on Partial Least Squares Path Modeling*. June 16–19, Seville, Spain.

MHED. 2018. *Laporan Dalaman Data Pengunaan MyHEALTH Portal*. Putrajaya: Unit Pendidikan Kesihatan.

Mokhtar, I. A., Goh, J. E., Li, K. J., & Tham, C. X. 2009. Medical and health information seeking among Singapore youths: An exploratory study. *Singapore Journal of Library and Information Management*, 38:49–76.

Monaghan, S., & Wood, R. 2010. Internet based interventions for youth dealing with gambling problems. *International Journal of Adolescent Medicine and Health*, 22(1):113–128.

National Audit Department. 2012. *Auditor General Report 2012*. Putrajaya: National Audit Department Malaysia.

Noor, A. D., Mohd, A. O., Ummi, N. Y., & Teh, C. H. 2014. *Burden of Disease Study: Estimating Mortality and Cause of Death in Malaysia*. Putrajaya: Institute of Public Health.

Pew Research Center. 2010. *Older adults and social media*. Available at: http://www.pewinternet.org/2016/08/27/older-adults-and-social-media/ (accessed January 21, 2017).

Singh, H. F., Sarah, A., Petersen, N. J., Shethia, A., & Street, R. L. 2009. Older patients' enthusiasm to use electronic mail to communicate with their physicians: Cross-sectional survey. *Journal of Medical Internet Research*, 11(2):e18.

Stake, R. E. 1995. *The Art of Case Study Research*. Thousand Oaks, CA: SAGE.

Staniszewski, A., & Wangberg, S. C. 2008. eHealth trends in Europe 2005–2007: A population-based survey. *Journal of Medical Internet Research*, 10(4):e42.

Tobi, S. N. M., Masrom, M., Kassim, E. S., & Wah, Y. B. 2018. Psychological influence towards health consumers intention to use a Malaysia national web based health information service. *E-BPJ*, 3(7):167–174.

Tobi, S. N. M., Masrom, M., & Mohammed, A. 2018. The 11th Malaysia Health Plan: Demand for investigation on health consumers' intention toward national web-based health information service. In F. Noordin, A. Othman, & E. Kassim (eds.), *Proceedings of the 2nd Advances in Business Research International Conference*. Singapore: Springer.

Tobi, S. N. M., Masrom, M., Mohammed, A., & Abdullah, M. N. 2015. The moderating effect of psychological factors towards the diffusion of a web-based health information service (WBHIS). *E-Proceeding of the International Conference on Social Science Research (ICSSR)*, June 8–9, 2015, Meliá Hotel Kuala Lumpur, pp. 616–628.

Tobi, S. N. M., Masrom, M., Rahaman, S. A. S. A., & Mohammed, A. 2017. MyHEALTH Portal: Malaysia national web-based health information service for public well-being. *Advanced Science Letters*, 23(4):2853–2856.

WHO. 2014. Health and ageing country factsheet. Available at: http://www.wpro.who.int/topics/ageing/ageing_fs_malaysia.pdf (accessed November 14, 2014).

WHO. 2015. *E-health*. Available at: http://www.who.int/trade/glossary/story021/en/ (accessed November 22, 2015).

Yin, R. 2003. *Case Study Research: Design and Methods*. (3rd ed.) Thousand Oaks, CA: SAGE.

Yun, E. K., & Park, H. A. 2010. Consumers' disease information-seeking behaviour on the internet in Korea. *Journal of Clinical Nursing*, 19(20):2860–2868.

Zuria, A. S., Noorsuraya, M. M., & Siti, K. M. 2013. Online health information seeking behavior among employees at two selected company. In *Proceedings of the 2013 IEEE Business Engineering and Industrial Applications Colloquium*, April 7–9,. Langkawi. Pp. 169–173.

Inclusive Development of Society – Lumban Gaol (eds)
© 2020 Taylor & Francis Group, London, ISBN 978-1-138-33476-2

Small innovative enterprises as a part of an organizational and economic mechanism supporting governmental economic security

L.P. Goncharenko, S.A. Sybachin & G.A. Khachaturov
SRI Innovative Economy, Plekhanov Russian University of Economics, Moscow, Russia

ABSTRACT: The innovative component of the Russian economy is not stable in the existing business environment, and within the framework of the current legal regulation system it cannot reach the level of feasible self-maintenance. Many attempts made and approaches tried to solve the current situation resulted in fluctuating development of the Russian innovative sphere. The strategy of the Russian economic safety for the period until 2030 presumes as one of the key elements of such an economy small innovative enterprises, set up under state educational and scientific institutions to incorporate the results of intellectual activity of academic science in economic development. The initiative to form such enterprises dates back to 1988 and by now has undergone many revisions. Now this line is facing a challenge of one more transformation and needs a vector to provide conformity with other elements of an innovative economy, consecutive transition to self-maintenance and further on to extended reproduction.

1 INTRODUCTION

Transition of the Russian economy to innovative development has been long and hard and even now we cannot say that it is close to completion.

Many attempts have been made to boost innovative activity of local market participants, but unfortunately to no stable self-maintenance effect. Based on accumulated foreign and Russian experience, we can draw a conclusion, that without efficient cooperation of the state, science, and business, extended reproduction of the economic innovative sphere is impossible. One of the first to raise the issue of an interaction system based on the model was Henry Etzkowitz in his work "Triple Helix. Universities– enterprises–state. Innovations in action" (Etzkowitz, 2008). The work describes the experience of such interaction in Silicon Valley as a leader of triple interaction in the sphere of innovations. In the "heart" of Silicon Valley there are three main universities, surrounded by a strong belt of small innovative and service companies. As noted correctly in the "Guidance on setting up innovative structures" (2012), every innovative center was formed in respective unique conditions and was called upon to solve a complex of unique, typical of a particular country or region, tasks; it has a unique control structure and organizational model of business processes. Quite often their practice is not compatible, like yellow versus flat. The Guidance contains a major analysis of successful and unsuccessful foreign experience in setting up macroeconomic innovative structures. A detailed review of each of them may prompt a conclusion that it acts as a macroaggregate for small venture and innovative enterprises.

Rich experience has been accumulated in Russia in building an innovative economy on the basis of small venture and innovative companies and start-ups, but here we are catching up with leading developed countries.

The first step along integration of science and business in the USSR was made in 1988 by passage of the law on cooperative societies. The first law legalized formation under different level educational institutions (from general education schools up to training workshops at enterprises) of cooperatives: "organization of USSR citizens, having voluntarily united together for joint execution of economic or other activity and on the basis of private property..." (USSR Law 8998-XI, 1988). Later in 1989 the Commission on Optimization of Economic Mechanism under the USSR Council of Ministers approved the Provision "On organization of small businesses" (Letter,of RSFSR Ministry of Education, 1989), accepting the notion of a "small enterprise."

This provision consisted of seven sections, setting out the following: General provisions; Procedure of formation, control, reorganization, and termination of small enterprises; Property and funds of small enterprises; Planning and accounting; Procurement; Sale of produce; and Finances and credit. The General provisions stipulated which enterprises were rated small and what they are created for; in particular we note that in the order of importance in p. 1.2. (Key purposes of creating small enterprises) we see in the second position: "faster implementation of scientific & technical achievements". Page 2 of the Provision says that such enterprises are formed upon decision of a founder, following which we can conclude that first small enterprises could be formed exclusively under legal entities. The Provision says that the founder and a small enterprise sign a contract, establishing their economic relations. Here it is worth noting that the funds of a small enterprise are formed fully by its founder, who, if necessary, may assign temporarily main assets and render temporary financial assistance to a small enterprise. Summarizing the information about the last three quoted paragraphs, we can conclude that small enterprises, regarding their economic activity, were actually structural units, as most issued were decided by a founder, while in other aspects of economic activity they were full- fledged legal entities. From the point of view of scientific research and innovative activity that arrangement was quite convenient, but it was only the first step along the transition to an innovative economy.

Subsequent steps were aimed at revising the form and establishing boundaries of a new entity: small innovative enterprise.

Simultaneously with formation of small enterprises under state educational and scientific institutions, a new system of legal regulation was created for the innovative component of the Russian economy. Here, placing an emphasis on important issues for small innovative enterprises, we should outline three states of formation: 1991–1996, creation of the legal regulation basis in the sphere of intellectual property; 1994–2009, optimization of the legal regulation basis in the sphere of intellectual property; from 2009 until now, rating small innovative enterprises as a special form of economic societies, which can be created under state educational and scientific institutions.

The beginning was laid on May 31, 1991 by passage of USSR Law No. 2213-1, "On inventions in the USSR" (1991). The document contained the following chapters:

1. Invention and its legal protection. The chapter outlines key moments, relating to inventions, first of all, criteria of invention and a list of (directions) characteristic subjects, as well as stating what is not considered as invention. The chapter regulates application of requests for patenting inventions and processing of requests and results in the form of issue of patent, refusal to issue a patent, and revocation of request.
2. Application of inventions. Reviewed here are issues of applying invention both under some sort of agreement and as an open license. In addition, this chapter sets out lines of state support and stimulation of invention activity.
3. Labor and other rights or privileges of inventors. The chapter's name characterizes the composition of information therein; for example, it shows an approach to stimulation of invention activity by granting privileges to inventors.
4. Organizational fundamentals of legal protection of inventions and of copyrights of inventors and patent holders. Presented for the first time as a single document are provisions on protection of inventors' copyrights and protection of inventions, covered by USSR patents.
5. Final provisions.

Forty days later, on July 10, 1991 the President signs USSR Law No. 2328-1 "On industrial samples" (1991), containing four chapters, similar to the aforementioned law "On inventions in the USSR." Passing these documents was a crucial moment for the Russian scientific community. Actually, these laws laid the basis for intellectual property, which is essentially the heart of any innovative economy. It is worth noting that these documents have the same template, testifying to the system approach toward forming legal regulation in the sphere of intellectual property.

A year later, on September 23, 1992, two laws were passed: No. 3523-1 "On legal protection of software for electronic computing machines and their data base" (1992) and No. 3520-1 "On trademarks, service labels and place of product origin" (1992), bringing about a list of results of intellectual activity, which can be granted legal protection by the state in the form of certificates and outlining state support for authors and inventors.

Then on July 9, 1993 Russian Federation Law No. 5351-1 was passed "On copyrights and adjacent rights" (1993), defining such notions as copyrights and adjacent rights, objects of such rights, and occurrence of rights on the objects, as well as personal property rights, occurring together with the aforementioned copyrights and adjacent rights. The second and third sections of the document stipulate the use of copyrights in covered publications and the form of copyright contract. The fourth section is about copyright management and the last section, No. 5, is about protection of copyrights and adjacent rights.

To regulate contract activity in the sphere of intellectual property and to motivate authors and inventors to conclude contracts on assigning exclusive and nonexclusive rights, as well as to manufacture products, using results of intellectual activity, the Russian Council of Ministers passes on July 12, 1993 Resolution No. 648 "On application procedure for inventions and industrial samples, protected in the Russian Federation by author's certificates on inventions and certificates on industrial sample as well as regarding author's remuneration" (1993).

Gradually the innovation sphere in Russia acquired its shape, but the process still has a long way to go owing to turbulent economic situation in the country. Revisions and transformations in priorities and development vectors of the innovative sphere occur much faster than adoption and implementation of regulating documents. Forecasting the development of the innovative sphere in Russia is lagging behind the actual requirements of the sphere, maintaining this lag, and having a negative impact on economic development, first of all on formation and development of small innovative enterprises on the basis of budget educational institutions.

Now, reviewing in detail the changes that have taken place, we can see several "economic waves" of the initiative rise and fall since 1988 until the present. These waves represent major modifications in the formation and development process of small innovative enterprises. The essence of these enterprises' existence remains the same. Their purpose is manufacture and sales of innovative products/technology and its optimization is based on information from the market and consumers.

Small innovative enterprises should not be assigned a task of improving innovative products/technology by scientific research in a particular sphere, of finding better solutions and implementing them through major modification of the current innovative products/technology (including its complete substitution).

Here it is important to note that in the present business conditions small innovative enterprises can be perceived as an economic entity, created to realize a particular project by marketing innovative products/technology, having a life cycle similar to that of the project and that cannot be restarted.

And the fact, skipping our attention, is that advanced inventions and technologies are often developed on experimental equipment, or at least with fine adjustment of mass production equipment, creating also prototypes for priority field research, which, considering modern technological progress, can be outsourced. This issue has been greatly promoted by 3-D printing with subsequent improvement of printing materials, printing accuracy, and larger printout volume.

The problem arising is that while a small innovative enterprise uses the production facilities of its parent educational institution, it is just short of potential to build full-fledged production and secure a place in the market. Sooner or later a small innovative enterprise will have to increase output of innovative products and have broader application of innovative technology, creating the necessity of complete equipping or re-equipping of its production and facing numerous issues, to be handled by the innovative enterprise together with its higher education institution. In addition, the form of such an enterprise existence, possessing the life cycle of an innovative product, does not make the problem easier. All dynamics, introduced originally by the state to small innovative enterprises, are rigidly tied to one result of intellectual activity, the right of using which is entered in the authorized capital of a small innovative enterprise. As a result, the attempt at revision calls for solution of the following issues:

1. Revision in the higher education institution's share in the authorized capital of a small innovative enterprise to change/correct the line of its activity.
2. Closing a small innovative enterprise on expiry of its life cycle, given to it to sell product/technology and to open another one.

Each of the aforementioned procedures is fraught with red tape and accompanied by substantiations of key solutions at several control levels. It is worth noting here that taking these decisions commits officials in charge to serious and sometimes nontransparent liability, creating obstacles and inert behavior.

Today to tackle most issues relating to small innovative enterprises, higher education institutions must obtain approval of the founder, which is quite a problem for state educational institutions.

To implement any revision, the following documents must be submitted:

1. Accompanying letter with data as per Order of the Russian Ministry of Education & Science of November 18, 2010 No. 1188 (2011), including basis, purpose of transaction, information about the parties in transaction, subject of transaction, deadline of execution of obligations in the transaction, sources of transaction funding
2. List of documents
3. Draft delivery and acceptance statement
4. Confirmation of transaction feasibility
5. Confirmation of ability to meet obligations in the transaction
6. Confirmation of forecast impact of transaction results on higher efficiency
7. List of especially valuable movable property, to be invested as deposit, with names of objects, inventory numbers, quantity of units, balance value (in soft and hard copies)
8. Documents confirming rating of objects as especially valuable movable property
9. Extract from the Federal Property Register
10. Extract from the property complex development program, containing detailed information about federal property objects, to be invested as deposit
11. Extract from MoM of the body that approved the property complex development program
12. Conclusion of commission at the university on feasibility of the proposed method of using and/or managing the federal property together with an extract from the MoM of organization, which approved the commission recommendations
13. Other documents relating to future transactions

Par. 7 and 8 of the list by the Russian Ministry of Education & Science present a simplified procedure. Order No. 800 of August 4, 2015 cancelled p. 3 of the Annex to Order No. 2261 of December 31, 2010 by the Russian Ministry of Education & Science, obliging us to consider all nonmaterial assets on university balance as especially valuable movable property. That is why this is important only for intellectual property, having balance value above 500,000 rubles. Note that this simplification has a dramatic impact on creation of small innovative enterprises under educational institutions, removing obstacles to their formation, although the timing was absolutely wrong. Now the trend of small innovative enterprise creation shows a steady slump.

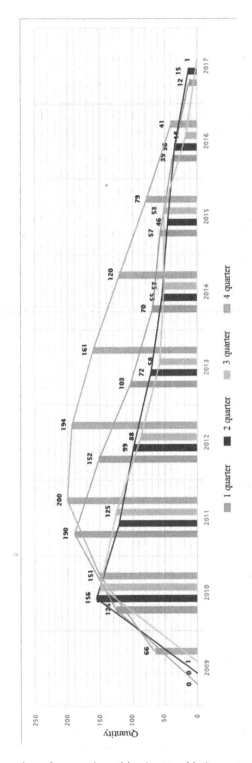

Figure 1. Dynamics of creation of economic entities (partnerships), quarters 1-4. (Information from www.mip.extech.ru).

The chart shows a steady slump, which will result by 2018 in the absence of newly formed small innovative enterprises, which calls for next transformation of this initiative; new approaches, especially considering improved financial sphere; and revised priorities of scientific and technical development.

Small innovative enterprises are one of the priority directions in the country's development within the framework of the Concept on long term social & economic development of the Russian Federation in the period up to 2020, being one of the deliverables ("Economy of leadership and innovations"). The main principles of economic innovative development are given in section 1, p. 4 of the Concept; important among them is small business development (Russian Federation Government Ruling No. 1662-p, 2008). Apart from the development strategy, shown also in the Strategy of Russian economic security until 2030, adopted on May 13, 2017 are issues of innovative development and specifically, development of small innovative enterprises (Russian Federation President's Decree No. 208, 2017). By 2019 we are to prepare legal regulation and an organizational basis to support the strategy implementation.

From some points of view development along these lines does not seem to be efficient because of low payback and high risks during creation and operation. Attempts have repeatedly been made to set up a system of small innovative enterprises under the budget of scientific and educational institutions, but to no avail, although resources had been spent. This is all not so straightforward. A small innovative enterprise should be commercially successful, but at the present stage of innovative sphere development in Russia most important is the factor of total social and economic efficiency.

According to the existing theory the efficiency of creation and activity of a small innovative enterprise includes positive (or nonzero) effects, split into two categories:

1. Economic effect
   The effect is calculated like for any investment project and is based on four main indices:
   a. NPV, net present value
   b. PI, profitability index
   c. IRR, internal rate of return
   d. DPP, discounted payback period
2. Social effect
   "Complex of social results, ..., projected to the quality of social environment and having both positive and negative values" (Ivushkina, 2001). Modern approaches to assessing this effect assume an expert cost evaluation of each separate effect and subsequent reference to economic indices:
   a. NPSV, net present social value (Shchekova, 2003)
   b. TSV, total social value (Shchekova, 2003)
   c. SV, social value of a separate element. This value can be estimated both through the value revision factor under the project's impact and average economic costs of alternative ways to reach the same level.
   d. KSR, social payback factor (Shchekova, 2003)

The key social effects, created by each small innovative enterprise are the following:

a. Interspace of higher education institutions:
   i Number of signed contracts on cooperation between higher education institutions, as well as with commercial companies
   ii Number of signed contracts on work execution/provision of services
b. Inner space of higher education institutions:
   i Involving teachers and lecturers of higher education institutions in activities of small innovative enterprises (more staff and higher revenues)
   ii Involving young scientists in work at small innovative enterprises (more staff, higher revenues, obtaining business experience (trying scientific research)
   iii Involving students in work at small innovative enterprises (obtaining business experience), support in search of own scientific or business lines of activity (this effect has considerable time lag)
c. Meeting goals within the framework of the Russian scientific and technical strategy:

i Number of small innovative enterprises

ii Additional budget funding of educational institutions through commercialization of intellectual activity results

iii Attracting additional external grants for fundamental and applied scientific research by inviting small innovative enterprises to participate in competitions

Note that there are quite a lot of spheres where small innovative enterprises can operate, and they can be more by efforts made by such enterprises. Depending on the sphere, where a small innovative enterprise and a budget educational institution operate, assessment methods will differ in threshold values, weight readings, assigned factors, etc.

In this view, to designate and confirm the importance of small innovative enterprises, and, consequently to substantiate the necessity in development along this line we will review as an example several different activities of small innovative enterprises under higher education institution (using publicly accessible data of the Tomsk State University as well as open data on small innovative enterprises related to the Tomsk State University).

For NPV calculation a discount rate was used, based on a modified CAPM model (incorporating extra specific risks). Assumed as risk-free assets for this model were Russian state bonds, using in the model their annual weighted average yield, and in the capacity of average market yield—weighted average yield by the PTC index. The specific feature of such a calculation is ignoring the cost of intellectual activity result, invested as deposit to the authorized capital from the calculation basis of original investment.

I. Small innovative enterprise, specializing in chemical and agrochemical production

Table 1. Economic efficiency.

Index/year	2013	2014	2015	2016	2017[a]
NPV	1.102 million rub.	1.103 million rub.	1.105 million rub.	1.101 million rub.	1.101 million rub.
PI	5.74	5.75	5.76	5.76	5.76

[a] Forecast value.

The payback period of the project was six months.

Based on the calculations we can say that this enterprise was economically efficient; at this moment the project has already passed its peak and is approaching the slump stage, so it has to be revived in the near future. We should also note its large life cycle, which confirms the importance of its innovative products.

II. Small innovative enterprise, specializing in production of non-foodstuffs

Table 2. Economic efficiency.

Index/Year	2013	2014	2015	2016	2017[a]
NPV	436.8 thousand rub.	591.3 thousand rub.	643.5 thousand rub.	0	0
PI	5.23	7.08	7.71	0	0

[a] Forecast value.

The payback period of the project was about three and a half months.

The calculations show that this enterprise was economically stable; its operation has been suspended, so the project needs to be revived in the near future. We should also note its large life cycle, which confirms the importance of its innovative products.

III. Small innovative enterprise, specializing in chemical production and technology

Table 3. Economic efficiency.

Index/Year	2013	2014	2015	2016	2017[a]
NPV	3.73 million rub.	4.4 million rub.	4.44 million rub.	5.28 million rub.	5.57 million rub.
PI	5.74	6.77	6.83	8.13	8.58

[a] Forecast value.

The payback period of the project was a little more than three months.

The calculations show that this enterprise was economically sustainable. As compared to the previous periods, in 2014 the volume of production and supplies plummeted, but even then the small innovative enterprise continued to enjoy a positive financial result, gradually increasing the volumes. These events could occur as a result of improvement of technology, bringing about revision of the technological cycle and, as a result the given economic result. We also note an extremely large life cycle of this innovative project, still ongoing, which demonstrates high market interest in the innovative products manufactured by this enterprise.

IV. Small innovative enterprise, specializing in production of medical appliances

Table 4. Economic efficiency.

Index/year	2013	2014	2015	2016	2017[a]
NPV	118 thousand rub.	34 thousand rub.	122.2 thousand rub.	394.4 thousand rub.	1 million rub.
PI	0.40	0.12	0.42	1.35	3.43

[a] Forecast value.

The payback period of the project was about three and a half years.

It took this small innovative enterprise quite a long time to stop suffering losses, first of all due to specific features of its innovative activity results, invested in the authorized capital, and consequently, specific features of the activity. Having the assistance of the higher education institution, however, it managed to gain a positive economic effect. In situations like this, support from a parent company is of great importance for a small innovative enterprise, since organization of production takes much effort (not only in production, but also in sales), which must be done simultaneously. Without the correct and timely support from the higher education institution this enterprise would have been doomed to closure.

V. Small innovative enterprise, specializing in research and development in the sphere of agriculture, as well as in agricultural production

Table 5. Economic efficiency.

Index/year	2013	2014	2015	2016	2017[a]
NPV	1.22 million rub.	2.41 million rub.	3.32 million rub.	4.06 million rub.	4.92
PI	1.11	2.19	3.02	3.69	4.48

[a] Forecast value.

The payback period of the project was about a year.

This was an extremely successful small innovative enterprise, which very soon gained payback and expansion. The calculations suggest that agriculture badly needs innovation and this small innovative enterprise accessed the market quite successfully. In this perspective it will expand and improve its technologies. This small innovative enterprise is a good example of solving issues, challenging all innovative enterprises.

Calculation of the social impact from activity of each small innovative enterprise under the Tomsk State University was quite complicated owing to a lack of many important data, which could allow us to assess the contribution of each enterprise. However, these data are sufficient to calculate the total social effect from activities of small innovative enterprises under the Tomsk State University.

Participating in the activity of small innovative enterprises were many employees and students of the higher education institution. In various years (2012–2016) the number of involved employees of the total scientific staff of the higher education institution varied from 2% to 3.5% (from 100 to 160 persons). Apart from the university employees, involved in the activity of small innovative enterprises were also students, postgraduates, and young scientists. However, over the past few years this practice has faded; for example, in 2013 the number of involved students, postgraduates, and young scientists was 227, while in 2016 it was only 29. There are many underlying causes, including effective barriers.

Today the main task for the coming transformation of small innovative enterprises is outlining their modification vector to make sure they meet the target of their existence, have more flexibility in daily activity, and maintain sufficient control on the part of educational institutions.

ACKNOWLEDGMENTS

This article was prepared as part of the project section of the government contract as requested by the Ministry of Education and Science of the Russian Federation on the subject formulated as 'Development of Methodological Principles and Organizational Economic Mechanism of Strategic Management of Economic Security in Russia" (Assignment No. 26.3913.2017/ПЧ).

REFERENCES

Danchenko, Ye. S. 2016. Assessment of social effect in investment project. *Issues of Economy and Management*, 5(1):4–6.

Decree of the Russian Federation Government No. 648. July 12, 1993. Available at: http://www.consultant.ru/document/cons_doc_LAW_2171/

Etzkowitz H. 2008. *The Triple Helix: University-Industry-Government Innovation in Action*. London: Taylor & Francis.

Fedosova, R., Kheyfits, B., & Ilyina, M. 2016. Formation of methodological approach to evaluation of the national innovative environment. *Knowledge, Service, Tourism and Hospitality*, 127–132.

Guidance on creation and development of innovative centres (technologies and principles), p. 144, 2012. Available at::

Ivushkina, N. V. 2001. Social effect of investment processes. Research paper. Candidate of economic science (08.00.01). Moscow. http://pressa.tomsk.gov.ru/files/2041/original/inno.pdf

Letter of RSFSR Ministry of Education No. 09-14/897. December 25, 1989. On provisions regarding activities of small enterprises" (together with "The provision…, approved by the Commission on Optimisation of the Economic Mechanism under the USSR Council of Ministers. June 6, 1989, MoM No. 14. Available at: http://www.consultant.ru/cons/cgi/online.cgi?req=doc;base=ESU;n=16696#0

Order of the RF Ministry of Education & Science No. 1188. November 18, 2010. (rev. September 13, 2011). Available at: http://www.consultant.ru/document/cons_doc_LAW_109183/

Order of the RF Ministry of Education & Science No. 800. August 4, 2015 Available at: http://www.consultant.ru/document/cons_doc_LAW_185395/

Order of the RF Ministry of Education & Science No. 2261. December 31, 2010. On determining especially valuable movable property. Available at:: http://www.consultant.ru/document/cons_doc_LAW_110099/

Russian Federation Government Ruling No. 1662-p November 17, 2008 (rev. February 10, 2017). On the concept of long term social & economic development of the Russian Federation till 2020.

Russian Federation President's Degree No. 208. May 13, 2017. On the economic security strategy of the Russian Federation till 2030.

Russian Federation Law No. 3523-1. September 23, 1992. On legal protection of software for electronic computing machines and their database. Available at: http://www.consultant.ru/document/cons_doc_LAW_1007/

Russian Federation Law No. 3520-1. September 23, 1992. On trademarks, service labels and places of product origin. Available at: http://www.consultant.ru/document/cons_doc_LAW_996/

Russian Federation Law No. 5351-1. July 9, 1993. On copyrights and adjacent rights. Available at:: http://www.consultant.ru/document/cons_doc_LAW_2238/

Shchekova, E. L. 2003. *Economy and Management of Non-commercial Organizations*. Saint-Petersburg: Lany.

USSR Law No. 8998-XI. May 26, 1988. On cooperation in the USSR, p. 1, clause 5, section 2.

USSR Law No. 2213-1. May 31, 1991. On inventions in the USSR. Available at: http://www.consultant.ru/document/cons_doc_LAW_18406/

USSR Law No. 2328-1. July 10, 1991. On industrial samples. Available at:: http://www.consultant.ru/document/cons_doc_LAW_18376/

. Zimin, A., Otto, V., Filimonova, N., Fedosova, R., & Kuznetsov, Y. 2016. New type of regions in the innovation economy. *Advanced Science Letters*, 22(8). Available at: www.scopus.com

Author Index

Ogannisyan, L.A. 277
Olisaeva, A.V. 182

Panakhov, A.U. 153
Panfilova, E.A. 187, 223
Pg Arshad, P.G.M.A. 292
Pirumyan, A.A. 196
Podchufarov, A.Y. 36
Podgajnov, D.V. 108
Pogorelova, T.G. 202
Pogosyan, N.V. 161
Polenova, A.U. 202
Ponomarev, V.I. 36
Prokopenko, Z. 131
Prokopets, T.N. 187
Pusparini, M.D. 256

Revalde, G. 230
Romanenko, M.A. 66
Rustamova, I.T. 196

Sagintayeva, S. 230

Salleh, N.Z. 44, 51, 59, 292
Semergey, S.V. 277
Senkov, R.V. 36
Sharko, E. 131
Sharkova, A.V. 176
Shkhagoshev, R.V. 202
Simionov, R.U. 153
Sinyuk, T.Yu. 187
Sokolnikova, I. 139
Sokolova, I.I. 73
Som, M.N. 51
Surov, A.S. 237
Sybachin, S. 131, 139
Sybachin, S.A. 339

Temirkanova, A.V. 313
Teslenko, I.B. 1
Tobi, S.N.M. 328
Tolstykh, T.O. 123
Tregub, I.V. 285
Tropinova, E.A. 161
Trousil, M. 244

Tukhkanen, T.N. 123

Vandina, O.G. 223
Volovik, M.E. 313
Voronina, T.V. 166

Yalunina, E.N. 187
Yatin, S.F.M. 328
Yatsenko, A.B. 166
Yevchenko, N.N. 166
Yudina, N.V. 321
Yudina, T.A. 176
Yusof, M.M. 328
Yussuf, A.A. 22

Zaidin, N. 44, 51, 59, 292, 298
Zainal , A.N. 44
Zakharchenko, E.S. 250
Zhanbayev, R. 230

Printed in the United States
by Baker & Taylor Publisher Services